Springer Series on Environmental Management

Robert S. DeSanto, Series Editor

Winter view of Lake McDonald, Glacier National Park. (Photo by Mel Ruder for *The Hungry Horse News*.)

Stephen R. Kessell

Gradient Modeling

Resource and Fire Management

With 175 Figures

Springer-Verlag

New York   Heidelberg   Berlin

Stephen R. Kessell
Gradient Modeling, Inc.
Box 2666
Missoula, Montana   59806
USA

Cover photo: Early summer view of Lake McDonald in Glacier National Park, Montana, looking toward the continental divide. Much of this book describes the development of a computer-based resource and fire management model for that park. (Photo by Mel Ruder for *The Hungry Horse News.*)

Library of Congress Cataloging in Publication Data

Kessell, Stephen R
Gradient modeling: resource and fire management.

(Springer series on environmental management; v. 1)
Bibliography: p.
1. Ecology—Simulation methods. I. Title.
II. Series.
QH541.15.S5K46   574.5′01′84   79-4371

ISBN 0–387–90379–8 Springer-Verlag   New York
ISBN 3–540–90379–8 Springer-Verlag   Berlin   Heidelberg

Dedicated to the folks whose encouragement and assistance
made this work possible:

> Willie Colony
> Jeff Johnstone
> Linc Brower
> Hugh Gauch, Jr.
> Charlie Hall
> Bob Whittaker
> Jim Lotan
> Meredith Potter
> Pete Cattelino
> Mel Ruder

Series Preface

This series is dedicated to serving the growing community of scholars and practitioners concerned with the principles and applications of environmental management. Each volume will be a thorough treatment of a specific topic of importance for proper management practices. A fundamental objective of these books is to help the reader discern and implement man's stewardship of our environment and the world's renewable resources. For we must strive to understand the relationship between man and nature, act to bring harmony to it and nurture an environment that is both stable and productive.

These objectives have often eluded us because the pursuit of other individual and societal goals has diverted us from a course of living in balance with the environment. At times, therefore, the environmental manager may have to exert restrictive control, which is usually best applied to man, not nature. Attempts to alter or harness nature have often failed or backfired, as exemplified by the results of imprudent use of herbicides, fertilizers, water and other agents.

Each book in this series will shed light on the fundamental and applied aspects of environmental management. It is hoped that each will help solve a practical and serious environmental problem.

Robert S. DeSanto
East Lyme, Connecticut

Preface

This book's origins can be traced back to a beautiful fall afternoon in 1969; I was tromping around the Holyoke Range just south of Amherst College with the excuse of collecting field data for Lincoln Brower's ecology course. I had started the afternoon as an astronomy major taking Ecology 41 because of a schedule conflict; I finished the day as the department's newest biology major. There just had to be ways to improve resource management by applying quantitative methods and computer technology to ecologic and environmental problems . . .

The summer of 1972 found me, after entering a doctoral program at Cornell University's Section of Ecology and Systematics, hiking in Glacier National Park to collect field data for a gradient analysis of the park's vegetation. Yet by summer's end, after many long discussions with Willie Colony, the park's Fire Management Officer, I realized that there are better uses for refined resource information than another dissertation collecting dust in some library. The next 18 months saw the formulation of what is now known as gradient modeling—that is, the linkage of gradient analysis vegetation models with site inventories and computer software to provide a natural resource information system.

When I left Cornell in 1974, gradient modeling was a nice idea still on the drawing board; I thought the approach would work, but many did not. I spent the next two years in West Glacier, Montana, developing and implementing the Glacier National Park gradient modeling system. It worked.

Since 1976, a similar system has been developed for southern California chaparral ecosystems, the method has found numerous other resource management applications, and recently it inspired the development of *FORPLAN,* a *FOR*est *P*lanning *LAN*guage and Simulator. *FORPLAN* links numerous models and data bases into a single package that is programmed in a conversational style using common English words; it and similar systems are currently being implemented as resource management tools in the U.S. and Australia, and are being evaluated for management implementation in several other countries.

I have written this book for three reasons. First, although various summary accounts of gradient modeling systems have been published, this is the first document to explain the detailed, step by step development of such systems. It is hoped that it will be of value to those considering the development of similar models for their own resource management problems and areas.

Second, there is still a paucity of good resource modeling texts available to the advanced undergraduate, graduate student, and field practitioner. C.

A. S. Hall's and J. R. Day's *Ecosystem Modeling in Theory and Practice* (1977) helps fill this gap by providing an (somewhat academic and esoteric) overview to numerous modeling approaches; to a certain extent it is a very good introduction to resource management modeling. I hope the present volume can further fill that gap by providing students with a more detailed view of how resource information is gathered, compiled, stratified, stored, and used to build resource management systems.

Third, I hope this book presents more than simply a description of gradient modeling systems. The work described here required the integration of numerous and seemingly unrelated methods and philosophies from biology, ecology, meteorology, combustion physics, remote sensing, statistics, and computer science; any development of a resource management information system from A to Z requires an open, interdisciplinary approach, a point too often ignored by overspecialized researchers and skeptical managers alike. Perhaps this book can also inspire other investigators to develop resource management models and systems, tools that we desperately need if we are to meet the ecologic future.

Canberra, A.C.T., Australia   Stephen R. Kessell
February 1979

Acknowledgments

The development of the gradient modeling method of constructing resource information systems represents the combined efforts of many people. Over 50 technicians assisted in this research effort during the period 1972–1978; many were underpaid and overworked. Significant professional assistance was provided by William M. Colony (Glacier National Park), Robert H. Whittaker (Cornell University), Hugh G. Gauch, Jr. (Cornell University), Jerry F. Franklin (U.S. Forest Service and Oregon State University), Meredith W. Potter (Rockford College and Gradient Modeling, Inc.), Charles A. S. Hall (Cornell University), James E. Lotan (U.S. Forest Service), Peter J. Cattelino (Gradient Modeling, Inc.), Richard C. Rothermel (U.S. Forest Service), and Jane E. Kapler (Glacier National Park). Richard Munro, Management Assistant in Glacier National Park, offered considerable administrative assistance in cutting through government red tape.

Numerous other U.S. Department of Agriculture Forest Service and U.S. Department of the Interior National Park Service personnel have also provided professional advice.

This work was supported by several agencies and institutions, including Gradient Modeling, Inc., an international nonprofit research foundation devoted to research in ecology and resource management applications; U.S. National Science Foundation grants to Amherst College and Cornell University; a research contract between the U.S. Department of the Interior National Park Service and Gradient Modeling, Inc.; several cooperative aid agreements between the U.S. Department of Agriculture Forest Service, Intermountain Forest and Range Experiment Station's Northern Forest Fire Laboratory (Fire in Multiple Use RD&A Program), and Gradient Modeling, Inc.; a cooperative aid agreement between the U.S. Department of Agriculture Forest Service, Pacific Southwest Forest and Range Experiment Station's Riverside Fire Laboratory (FIRESCOPE Program), and Gradient Modeling, Inc.; and a research contract between the National Parks and Wildlife Service of New South Wales (Australia) and the author.

Some of the research described here was first reported in Springer-Verlag's journal *Environmental Management*. Articles cited include:

Kessell, S. R. 1976. Gradient Modeling: a new approach to fire modeling and wilderness resource management. Environmental Management 1(1): 39–48.

Kessell, S. R. 1976. Responsible management of biological systems? Environmental Management 1(2):99–100.

Kessell, S. R., and P. J. Cattelino. 1978. Evaluation of a fire behavior information integration system for southern California chaparral wildlands. Environmental Management 2(2):135–157.

Kessell, S. R. 1978. Perspectives in fire research. Environmental Management 2(4):291–294.

Kessell, S. R., M. W. Potter, C. D. Bevins, L. Bradshaw, and B. W. Jeske. 1978. Analysis and application of forest fuels data. Environmental Management 2(4):347–363.

Kessell, S. R. 1979. Fire modeling, fire management, and land management planning. Environmental Management 3(1):1–2.

Kessell, S. R. 1979. Phytosociological inference and resource management. Environmental Management 3(1):29–40.

Cattelino, P. J., I. R. Noble, R. O. Slatyer, and S. R. Kessell. 1979. Predicting the multiple pathways of plant succession. Environmental Management 3(1):41–50.

Potter, M. W., S. R. Kessell, and P. J. Cattelino. 1979. *FORPLAN: A FORest Planning LANguage and Simulator.* Environmental Management 3(1):59–72.

Contents

SECTION I

RESOURCE MANAGEMENT INFORMATION SYSTEMS

1

Introduction

Historically, resource management has entailed the preservation of the status quo. Resource demands, legal constraints, economic considerations, and conflicting public pressures taxed neither the land itself nor its stewards. Passive management, which minimized negative interferences (fires, insects, overcutting, or overgrazing), was sufficient; active management, which would have maximized the output of nature, was unnecessary. For many nations, that chapter of history has come to an inevitable close.

Today's manager faces a demanding and informed public that voices conflicting concerns, a mountain of legal mandates and administrative requirements, skyrocketing management costs despite fixed budgets, and conflicting demands for preservation, conservation, and exploitation. "Sustained yield" once implied continued timber output—now we require a sustained yield of recreation opportunities, wildlife, watershed quality, and even wilderness solitude. "Multiple use" is no longer simply reserving one watershed for timber production, another for wildlife and recreation. "Wilderness" is no longer an inaccessible area where no one has yet bothered to harvest trees or remove minerals. "Recreation" no longer implies a few campsites for the occasional hiker or family outing. We have reached the point where active, aggressive, positive land management planning is required by public pressure, economic reality, legal mandate, and ecologic necessity (Kessell 1978, 1979c).

Nowhere is this change more evident than in the development of simulation models and information systems for resource management, especially in the integration of fire management considerations into the resource management and land management planning processes. Why are land managers increasingly involved in such activities as modeling, simulation, information systems analysis, and data processing? Franklin (1979) responds to this question with the following commentary:

> Intuition and knowledge, developed from experience and simplistic models, based upon one to a few factors (and often not recognized as models), have served us well in the past. Why involve ourselves in these complex computer simulations? . . . I believe the reasons for our increasing involvement with simulation models are found in the types and numbers of questions land managers are asking. Let's consider a few examples.
>
> On a hot afternoon in July a fire control officer in Glacier National Park receives a report of a fire. What suppression activity will be undertaken? It is an area where there is the option of letting the fire burn unimpeded. Key questions concern rate and direction of fire spread. How big will it be in an hour, six hours, 24 hours from now? Many factors enter into any predic-

tion of this type including the fuels in the vicinity of the fire, current and predicted weather, topography, etc. . . .

Staff professionals in wildlife and silviculture on a western National Forest consider long-term needs for large, standing dead trees or snags, critical for many wildlife. What prescriptions will provide the necessary continuous flow of these structures as most natural stands are converted to a managed state, as the currently reserved snags decay and fall, as shorter rotations reduce average tree size?

In a western National Park the superintendent considers alternative fire management policies. If a policy of essentially complete fire suppression is continued what probable appearance will the park landscape have 50, 100, 200 years from now? Will persistence of individual species be endangered? What will be the impact on insect epidemics, visitor safety, pre-suppression activities?

Restoration of the natural vegetation is adopted as the long-term management objective within a small Oregon estuary. Most of the original salt marsh has been diked for agricultural use for many years. How should the responsible manager implement the restoration policy? Are extensive mudflats a necessary early result of dike breaching? What timing and techniques will lead to the desired result with minimal ecologic and aesthetic damage? (Franklin 1979).

Franklin notes that the land management problems represented by his examples are diverse in subject, scale, and relevant research methods; nevertheless, they not only share several important features but also represent a major class of questions being asked by today's land manager.

1) They are complex questions, deal with the multiple ramifications of management activities, and impact whole ecosystems. When economic and social consequences are added to the biologic and physical, complexity increases further; large amounts of diverse information are required in the formulation of a solution.
2) Alternative strategies need to be explored. Managers must evaluate their options, seek optimal solutions in resource conflicts, and consider tradeoffs.
3) There is a need to predict the short- and long-term outcome of the various alternatives, often under conditions involving considerable uncertainty.

The sheer volume and complexity of today's resource management questions and the information required for their solutions dictate the use of resource information systems, automatic data processing (ADP) techniques, and simulation modeling. These techniques further allow us to handle vast quantities of complex data, to evaluate the consequences of various management options without incurring the costs or consequences of implementing each option, and to predict the outcomes of various alternatives. These are tools that allow us to organize, abstract, stratify, simplify, interrelate, store, retrieve, and use large amounts of information.

They show us the implications of our assumptions, help us predict and compare the outcomes of various actions (or inactions), and provide us with methods of testing our goals and our strategies. They are no panacea; models are only as sound as the understanding and data that go into their construction. They allow us to manipulate and use current knowledge; they can produce no new knowledge—only improved understanding. Quite simply, they help us efficiently meet our exploding need for information applied to real world problems (Franklin 1979, Kessell 1976b, 1977, 1978, 1979b,c).

Section I of this book reviews the information and simulation needs of resource management and land management planning (Chap. 2), and the basic concepts and specific systems that have been advanced to meet these needs of both activity (project) management and long-range land management planning (Chap. 3). This section thus attempts to integrate two basic problems: What do land managers need, and how have researchers attempted to meet these needs?

In 1973, I proposed a new approach to resource information systems and simulation which linked gradient analysis vegetation and fuel models with remote site-specific inventories and computer software. Called *gradient modeling,* the method attempted to better meet management information system and simulation requirements within specified accuracy, resolution, and cost constraints. The system was originally designed to provide a resource management—fire management information system for Glacier National Park, Montana, USA (Kessell 1973, 1976a) and was implemented in that park in 1976 (Kessell 1976b, 1977). Subsequent applications of the gradient modeling method, with varied accuracy, resolution, and cost constraints, have been made for fire management modeling in southern California chaparral ecosystems (Kessell and others 1977a, Kessell and Cattelino 1978) and for multiple-use management modeling in the Pacific Northwest of the United States (Kessell and others 1977b, 1978, Cattelino and others 1979). The basic gradient modeling method inspired the development of *FORPLAN,* a *FOR*est *P*lanning *LAN*guage and Simulator that is programmed in common English words, incorporates unique characteristics of previous systems, links numerous models and data bases, allows selection of variable resolution levels, and permits discrete-time simulation of disturbances on plants, fuels, and animals (Kessell and others 1977a, Potter and others 1978, 1979). Land management agencies from several countries are now evaluating this system for application to forest, grassland, brush, and semitropical ecosystem management. Our latest system, *PREPLAN* (*PR*istine *E*nvironment *P*lanning *LAN*guage and Simulator), combines basic gradient modeling capabilities with FORPLAN's English language vocabulary. I am currently implementing this system in Kosciusko National Park, Australia, under contract with the National Parks and Wildlife Service of New South Wales.

Section II provides a detailed description of the research conducted in

Glacier National Park from which we have built the gradient modeling information system. An introduction to the park and its communities (Chaps. 4 and 5) is followed by description of the field methodology (Chap. 6), analytical methodology (Chap. 7), and the development of the gradient models (Chap. 8). The results of this research are presented in Chaps. 9–12; Appendix 2 includes habitat utilization population nomograms for all overstory species and for common understory species.

Section III describes how the results of this extensive terrestrial gradient analysis have been used to build the information system. Chapter 13 describes the basic components of the gradient modeling system and Chapter 14 presents the detailed methodology showing how components have been selected, constructed, and linked. Chapter 15 discusses our testing and evaluation efforts and Chapter 16 describes the computer program and software package.

Section IV highlights gradient modeling applications subsequent to the implementation of the Glacier National Park system. It includes a discussion of the accuracy and resolution requirements posed by different management applications and needs (Chap. 17), recent work on modeling vertebrate habitat utilization (Chap. 18—based largely on Thomas and others 1977), and the recent application of gradient modeling techniques to the development of *FORPLAN* and related systems (Chap. 19). Conclusions (Chap. 20) center on the pressing need for better communications and for the transfer of new technology between researchers and resource managers.

2

The Needs of Resource Management and Land Management Planning

Sound resource management requires knowledge of the land area, an accurate and quantitative inventory of the biota, an understanding of the interactions among the various components, and the ability to predict or simulate changes to the system resulting from both natural disturbances and management actions. In addition to the broad land management planning problems cited in the last chapter, the manager daily must face such questions as:

> What are the distribution and relative abundance of rare or endangered species in an area proposed for development?
>
> What buildings or trails in a national park are in areas subjected to frequent flooding?
>
> What are the fire behavior predictions for a prescribed or natural fire needed for a decision that must be reached in a few minutes? What could the fire destroy? How much fuel is present now? How much will be present after the fire? What will be the postfire successional communities? What is the anticipated increase in natural fuels, and thus the fire hazard, if the fire is suppressed?
>
> Where can stands with abundant huckleberries or other prime grizzly bear habitat be found? Are these areas adjacent to popular hiking trails?
>
> What management actions can help restore an endangered wolf population?
>
> If prescription burning is used to reduce fuel loads and fire hazard, will it impact the grizzly bear and wolf populations?

Answers to such questions are becoming increasingly important to the resource manager. All of these questions demonstrate the need for a resource information system and the ability to synthesize component data to provide specific answers. In the absence of such a system, the manager must make a decision without the aid of all the available and necessary information. Working without the necessary data, even the best, most experienced, best intentioned manager is bound to guess wrong on too many occasions.

Inventory, modeling, and resource information systems that attempt to meet the needs of resource management must be (Kessell 1976a):

1) *Comprehensive:* include all types of data and parameters necessary to understand the problem

2) *Accurate:* include detailed ecologic descriptions showing the dynamic ecologic processes within an area with the required resolution
3) *Flexible:* applicable to a wide range of problems by easily manipulating and integrating various kinds of information
4) *Synthetic:* capable of drawing together diverse information to make predictions, such as integrating site, fuel, vegetation, animal, weather, and cultural information to simulate the effects of a forest fire
5) *Problem oriented:* permitting selective retrieval of only that information needed to solve a particular management problem
6) *Accessible:* available to the manager in an easily and quickly interpretable form when it is needed
7) *Economical:* both development and application costs must be within the manager's budget.

The availability of resource management systems that meet all of these requirements is limited; in fact, they have slowly become available only in the past few years. Furthermore, many of these goals for resource information systems are conflicting. For example, systems that are expanded to make them more comprehensive, more accurate, more flexible, and more synthetic often become so only along with more limited accessibility and increased development and operating costs.

Additional requirements are posed, and constraints placed, by the management goals for each area. Management of the natural resources within our national parks and wilderness areas requires action to preserve, maintain, and often restore pristine ecosystems; however, the legislation that has established these areas may preclude the use of certain desired management actions (such as prescribed burning). As noted in the last chapter, the multiple-use areas of our national forests often pose conflicts among the various uses; a management action aimed at optimizing timber production can have devastating impacts upon wildlife (Thomas and others 1977). Interfaces between wildlands and suburban areas pose the still different problem of dual encroachment. The public's proximity, use, and pollution often impact the wildlands, whereas natural agents of the wildlands, such as fire, are unacceptable when they threaten the urban area (Phillips 1977; cf. Storey 1972, Kessell and Cattelino 1978).

In addition to being a required input to the formulation of management action decisions, information systems and simulation models are prerequisites to the long-term land management planning process. Public land managers are required to balance various land uses, to minimize conflicts, to provide for a sustained yield of various outputs (timber, rangeland, recreation, wildlife, watersheds, etc.), and to protect the public's interests. This is accomplished by numerous management activities and action decisions, which interact with each other and frequently cause environmental change. In the United States all of these activities and their interactions must be investigated before they are implemented (National Environmental

Policy Act of 1969). This evaluation is ideally accomplished through the land management planning process (Egging and Barney 1979).

As described by Egging and Barney (1979) and shown in Fig. 2.1, land management planning includes four basic inputs:

1) Public demands for goods and services
2) Administrative and legal constraints
3) Fiscal constraints
4) The resource potential (i.e., ecologic constraints).

In the planning process, the manager must balance the public demands with the administrative, legal, economic, and ecologic constraints; determine the use allocations and management activities that will accomplish this balance; and direct the quality, quantity, and timing of the management activities. This process requires an intimate understanding of the ecosystems being managed. The process is successful only when the manager can predict the outcome of each allocation and its included management actions and can superimpose the potential effects of unplanned, uncontrollable impacts, such as fire, insects, unusual weather, and the like. The planning process requires estimating the answers to numerous "What if?" questions and will not succeed if such questions cannot be answered. Clearly, resource inventories, information systems, and especially the ability to predict the consequences of management actions are extremely critical inputs to the land management planning process.

Answering such "What if?" resource management and land management planning questions becomes especially critical in the area of fire management. Barney (1975) defined fire management as "the integration of fire related biological, ecological, physical, and technological information into land management to meet desired objectives," or simply "the integration of fire considerations into land management planning" (Kessell 1979c, Lotan 1979). Barrows (1977) proposed that:

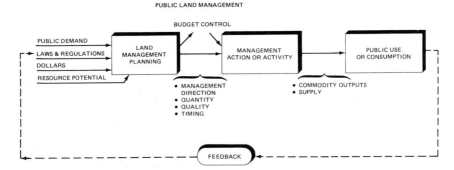

Figure 2.1. Simplified flowchart of land management planning, action, and use. (Egging and Barney 1979)

1) Fire management activities be governed by resource management objectives;
2) Both beneficial and damaging aspects of fire be considered;
3) Fire effects be evaluated for various fire management activities;
4) A systems approach be involved linking the various fire management activities.

Lotan (1979) further discusses why we should integrate fire into land management planning:

> We will not have adequate fire management until all activities of a fire organization—including fire prevention, control, and beneficial uses of fire—are directed by land management objectives. Fire is a complex and powerful force in natural ecosystems. Its impacts are varied and affect the ecosystem over long periods of time. The soil, water, air, and vegetation are all affected. Hence, many products and uses of the ecosystems are affected. Because it cuts across so many resource management boundaries and affects both short-term and long-term resource outputs, the only effective way to deal with fire is on a multi-resource, multi-objective basis in those countries where multi-resource objectives are sought. Further, because fire does not respect property boundaries, planning must consider objectives of all land-owners in the process.
>
> From a planning viewpoint, we must be able to predict the consequences of various fire management activities. To predict fire effects is important not only from the ecological perspective but also to judge the trade-offs involved in meeting complex arrays of resource demands.
>
> Our ability to predict the effects of fire requires a knowledge of preburn conditions, the particular kind and amount of fire involved, and the response of the ecosystem over time. To be meaningful in a planning context this needs to be done for some stratification of land and vegetation (whether by classification methods or environmental gradients—see Kessell 1979b and Franklin 1979) together with some measure of distributions of the consequent results. . . .
>
> It is essential that fire considerations be included in the planning process and that objectives of fire organizations be established in the process where demands of society are addressed. . . . The premise here is that it is in the land management planning context that the need for fire management expenditure should be made, the economic and ecological implications be evaluated, and more specifically that fire management alternatives be evaluated. . . . We must be able to predict consequences of various fire management activities. We must know in advance what various fires will do to resources or values for which we are managing. . . .

This recent increased interest in fire management in the United States, Canada, Israel, Australia, and elsewhere demands:

A better understanding of fire behavior, fire ecology, and fire effects
A better understanding of the short- and long-term effects of fire exclusion

> The ability to predict these consequences in advance of a fire
>
> An increased responsibility on researchers to communicate their findings to managers, and on managers to communicate their needs to researchers
>
> Most importantly, the mandate to integrate these considerations into the land management planning process

This, in turn, has spurred fire research during the past few years, especially in the development of resource management simulation models and systems that assist the integration of fire considerations into land management planning (Kessell 1978, 1979c). In addition to meeting their primary goal, such systems also have provided numerous other benefits, for the understanding required to predict fire effects also aids us in predicting the effects of other natural disturbances and management activities. Furthermore, because of the exponential increases in fire suppression costs (Lotan 1979), these developments have helped demonstrate the cost-effectiveness of resource information systems.

The following chapter reviews some of these modeling methods and information systems, shows how such components are used to construct integrated systems, and suggests how disturbance modeling techniques may be applied to simulate the ecologic effects of fires, other natural disturbances, and various management activities.

3

Review of Some Methods, Information Systems, and Models

Chapter 1 highlighted the extensive problems facing land managers and suggested that modeling, simulation, and resource information systems were valuable tools for solving some of these problems. Chapter 2 continued that discussion of management problems, pointed out the desirable features of management models and systems, and noted that different geographic areas, management priorities, and planning requirements also affect basic information requirements. An unstated assumption of those discussions was that systems are available, or at least can be constructed, to provide the information needed for resource management and land management planning. I believe that assumption to be true and I review in this chapter some of the available methods, models, and systems that have been advanced to meet these management needs. First, however, a brief description of the modeling process is in order.

Modeling and Simulation

Mathematics provides us with ideas and tools that enable us better to describe and understand the real world around us. A useful application of mathematics in real world studies is the process of data collection and analysis, theory construction, and event prediction, or, in other words, the construction, development, study, and application of models (Maki and Thompson 1973).

A model is simply a substitute expression of reality. It is usually simpler than the real world object or process it represents. I built my first model at an early age; it consisted of a fishhook, some silk thread, some yarn, and a hackle and was intended to convince a brook trout that it was a tender, juicy mayfly. Other modeling attempts included mimics of a P-51 Mustang fighter aircraft and a Canadian Northern railroad train. Many years later I tried to model plant communities, succession, and forest fires.

Table 3.1 gives some basic steps in the modeling process. They include the collection and analysis of data, the formulation of theories, the construction of the "real" (physical) model, the formulation of a mathematical expression of that model, the study of the system or model produced, and finally the very important comparison of the results predicted by the model with the real world. Interestingly, the basic steps in modeling a small insect by tying a dry fly and in modeling a forest fire's effects on plant communities by writing a computer program are quite similar, as shown in Table 3.2.

Table 3.1. The Basic Steps of Modeling.

1. Collection and analysis of data
2. Formulation of conjecture about system to be studied
3. Construction of the "real" model:
   Making the problem as precise as possible by
   a. Identification
   b. Approximation
   c. Idealization
4. Formulation of the mathematical model   (expression of "real" model in symbolic terms)
5. Study of system using mathematical ideas and technique
6. Comparison of results predicted by mathematical model with real world

Table 3.2. The Basic Steps of Modeling Applied to Predicting Postfire Successions and Catching Trout on an Artificial Fly.

Modeling a forest fire for the purpose of predicting ecologic effects	Modeling a mayfly for the purpose of catching a trout
1. Collection and analysis of data Field work to determine the distribution and abundance of species, their successional roles, their adaptive traits, their response to fire	Field observation that trout are eating mayflies and on the size, shape, color, and appearance of a mayfly
2. Formulation of conjecture about system to be studied Analysis of data collected; formulation of theories about postfire successional changes	Analysis of mayfly appearance; formulation of theory that a good replica will lure trout
3. Construction of the "real" model Determination of successional pathways following fires of different intensities	Construction of mayfly replica from fur and feathers
4. Formulation of the mathematical model Express the successional relationships and pathways as mathematical functions in a computer program	Not necessary
5. Study of the system What traits does the model exhibit; i.e., what are the implications of its assumptions?	Study the artificial mayfly; does it reasonably replicate the real thing? (no mathematics necessary)
6. Comparison of results Is it realistic; does it mimic postfire succession and produce accurate predictions? Field test in Glacier National Park, Montana	Is it realistic; does it mimic the real mayfly and lure trout? Field test in Rangeley Lake, Maine

These basic steps of modeling, based on Maki and Thompson (1973), are not "cookbook" instructions, but a simplification and idealization of how models are constructed—a "model" of the modeling process, if you like. Using the postfire succession example given in Table 3.2, let us briefly review this process.

Once we state our real world problem and the purpose of the model (e.g., to understand and predict postfire successions), we collect and analyze field data and observations; perhaps some statistical methods are applied at this point. Next, we use these results to formulate conjectures and theories—to speculate on what our data appear to tell us about the successional process. In doing this, we begin to construct the "real" model; we will make some approximations and idealizations to keep our real model manageable. We may also attempt to generalize our findings and perhaps make some assumptions [e.g., that succession works as described by Clements (1916) or, alternatively, that a monoclimax, single-pathway view of succession is unrealistic]. Next we conduct the often creative process of constructing the mathematical model—the entire situation is expressed in symbolic terms. Study of the mathematical model can then produce:

> . . . theorems, from a mathematical point of view, and predictions, from the empirical point of view. The motivation for the mathematical study is not to produce new mathematics, i.e., new abstract ideas or new theorems, although this may happen, but more importantly to produce new information about the situation being studied. In fact, it is likely that such information can be obtained by using well-known mathematical concepts and techniques. The important contribution of the study may well be the recognition of the relationship between known mathematical results and the situation being studied (Maki and Thompson 1973).

The final step of modeling is the comparison of the results predicted by the model with the real world.

> The happiest situation is that everything actually observed is accounted for by the conclusions of the mathematical study and that other predictions are subsequently verified by experiment. Such agreement is not frequently observed, at least not on the first attempt. A much more typical situation would be that the set of conclusions of the mathematical theory contains some which seem to agree and some which seem to disagree with the outcomes of experiments. In such a case one has to examine every step of the process again. Has there been a significant omission in the step from the real world to the real model? Does the mathematical model reflect all the important aspects of the real model, and does it avoid introducing extraneous behavior not observed in the real world? Is the mathematical work free from error? It usually happens that the model-building process proceeds through several iterations, each a refinement of the preceding, until finally an acceptable one is found. (Maki and Thompson 1973).

This iterative process of modeling is shown in Fig. 3.1.

A final note on model construction is the choice between a deterministic and a stochastic process. A model is called *deterministic* when it predicts that if sufficient information is given for one point in time, then the entire future behavior of the system can be predicted. Alternatively, a model is called *stochastic* if it incorporates probabilistic behavior; i.e., no matter how much we know about the system at any time, it is impossible to predict with absolute certainty its future behavior. Note that the gradient modeling process described in this book is essentially a deterministic model; however, certain stochastic elements are introduced in subsequent chapters.

Some authors use the terms "modeling" and "simulation" interchangeably; I do not. A simulation, or a simulation model, is simply a model of a system through time. For example, the toy airplane and the trout fly are models, but when I throw the plane through the air or retrieve the fly across a stream, I am simulating. Two basic types of simulation are *discrete-time simulation*, where the predictions are for specific, fixed points in time, and *continuous-time simulation*, which predicts over the entire continuum of time. Conceptually different, the two methods often converge in real modeling applications, as shown in Chapter 14.

Virtually all of the models discussed in this book are simulators, for a model that did not predict events through time would have (with a few special exceptions) little application in resource management. The latter part of this chapter views some current resource management simulators, with emphasis on fire simulators, succession simulators, plant and animal habitat utilization models and simulators, and integrated systems. Before we discuss these specific tools, however, let us view some more basic considerations: How are community data obtained, abstracted, stratified, generalized, stored, retrieved, and used in the resource management–land management planning decision-making process, and how does this process place accuracy and resolution constraints upon the data, their abstraction, and their ultimate use for simulation?

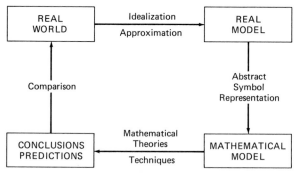

Figure 3.1. The basic process of modeling, showing comparison to and feedback from the real world situation. (Based on Maki and Thompson 1973, p. 4)

Community Stratification and Abstraction

Plant ecologists and resource managers seek descriptions and understanding of complex vegetation patterns and the changes in these vegetation patterns that result from natural processes and management actions. Various applications pose widely varying requirements for accuracy and resolution. Furthermore, while many are interested in vegetation in its own right, others are concerned with using the information to infer other ecologic properties. Flammable fuels, animal habitat, potential productivity, and watershed quality are all important management concerns that are frequently inferred from phytosociological data.

In his introduction to *Ordination and Classification of Communities,* Whittaker (1973a) emphasizes that understanding is based on abstraction:

> From phenomena that are often intricate and obscure, significant relationships are to be detected, embodied in concepts, related to one another, and tested and revised in the systems of abstraction that are science. As much as any other phenomena, plant communities offer complexities that challenge our efforts at abstraction and understanding.

Historically, two broadly conceived research methods have evolved to allow stratification and abstraction of plant communities. The first method is *classification,* in which a number of samples representing communities are grouped together on the basis of shared characteristics into an abstract class of plant communities. Such a grouping of communities, on any level, by any definition of shared characteristics, is referred to as a *community type* (Whittaker 1973a, b). The second method, *gradient analysis,* deals not with discontinuous classes but with continuity and gradient relationships. Samples from communities are arranged along one or more gradients or continua of environment and/or community characteristics. When the arrangement is along a predetermined environmental gradient, the method is termed *direct gradient analysis.* Alternatively, *indirect gradient analysis* is the arrangement of samples along abstract axes that may or may not correspond to environmental gradients. The process of arranging samples along one or more environmental gradients is called *ordination* (Goodall 1954). The following subsections review classification and gradient analysis in view of their application to resource management problems (based on Kessell 1979b).

Classification

Traditionally, plant ecologists and resource managers have viewed natural communities in terms of categories or community types. Each unit of the landscape is assigned membership within a class and is then viewed as possessing the attributes of that class. The definition and description of each community type depends on the interests, training, and prejudices of the classifier; the anticipated use of the system; and the classifier's requirements for community-type accuracy and resolution.

> If he is most interested in vegetation structure, he may state a definition of a formation: Forests dominated by needle-leaved trees in subarctic and subalpine climates are members of community-type (formation) *A*. If his interests center on floristic composition, he may define an association: Plant communities in which several of the following character-species . . . occur together belong to community-type (association) *B*. If he is concerned mainly with dominant species, he may define a dominance-type: Grasslands in which the two species *a* and *b* are more important than any others represent community-type (dominance-type) *C*. Such statements of class-concepts are also intensional definitions of community-types. The particular communities in the field for which a defining statement becomes true, which therefore conform to the class-concepts, constitute its extensional definitions. (Whittaker 1973b)

Obviously, the utility of such a system depends both on the resolution of the classification method (number and refinement of community types recognized) and on the correspondence between the real landscape and the intensional definition of the classes into which landscape units are assigned.

Despite the American preoccupation with canopy dominance and "habitat types" (a community type that indicates community potential through a "sociation" of a canopy and understory species) (Cowles 1899, 1901, Clements 1905, 1916, 1936, Clements and Shelford 1939, Daubenmire 1952, 1954, 1966, 1968, Pfister and others 1977; reviewed in Whittaker 1962, 1973b, c), five major historical traditions have led to the development of 12 major schools of vegetation classification (Whittaker 1962, 1973b). These include:

1) The physiognomic, structural, or formation-type approach (reviewed by Beard 1973)
2) The environmental or biotope-type approach (Merriam 1894, Emberger 1936, 1942, Holdridge 1947, 1967, Elton and Miller 1954, Dansereau 1957)
3) The landscape-type or biogeocoenose-type approach (including both landscape and microlandscape levels, as reviewed by Whittaker 1973b)
4) The biotic area or province approach (Whittaker 1973b)
5) The life-zone and segments of community gradients approach (Merriam 1890, 1898, 1899, Kendeigh 1954, Beard 1955, 1973, Whittaker 1973b, Aleksandrova 1973)
6) The dominance-type approach (reviewed by Whittaker 1962, 1973c),
7) The vegetation dynamics approach (Tansley 1911, 1920, 1939; Clements 1916, 1928, 1936, Weaver and Clements 1929, Whittaker 1953, 1974)
8) The stratal or lifeform approach (Gams 1918, 1927, Lippmaa 1933, 1935, 1939, Barkman 1958; reviewed by Barkman 1973)

9) The stratal combination or sociation approach (Fries 1913, Du Rietz 1921, 1932, 1936, Daubenmire 1952, 1966, Aleksandrova 1973, Trass and Malmer 1973)
10) The northern European site-type (understory) approach (Cajander 1909, 1949, Cajander and Ilvessalo 1921; reviewed by Frey 1973)
11) The numerical classification approach (reviewed by Goodall 1973)
12) The floristic units approach, best represented by the school of Braun-Blanquet (1913, 1921, 1932, 1951, 1964; reviewed in English by Westhoff and Maarel 1973)

Obviously, there is no single "natural," "objective," or "correct" method of classifying vegetation. The various methods of defining community types imply different classifications of the same vegetation (Kessell 1976a, Whittaker 1962, 1973b). An excellent example is provided by Ellenberg's (1967) comparative classification/mapping project. Particular classification systems can be evaluated only in terms of how well each provides the information needed by the resource manager for a particular application. For example:

1) An excellent physiognomic classification is of little value to a manager concerned with the preservation of specific endangered herbaceous species.
2) An excellent habitat-type classification offers little insight into the details of species importance changes through postfire succession.
3) A low-resolution landscape-type classification cannot meet the needs of a manager attempting to infer fuel properties needed by a real-time fire behavior model.
4) A high-resolution numerical classification would be overly detailed and prohibitively expensive for a planner dealing with a 100 000 km² area.

Probably every classification method reviewed above, when built from an appropriate data base, designed at an appropriate resolution level, and stored in an appropriate form, could be very valuable in solving some pressing resource management problem. However, before I describe the linkage among classification systems, data sources, and storage/retrieval methods, let us also review another approach to stratifying vegetation information: gradient analysis.

Gradient Analysis

An alternative to classification is to view and describe plant communities in terms of the spatial and temporal environmental gradients that affect their composition and structure. Originated independently by Ramensky (1926, 1930) and Gleason (1926, 1939), gradient analysis has achieved its present sophistication as a result of extensive independent development in the United States by Bray and Curtis (1957, Bray 1956, 1960, 1961) and Whittaker (1956, 1960, 1967, 1970, 1973d, e, Whittaker and Niering 1964,

1965). Recent years have also seen the accelerated development of new mathematical tools for arranging samples along environmental axes; some of these more important ordination techniques are reviewed below. Despite the attractive theoretical and quantitative qualities of gradient analysis, however, until recently it had found little application to resource management problems. Most managers unfortunately dismissed it as an esoteric exercise performed by theoretical phytosociologists, while too many of its proponents did little to dispel this view.

In 1973, I introduced the concept of gradient modeling (Kessell 1973, 1976a, b, 1977), the first extensive application of gradient analysis to the needs of resource management information systems. Gradient modeling is nothing more than the linkage of a multidimensional gradient analysis with a site-specific stand inventory and appropriate computer software. Yet this linkage provides capabilities that vastly exceed those of either a gradient analysis or a stand inventory alone. The linkage works in this fashion:

1) The *gradient models* (from the gradient analysis) can provide quantitative community inferences (species importance values, fuel properties, and so forth) if the location of each stand within the gradient matrix is known. To locate an individual stand within this gradient matrix, one must know its location on each gradient (for example, its elevation, its aspect, and the number of years since the last disturbance).

2) A *remote site-specific inventory,* determined from topographic maps, aerial photography, and disturbance history, can provide the location of each stand on each gradient. This inventory is then stored on a digital computer.

3) An appropriate *computer program* can link these two components.

Thus, the completed system permits the user to enter the geographic coordinates of a stand and obtain from the program species importance values, fuel parameters, and so forth.

The original application of gradient modeling, described fully in Sections II and III, was in Glacier National Park (Montana), USA and included additional modules to infer fuel moistures, fire behavior, and the postfire succession of plants and fuels. The system possessed very high resolution (10–20 m) but was moderately expensive to develop and implement ($1.50–1.00 per hectare). A subsequent application of gradient modeling in southern California, USA, demonstrated the applicability of the method at lower resolution levels (100–1000 m) and correspondingly reduced costs ($0.20–0.02 per hectare, respectively) (Kessell and others 1977d, Kessell and Cattelino 1978). Recently the method has been applied to the construction of fuel inference models (Kessell and others 1978), multiple-pathway succession models (Cattelino and others 1979), and integrated forest management simulators (Kessell and others 1977a, Potter and others 1979) in the United States and Australia.

Just as the union of an appropriate classification technique, data base, resolution level, and storage and retrieval mode is the key to successful application of community classification to resource management problems, the correct construction and linkage of gradient model components is vital to a successful application of this method. Furthermore, the development of a gradient model often requires an intricate interplay of direct gradient analysis and ordination techniques, the applicability and usefulness of which are dictated by data quality and quantity, species and communities properties, and the required resolution level. A brief review of some of these techniques follows.

Numerous ordination techniques have been introduced, reviewed, and evaluated in recent years. Let us consider four of the more important methods: Bray–Curtis (Wisconsin comparative) polar ordination (PO), principal components analysis (PCA), Gaussian ordination (GO), and reciprocal averaging (RA). More extensive reviews are provided by Gauch and Whittaker (1972a, Whittaker and Gauch, 1973), Kessell and Whittaker (1976), Gauch and others (1977), and Noy-Meir and Whittaker (1977).

Bray–Curtis PO (Bray and Curtis 1957; also see Cottom and others 1973, Gauch 1973a) arranges samples by their similarity to two endpoint samples using a pythagorean algorithm; the user may select the choice of sample similarity measurements, endpoints, and relativization method. Endpoints may also be selected by objective criteria (Bray and Curtis 1957). Despite its simplicity, evaluation shows PO to be one of the most reliable and flexible techniques available (Gauch and Whittaker 1972a, Kessell and Whittaker 1976).

Principal components analysis (Orloci 1966, 1973, 1975) uses eigenanalysis to extract axes of variation. Although it objectively selects axis directions, it assumes a linear relationship between species importance values and the axes and so performs very poorly with the typical curvilinear and polytonic species distributions observed in the field (unless the samples are very homogeneous and show low beta diversity [species turnover rates] along the axes) (Whittaker and Gauch 1973, Kessell and Whittaker 1976).

Gaussian ordination (Gauch and others 1974) assumes Gaussian (bell-shaped) species distributions instead of linear relationships. Both formal testing (Gauch and others 1974) and application to field data (Chap. 7) show it to be an effective technique if the majority of the species distributions are approximately Gaussian. However, results can be ambiguous or even misleading if it is applied to data sets that contain several bimodal or polymodal species distributions (Chap. 7).

Reciprocal averaging (Hill 1973a, Gauch and others 1977) is described as a weighted-average ordination effected by successive approximations. Species are weighted by positions along a rough initial gradient; the weights are used to calculate sample scores. These sample scores as weights are then used to calculate an improved calibration of species. After numerous iterations, a stable, optimal solution is obtained. The initial configuration does not affect the final result but does affect the number of iterations required. Evaluation of RA (Gauch and others 1977), using simulated community gradients (from Gauch and Whittaker 1972b), simulated multi-dimensional community patterns (from Gauch and Whittaker, 1976), and field data sets, shows it to be one of the more powerful ordination techniques available.

The results of numerous evaluations of these and other ordination techniques, using simulated data primarily, reveal PO and RA to be the best available techniques. Principal components analysis performs poorly with noisy data and/or high beta diversity (rapid turnover of species along a gradient). Gaussian ordination has not, in my opinion, received sufficient testing. For that matter, none of these techniques has received adequate testing with real field data. Too often, the results of tests with simulated data reflect the assumptions of the data simulation as well as the effectiveness of the ordination methods. The reader is referred to the reviews and evaluations cited above and to Chapter 7 for further details.

Integrated Methods

As noted above, the successful construction of a gradient modeling system requires an intricate interplay between direct gradient analysis and ordination techniques. To effect a direct linkage between gradient models of the vegetation and a site-specific stand inventory, species distributions must be expressed along direct, measurable gradients so that a stand's location on each gradient can be determined by remote methods. Often the researcher is unaware of all the important gradients in the study area and must rely on ordination techniques to identify them (Chaps. 7 and 8). However, the research cannot stop there; once new axes are identified, they must be expressed in terms of measurable gradients before the system can be completed.

Similarly, classification can have a significant role in the construction of gradient models. The researcher is likely to encounter some discontinuities in the data sets that are best treated as classification units (community types). Alternatively, he or she may purposely impose community types on a continuum or near-continuum to simplify and streamline the system.

In this same fashion, the development of a sound classification system often requires the use of ordination methods. Discontinuities in the natural vegetation are sought for the purpose of determining the boundaries of the community types recognized. These are often best determined objectively by employing the methods of gradient analysis and ordination. The development of the Montana habitat-type system (Pfister and others 1977) is a good example of the successful use of ordination in developing a classification system.

One should therefore view classification and gradient analysis as complementary philosophies and methods. They need be in no way antagonistic or mutually exclusive methods; rather they can and should be combined in various ways for an effective study of natural vegetation and application to resource management problems. An understanding of existing, potential, and successional vegetation requires that the stand be described in both an environmental context and a temporal (successional) context (Franklin, 1979). Whether this is accomplished with a gradient model or by stratification of community types along a successional sequence is quite irrelevant if the system accomplishes its management purpose. More important considerations in the selection of stratification methods are the source of the vegetation data, the intended storage and retrieval mode, the accuracy and resolution requirements, and the intended use of the system. These considerations are discussed in the following sections.

Storage and Retrieval of Vegetation Information

Effective utilization of a vegetation information system for resource management purposes requires an efficient and appropriate method for data storage and retrieval. Two primary methods are available: graphic representation (mapping) and digital inventories (usually stored on a digital computer). Although mapping is usually associated with a classification system, and gradient models with digital data bases, this need not be the case, as shown below.

Mapping

As noted by Mueller-Dombois and Ellenberg (1974), the contents and scales of both existing vegetation maps and site potential maps depend primarily on three factors:

1) The objectives for which the map is prepared
2) The detail and accuracy of the underlying information
3) The funds available to the project.

The following discussion of vegetation mapping is based in part on reviews by Mueller-Dombois and Ellenberg (1974) and Küchler (1967). An excellent worldwide bibliography of published vegetation maps has been compiled by Küchler and McCormick (1965, Küchler 1966, 1968 1970).

Vegetation maps are constructed from direct field samples and/or remote sensing materials, such as aerial photographs and satellite imagery. Mueller-Dombois and Ellenberg recognize five major scales of vegetation maps:

1) Small-scale maps for a general overview at a scale of 1:1 000 000 or smaller. Such maps usually present major formation (cover) types. Examples include Küchler's (1964, 1965) 1:7 500 000 map of the "Potential Nat-

ural Vegetation" of the United States, Rowe's (1959, 1972) 1:6 400 000 map of the "Forest Classification" of Canada, and Schmithusen's (1968) 1:25 000 000 map of world vegetation.

2) Intermediate-scale maps for regional orientation at a scale of from 1: 1 000 000 to 1:100 000. Such maps permit the generalized representation of floristically defined vegetation units. Examples include Gaussen's 1:200 000 map of the "Vegetation of France" (cited by Mueller-Dombois and Ellenberg 1974) and Mueller-Dombois' (1972) 1:140 000 "Generalized Vegetation Map" of Ruhuna National Park, Ceylon.

3) Large-scale maps at a scale of from 1:100 000 to 1:10 000. Such maps permit representation of most of the vegetation units defined through dominant species or other floristic definitions. Examples include Küchler's and Sawyer's (1967) 1:30 000 map of the "Vegetation West of Maenam Ping in Thailand," the comparative 1:10 000 ecologic maps prepared for the International Methods comparison in Switzerland (Ellenberg 1967), and the numerous 1:24 000 scale overlay maps (to the U.S. Geological Survey 7½ minute topographic maps) produced for various purposes in the United States.

4) Detail maps of very large scale (on the order of 1:5 000 to 1:1 000). Such maps are prepared for special purposes, for example, the documentation of a nature preserve or a critical fire-prone wildlands–suburban interface. Examples are provided by Ellenberg and Klotzli (1967) and Mueller-Dombois and Ellenberg (1974).

5) Chart quadrats for mapping all important species (on the order of 1:100 to 1:10). Primary applications are for recording permanent sample plots and in landscape architecture.

The reader will see that the large number of classification methods described above, when combined with the huge variation of possible map scales, permits the construction of an almost infinite combination of vegetation maps for any given area. Before the initiation of any mapping project, therefore, it is extremely important to evaluate carefully the source and quality of vegetation data, the application's resolution and accuracy requirements, the cost constraints, and the primary (and potential secondary) applications of the map. Neglect of any of these considerations will often produce a less than adequate and/or flexible return on the manager's investment.

A relatively new area of mapping is the computer-aided construction of maps from gradient modeling systems. Such maps, on a scale of from 1:100 000 to 1:5000, make use of the information capabilities of gradient models, the digital data base of the remote site inventory, and the digital computer's speed to print maps of requested features. For example, a manager might request a map at 1:5000 that shows, for 0.25-ha grids, the total fuel loadings in tons per hectare, or a map at 1:24 000 showing areas where potential bear habitat (floristically defined) occurs within 1 km of a maintained trail.

Alternatively, we must also recognize the limitations of vegetation mapping. While small- to moderate-scale maps are often useful for regional or area planning purposes, the development of high-resolution, large-scale maps often taxes the managers' comprehension, patience, eyesight, and budget. In addition, a refined large-scale map designed for one purpose, such as fire management, may be less than adequate for other resource management purposes, such as trail construction, regulation of backcountry use, minimizing conflicts between bears and people, or understanding ungulate–habitat relationships (Kessell 1976a). For example, Colony (1974) described an ill-conceived vegetation map of Glacier National Park, USA at a scale of 1:125 000 which:

> has been simplified to the extreme (but) the map is virtually unusable to the manager. A melange of 21 colors and 114 symbols are crowded onto this chart in an attempt to maintain at least a pretense of accuracy. As a result the map gives the overall impression of a plate of chop suey. There is a certain amount of basic information displayed if the user is willing to learn the symbol table and study the map long enough to discern the distribution patterns. In the same manner the patron of a Chinese restaurant can pick out the bean sprouts and mushroom pieces if he has a pair of tweezers, enough time, and sufficient motivation. Complex as the Glacier map is, it does not have good enough detail to be used for precise decision making. Questions of wildlife habitats and range are difficult to answer, and fire management problems cannot be related to this map with an acceptable margin of safety.

Digital Inventories

Digital inventories record vegetation or other management information in geographically referenced numerical (rather than cartographic) form, usually in a computer-accessible code. We can distinguish two basic types of digital inventories: *basic digital inventories,* which serve as an alternative storage mode to mapping, and *gradient modeling remote site inventories,* which may include basic cartographic information but also include parameters required for linkage to gradient models of the vegetation, fuels, and/or animals (Kessell 1976b, 1977, Kessell and Cattelino 1978).

Basic digital inventories include the same information as maps, but with three additional capabilities:

1) Graphic complexity resulting from large scales (high resolution) is eliminated. The smallest feature may be included in the data base without encumbering the user with impossible visual detail.
2) Numerous kinds of information may be included in a single inventory without undue visual complexity. In addition, computer-readable inventories allow correlations among stored variables (such as between floristics and bear habitat utilization) that are not readily available from simple graphic records.

3) Computer-readable files permit statistical summarization, stratification, and simplification of the inventory information. Generally, this increases system flexibility at very low cost.

As noted above, gradient modeling remote site inventories include the capabilities of a basic digital inventory, along with the added ability to link with gradient models. For example, the Glacier National Park inventory coded site, cultural, and hydrologic features directly and, through the gradient models, provided quantitative information on the floristics and fuels (Kessell 1976b, 1977; see also Section III). The southern California *Fire Behavior Information Integration System (FBIIS)* provided similar output plus the ability to summarize fuel parameters for large areas by calculation of a sample mean and variance (Kessell and Cattelino 1978; see also Chap. 17). As a final example, *FORPLAN* (*FORest Planning LANguage* and Simulator; Potter and others 1979; see also Chap. 19) stratifies vegetation, fuels, and terrestrial vertebrates (from Thomas and others 1977; see also Chap. 18) by the same two-axis (community type and stand age) system; a digital inventory that recorded these two parameters for each stand would permit automatic inference for the communities' plants, animals, and fuels.

Sources of Vegetation Information

The development of a vegetation information system, regardless of stratification technique or storage method, requires data on the vegetation. Although some studies are conducted almost totally from field samples ("ground truth" studies) and others from remote methods (aerial photography and/or satellite imagery), most systems use both methods. That is, detailed ground truth data are used to derive the community types or gradient models, whereas aerial photographs (or imagery) are used to infer the vegetation of unsampled areas. Let us briefly review some basic requirements of vegetation samples.

Whittaker (1973d) offers an excellent summary of the requirements of field vegetation samples:

> There are a wide range of samples that can be applied to plant communities. . . .Some considerations affecting choice and kind of sample are: (i) The sample should be large enough in area, or should include counts of a sufficient number of plants, to represent effectively the composition of the plant community. If the sample is too large, however, difficulties are encountered in meeting the two following considerations. (ii) The sample should be homogeneous—there should be no trend of change in community composition or structure from one edge of the sample to another. (iii) The sample should be efficient. Since considerable numbers of samples

may be needed, the samples should be designed to obtain and record rapidly the kinds of information regarded as most important. (iv) The sample should be appropriate. Among the many kinds of information that might be gathered on a plant community, some are more interesting or significant, appropriate to the character of the community, and informative in relation to time spent and the purposes (of the study).

The required sampling intensity (in terms of type of samples, number of samples, and sample detail) will be almost totally dictated by the floristics and diversity of the study area and by the manager's required accuracy and resolution.

Once ground truth information is obtained, stratified, and stored, most management applications require extrapolation of those data to unsampled stands through inference. The most frequently used materials for both community-type mapping and gradient model remote site inventories are aerial photographs. A wide range of photographic films (black and white, color, false-color infrared), photographic methods (vertical, oblique), and scales can be obtained. Choices will usually be determined by availability of existing photographs, management purposes, and funding. For example, the Glacier National Park gradient model used both 1:16 000 vertical black and white photos (already available) and oblique false-color infrared photos taken from a light aircraft with a hand-held 35-mm camera (Chap. 14). The southern California *FBIIS* (Kessell and Cattelino 1978) used high-quality 1:100 000 vertical false-color infrared transparencies taken from a high-altitude U2 aircraft. Good reviews and bibliographies of materials and techniques are provided by Salerno (1976) and in various Kodak publications (Eastman Kodak Co. 1970, 1971, 1972).

A recent and promising source of remote vegetation information is satellite imagery. Multiband sensing and digital picture processing techniques are advancing rapidly and should offer increased capabilities in the years ahead (Heaslip 1976, Schanda 1976, Hartl 1976). Evaluations of LANDSAT imagery applications to vegetation and fuels mapping have produced varying results. D. L. Williams (1976) reports good results for southern pine forests in the United States. Single-pixel (approximately 0.4 ha) classification, which stratified the forests into six major canopy types, provided accuracy ranging from 54% for clear cuts (unacceptable) to 94% for hardwoods (excellent). Similar success is reported by Kan and Dillman (1975) and Kalensky and Scherk (1975). Alternatively, Nichols' (1975) imagery classification of southern California fuel and vegetation characteristics appeared unsatisfactory as a site-specific data base for the higher resolution levels (4 ha or better) of the *Fire Behavior Information Integration System* (Kessell and Cattelino 1978), yet Kourtz (1977) indicates good potential from an imagery fuel mapping study conducted in Province of Quebec, Canada. These results demonstrate the potential capabilities of imagery but also urge continued local evaluation of each new imagery application.

As with field data, choices of remote sensing materials and scales must depend upon management purposes, on funding, and especially on the required accuracy and resolution.

Accuracy and Resolution

The above sections strongly emphasize the need to consider accuracy and resolution requirements before vegetation information is collected, stratified, or stored. This can be done only within the framework of the system's intended application. As Goodall (1977) notes: "The question here of course is, How imperfect may it be? and this can only be answered after the other question, For what? An objective is implied."

For every application, moreover, one must consider: (1) How are accuracy and resolution requirements determined? and (2) once determined, how do they affect the development of vegetation information systems?

Accuracy and resolution are two very different concepts. Furthermore, accuracy is rather meaningless unless it is accompanied by the specification of resolution level. *Accuracy* is defined as "exactness, having no error, correctness," whereas *resolution* is "the act or process of separating or reducing something into its constituent parts." In the context of plant communities, resolution implies a community complex that may be further differentiated into varying constituent communities. Any resolution level that is less than perfect therefore implies a community type that is really composed of two or more different community types.

For example, consider a 100-ha community type of "coniferous forest." This would be an "accurate" characterization of the entire 100-ha area. However, this 100-ha community may be divided into a 35-ha lodgepole pine *(Pinus contorta)* forest, a 45-ha Douglas-fir *(Pseudotsuga menziesii)* forest, and a 20-ha western larch *(Larix occidentalis)* forest. Each of these three new, more refined community types may be further divided among different stand ages (120-, 160-, and 225-year-old Douglas-fir), different topographies and aspects, and different substrates. Each additional environmental factor considered by the more refined subdivisions affects community and plant population characteristics. A community type of "coniferous forest" is therefore an accurate characterization at the 100-ha resolution level but is less than adequate for characterizing a 160-year-old Douglas-fir subset. It has very little value indeed (that is, it is terribly inaccurate) as a community description for a 2-ha subset that includes only 160-year-old Douglas-fir on west aspect ridges and a shallow, rocky substrate.

Similar problems are encountered by managers attempting to describe forest fuels. A broad fuel model, such as the National Fire Danger Rating System H (coniferous forest) model (Deeming and others 1977), may be very accurate for characterizing a 100 000-ha forest but terribly inadequate when applied to a 1-ha stand (Kessell and others 1978).

A manager's selection of an appropriate accuracy and resolution level often requires experimentation. In the development of both the Glacier National Park and southern California gradient models, we tested inventory system resolution levels that ranged from 10-m to 1000-m linear resolution (0.01-ha to 100-ha area resolution—see Chap. 17). In each location, representative areas were inventoried at numerous resolution levels; this allowed determination of errors incurred (and money saved) at each successively lower resolution level. Both studies measured vegetation and fuel inference errors. The Glacier National Park work went an additional step and predicted fire behavior from the fuels information inferred at each resolution level (Chap. 17). In this fashion, the manager could see the loss of modeling accuracy incurred at each successively lower resolution level. Interestingly, the Glacier results showed that whereas "typical" vegetation, fuels, and fire behavior could be predicted accurately at a resolution of 25 ha, atypical communities and extreme fire behavior predictions required a minimum resolution of about 4 ha.

Once a manager determines his or her accuracy and resolution requirements, it is possible to tailor a vegetation information system to these needs. These constraints will determine the sources of vegetation information, the stratification methods, and the storage/retrieval options that are to be used.

For example, a system requiring only 2000-m (400-ha) resolution can be developed from low-resolution satellite imagery or aerial photography and conveniently stored as a vegetation map. Resolution requirements in the range of 1000 m (100 ha) can be met in a similar fashion; alternatively, this resolution level can also be effectively achieved by basic digital inventories. This is also the lowest cost-effective resolution level for gradient modeling systems.

For resolution on the order of 100–500 m (1–25 ha), the options are more limited. Field sampling augmented by moderate- to high-resolution aerial photos can provide: vegetation maps of community types on a scale of 1:24 000; basic digital inventories based on community types; or a gradient modeling system. Satellite imagery should be considered as an alternative to aerial photographs for resolution requirements of 200–500 m.

At resolution levels finer than 100 m, graphic vegetation maps are not a good choice. Down to about 50 m, basic digital inventories (based on cover types) are feasible (but not as cost-effective as gradient models). Gradient models are the best alternative at resolution levels of 10–50 m; below 10 m, the only feasible alternative is (very expensive) ground truth mapping on a scale of about 1:100 or larger.

A real concern in selecting accuracy and resolution levels is system cost. For example, Table 3.3 shows comparative costs (per hectare) of gradient modeling systems at six different resolution levels. Suppose a manager

Table 3.3. Comparative Costs of Developing and Implementing a Gradient Modeling System at Six Different Resolution Levels. (Based on Kessell 1979b)

Resolution level		Cost per hectare (U.S. $, 1976)
Area (ha)	Linear (m)	
100	1000	0.02
10	333	0.06
4	200	0.10
1	100	0.20
0.1	33	0.60
0.01	10	1.50

would like 1-ha resolution in the development of a fire effects model. Is the improved accuracy of a 1-ha system over a 4-ha system worth the 2:1 cost difference? Which is a better option, a 1-ha system for half of the management area or a 4-ha system for all of it? Given cost constraints, would it be wise to develop a 1-ha resolution system for critical areas and apply a 10-ha system to the remainder? Once again, the manager must first define the system's purposes explicitly and then evaluate the potential management impacts, before he or she can reach a sound decision.

Fuel and Fire Modeling

Until about 1970, there was little need for or efforts directed toward systematic collection, description, or application of quantitative data on flammable forest fuels. Because of a paucity of mathematical tools, models, and simulation methods, fuel loads (dry weight per unit area) were described as "light" or "heavy," fire spread was characterized as "moderate" or "high," and fire danger was rated "low," "moderate," or "extreme." Although the quantity, arrangement, and condition of flammable fuels on the forest floor were known to exert a significant effect on fire behavior, hazard, and danger, there was as yet no way to use detailed quantitative data efficiently. Too often, an experienced Fire Control Officer's cursory examination of a fuel array provided all the information available for fire management decisions. Several developments during the past few years have drastically changed this situation.

First, Rothermel (1972) developed a mathematical model of fire spread rate and intensity for wildland fuels. Unlike previous models, which required inputs that could not readily be determined before combustion occurred, the Rothermel model predicted fire behavior from such parameters as fuel loads, moisture, wind, and slope steepness, which describe the fuel array and the conditions under which it burns. These parameters can be measured or estimated before a fire occurs. As we shall see, most new fire management systems either use the Rothermel model directly or link it with other simulation models and information systems. The Rothermel model is described in detail in Chapter 14.

Concurrent with the development of the Rothermel model, Brown (1971, 1974) developed the planar intersect technique for obtaining quantitative estimates of the loadings and packing ratios of natural forest fuels. That method has been widely used and frequently modified for various applications in different habitat and fuel types (Jeske and Bevins 1976, Kessell and others 1978; see also Chaps. 6 and 7).

During this same period, the construction of stylized "fuel models" (basic models for "grass," "coniferous forest," "chaparral," and other major vegetation types) and application of the Rothermel fire behavior model permitted the development of the National Fire Danger Rating System (NFDRS) by Deeming and others (1972). NFDRS is a system that integrates meteorologic inputs with fuel and fire behavior models to generate relative scales (indices) of fire danger. With it, and from a historical weather data set, a user can derive estimates of relative fire danger for a stylized fuel model and any historical weather day. By repeated iterations through a complete weather set, the user generates cumulative probability distributions of fire danger. These are displayed through computation of a "spread component," an "ignition component," and a synthetic "burning index." The most recent edition of the NFDRS (Deeming and others 1977) allows the user to obtain actual estimates of fire behavior for the stylized fuel models under "worst possible" fire conditions.

The cumulative probability distributions generated by NFDRS allow the user to assess the current fire danger in the context of the historical record. The user knows not only that today is "dry," but that it is among the driest 5% of the past 10 years' summer fire season days for that station.

The NFDRS uses the Administrative Forest Fire Information Retrieval and Management System (AFFIRMS) for forecasting fire danger over a 2-day period. AFFIRMS is a computer system that allows National Weather Service forecasters to integrate their fire weather forecasts with NFDRS. The results are both a regular daily fire weather forecast and a forecast for the next day's fire danger indices.

The NFDRS and its family of computer routines (AFFIRMS, FIRE-DAT, and others) are now widely used in the United States. The 1978 version of the NFDRS expands the system from 9 to 20 stylized fuel models and offers other significant improvements to the 1972 system. Other fire danger rating systems are in use in Canada, Australia, France, and Spain; an international system has been proposed to the World Meterological Organization (Reifsnyder 1976).

Following the development of these tools and models, the White Cap

Fire Management Plan was developed for a portion of the Selway–Bitter-root Wilderness in western Montana (Aldrich and Mutch 1972). The White Cap study assessed landforms, plant communities, fuels, and historical fire occurrence for a 25 000-ha portion of the wilderness. From these data, fire prescriptions were developed that specified, for each major habitat (called "ecological land units"), the conditions under which a fire would be suppressed, contained, or permitted to burn under observation. Although somewhat outdated by more recent and sophisticated approaches, the White Cap Plan was a milestone because it represented the first sound natural fire management plan derived from quantitative estimates of the natural communities and potential fire behavior.

Near the completion of the White Cap program, the gradient modeling approach was developed. Gradient modeling offered improved land management simulation capabilities but in turn demanded greater resolution of fire modeling input parameters (primarily fuels and weather data); some of this work is described in Section II (Chaps. 6 and 7) and Section III (Chaps. 14 and 15). The resulting increase in plant community, fuels, weather, and fire effects simulation resolution in turn placed greater demands upon the fire model itself; the response was refinement of the existing model (Frandsen 1974; Bevins 1976; Kessell and others 1978) as detailed in Chapter 14.

As noted earlier in this chapter, model construction is an iterative process; we build models, test them, identify the problems and limiting factor, improve them, test them again, and so forth. We have already completed the cycle shown in Fig. 3.1 several times, and may well have to complete it several more times before our simulation capabilities produce consistent satisfactory agreement with real world fire behavior and effects.

Animal Habitat Modeling

Much of this chapter has been devoted to problems associated with the collection, analysis, and simulation of plant communities; similar problems occur with fuel information, as noted above, and animal habitat utilization. Ideally, data collection, stratification, simulation methods, and information systems analysis appropriate to plant communities and their component fuels can be used for animal habitat modeling.

This was assumed in the Glacier National Park gradient modeling effort, and a modification of the phytosociologic gradient scheme was used by Singer (1975a, b) to model stand utilization by ungulates. We assumed that the environmental gradients affect the vegetation, which in turn affects the animals, and we were content to infer animal habitat utilization directly from the gradients. Subsequent work in western Montana attempted to obtain a better understanding of these relationships: The quantitative composition of plant communities was inferred from the gradients; indexes of relative palatability to a few common vertebrate species were assigned to each common plant species; and these combined data were then used to infer community utilization by the vertebrates (Kessell and others 1977b). This approach seemed conceptually sound, but because of a paucity of information on both the relative palatability of different plants to different vertebrates, and the unknown extent to which food sources are limiting factors of each vertebrate species' distribution, we decided that the approach was simply not practical.

In the meantime, Thomas and others (1977) completed an extensive study of vertebrate habitat utilization in the Blue Mountains Province of northeastern Oregon, USA, for all 327 terrestrial vertebrate species present in the area. Conceptually similar to Singer's work in Glacier National Park, their study stratified the area into major plant community types and further stratified each community type into stand age classes. Then, for each vertebrate species, they determined the combinations of community types and stand ages that the species could utilize for either reproduction or feeding. Their methods are now being applied in several other areas and are, I believe, a real breakthrough in modeling habitat utilization by vertebrates.

Details of Singer's work in Glacier Park and of the Thomas and others Blue Mountains study are given in Chapter 18; the inclusion of these methods and concepts into the *FORPLAN* Simulator is discussed in Chapter 19.

Succession Modeling

As noted early in this chapter, simulation is modeling through time; in the context of natural communities, many of these changes through time are the directional changes known as succession. For the purpose of resource management simulation, we wish to understand, describe, and predict these changes for plant communities, their component fuels, and their utilization by animals.

Earlier Concepts of Succession

Throughout this century, succession has been a prime factor in the study of communities following various disturbances. Early concepts (Clements 1916) assume that following a disturbance, a community regenerates a semblance of itself by an orderly and predictable series of species replacements. This assumption allows a deterministic, "single-pathway" prediction of succession. More recently, others have developed different perspectives. The JABOWA model (Botkin and others 1972) observes successional changes on an individual tree basis by considering competition, environmental variables, growth rate, and survival characteristics. A slightly different approach was taken by Horn (1974). For each individual tree in the canopy, he observed the understory species that might replace it and assigned a probability to that event. The replacement probabilities were then used to predict successional changes. Alternatively, Whittaker (1975a) characterizes succession by developmental trends; that is, characteristics of the ecosystem which change through succession. These include species diversity, productivity, biomass, and nutrient stock. Odum (1969) similarly views succession as a process of ecosystem development. The three models of Connell and Slatyer (1977) describe succession in terms of species facilitation, tolerance, and inhibitive properties.

These models deal with disturbances in general, with emphasis on replacement sequences (either species populations or community characteristics) observed in the plant communities. They address neither the adaptive traits of a species persisting through a disturbance (Gill 1977) nor the periodicity of the disturbance (Vogl 1977). Many communities follow a single regeneration pathway under "normal" disturbance periodicities, yet widely depart from it when subjected to very short or very long interdisturbance periods (Noble and Slatyer 1977, Cattelino and others 1979). Since resource management—especially fire management—often affects natural disturbance periodicities and interdisturbance periods it is highly desirable to include these effects in succession models.

While constructing the Glacier Park gradient model in 1975–1976, we dealt with this problem at length. Community data of sufficient resolution to apply the JABOWA or Horn techniques were simply not available, and as a result a Clements "single-pathway" assumption was made. This proved adequate for many community types when subjected to normal interfire periods, but for some communities it was simply unsatisfactory; the comparison of the model's predictions to the real world showed large disparities. No solution was apparent.

The Noble and Slatyer Model

Then in 1977 Noble and Slatyer proposed new concepts of replacement patterns following disturbance. They have developed a successional scheme that has both descriptive and predictive capabilities under various interdisturbance periods. This system defines a set of species attributes that are vital to the reproduction and survival of a species within the community.

Noble and Slatyer define the following attributes as most important to a species' success on a site subject to periodic fires:

1) The method of arrival or persistence of propagules at the site following a disturbance
2) The conditions in which the species establish and grow to maturity
3) The time taken to reach critical stages in their life history

If some indication of relative importance of species is needed, a fourth attribute may be added:

4) The size, growth rate, and mortality of the species.

Method of Persistence

This attribute describes the method of persistence through the disturbance. Four mechanisms are recognized by Noble and Slatyer:

1) Persistence by the arrival of widely dispersed seeds (D species). Seed are dispersed from the surrounding undisturbed areas and are always available at the site.
2) Persistence by seeds with long viability that are stored in the soil (S species). Not all of these seeds germinate after the first disturbance and the seed pool may persist through a series of disturbances without replenishment.
3) Persistence of seeds with short viability that often survive the disturbance within protective fruits or cones that are stored in the canopy (C species). In this case, seeds are available only if the adults have been present before the disturbance.
4) Persistence by part of the individual surviving the disturbance and recovery by vegetative regrowth (V species).

Conditions for Establishment

This attribute describes the conditions necessary for establishment. Three types of species and mechanisms are recognized:

1) Species able to establish at any time, with adults of the same species and of other species occurring at the site. These species can tolerate competition (T species). As used here, tolerance can apply to light, soil nutrients, or other physiologic requirements.
2) Species able to establish only immediately after a disturbance, when competition is usually reduced. These species are intolerant of competition in established communities (I species).
3) Species unable to establish immediately after a disturbance but able to become established once mature individuals of either the same or another species are present. These species have some requirement that is only provided by established communities (R species).

Life History

This vital attribute plays an important part in the replacement sequence and in describing the role of a species after a disturbance. There are four critical events within the life history of a species that are arranged in order of occurrence following a disturbance (time 0):

1) The replenishment of sufficient propagules to survive another disturbance (p).
2) Maturity (m) is the time at which individuals will have recovered or
   grown sufficiently to be regarded as established. Maturity by this definition is therefore the time when the individual is able to contribute
   propagules into the propagule pool that will enable the species to persist
   through another disturbance.
3) The senescence and loss of the species from the community (l).
4) Loss of propagules from the site so that the species is extinct (e).

By combining the first two vital attributes of a species (method of
persistence and conditions for establishment) with its critical events, the
life history characteristics can be represented as shown in Fig. 3.2.

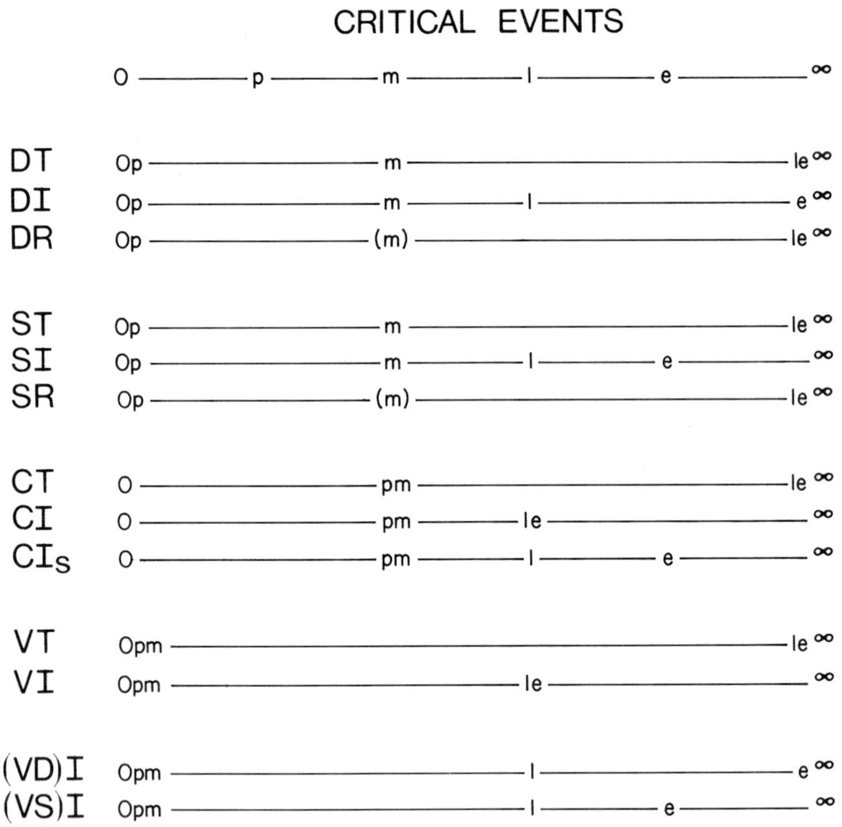

CRITICAL EVENTS

Figure 3.2. Life history characteristics for various combinations of mechanisms of
the first two vital attributes. The critical events are described in the text. The
parentheses around the m for the R species indicates that their reaching maturity
depends on the presence of other mature individuals. (VD) and (VS) species are
those which can reproduce either vegetatively or from seeds. (Noble and Slatyer
1977)

Application of the Model

With this knowledge of a species occurring on a site of periodic disturbance, the replacement sequence can be predicted for that species. To
illustrate, consider a *Callitris*–mallee community in western New South
Wales (Noble and Slatyer 1977). The mallee species are typical of the
majority of eucalypts, with a shrublike multistemmed form, and have a
long-lived underground lignotuber from which they can resprout after a fire
or mechanical damage. Seedlings are rarely observed. The mallee therefore
can be described as a VI species (Fig. 3.2), and the only information
required for their life history is their longevity (l); all other critical events
either occur at time zero or coincide with time l. Although individual plant
longevity is not known, the mallee lignotuber may live for a century.

Callitris columellaris is a slow growing conifer that can attain a height of
25 m. Generally mallee and *Callitris* communities are distinctly segregated,
but in some cases they occur in closely interwoven patches. *Callitris*
regenerates from seed, can reach a reproductive age in as few as 6 years,
and is classified as a CT species (Fig. 3.2). No futher information is needed
regarding its life history because events p and m coincide and events l and e
occur at infinity.

Now let us consider a site that is currently supporting both mature
mallee and *Callitris* (that is, VI + CT). If the site is burned by a hot fire, the
above-ground tissues of both species will be killed The VI mallee will
regenerate immediately and the first stage of the replacement sequence will
be occupied by the mallee. Seeds of the CT *Callitris* will germinate when
conditions are favorable but the mallee will remain the only mature species
in the community until the *Callitris* seedlings have matured (6–8 years) and
the site returns to a VI + CT community. If this community remains
unburned for a long period, the mallee will die off, leaving a *Callitris* (CT)
community. This succession sequence is shown in Fig. 3.3a.

However, fire periodicity does not always follow this sequence. If a fire
occurs during the VI stage, within 6–8 years after a previous fire, then the
young, immature *Callitris* will be destroyed and lost from the community.
Similarly, if a CT community burns after a long period without fire, the
Callitris regenerates from its seed source but the VI mallee is lost because
of its V mechanism and its relatively short life span. These critical events
for both species are shown in Fig. 3.3b; the overall successional possibili-
ties are shown in Fig. 3.3c (where the solid lines indicate no disturbance
and the broken lines indicate fire recovery).

The Noble and Slatyer succession model has provided valuable insight
to the replacement patterns following disturbances for Mediterranean climate ecosystems in Australia. Although it was not developed until after the
completion of the Glacier Park gradient modeling effort, we have since
applied their model to many of the communities from Glacier Park for
which the Clements method proved unsatisfactory. The Noble and Slatyer

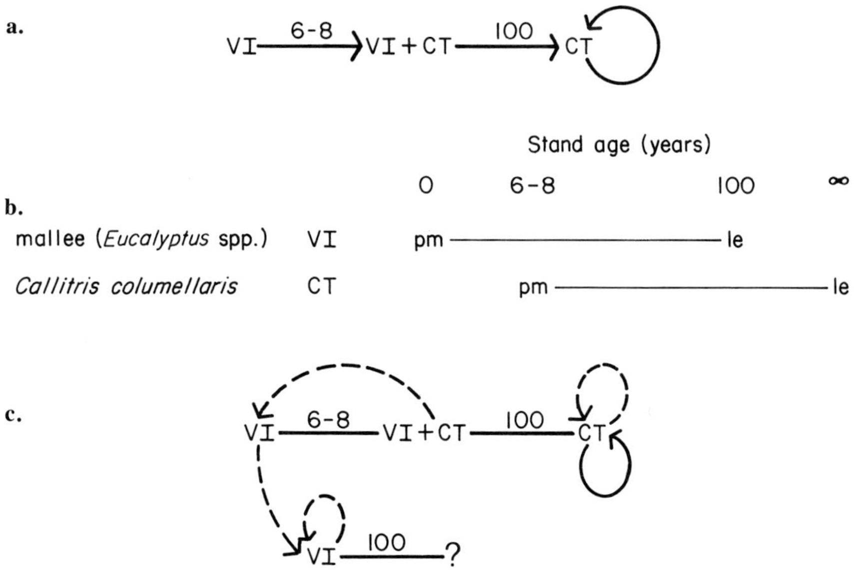

a.

$$\text{VI} \xrightarrow{\;\;6\text{-}8\;\;} \text{VI}+\text{CT} \xrightarrow{\;\;100\;\;} \text{CT} \circlearrowleft$$

Stand age (years)

		0	6-8	100	∞

b.

mallee (*Eucalyptus* spp.)   VI   pm ———————————————— le

Callitris columellaris   CT   pm ———————————————— le

c.

$$\text{VI} \xrightarrow{\;\;6\text{-}8\;\;} \text{VI}+\text{CT} \xrightarrow{\;\;100\;\;} \text{CT}$$
$$\text{VI} \xrightarrow{\;\;100\;\;} ?$$

Figure 3.3.a–c. a Mallee (VI)–*Callitris* (CT) succession without disturbance. **b** Life history characteristics for mallee and *Callitris*. **c** Species replacement sequences in a mallee–*Callitris* community. The approximate time (in years) for transitions in undisturbed communities is shown by solid line arrows; the broken line arrows show the transitions caused by fire. (Noble and Slatyer 1977)

model gave significantly better predictions of successional patterns and greatly improved understanding of the successional process. Those results have been reported by Cattelino and others (1979) and are highlighted in Chapter 10. Furthermore, although the *FORPLAN* Simulator originally used the deterministic Clements predictions, the succession module is being converted to the Noble and Slatyer method, community by community, as the alternative pathways are established (Potter and others 1979; see also Chap. 19).

Simulation as a Managerial Tool

As noted in Chapter 1, and reviewed by Franklin (1979), Kessell (1976a, 1978, 1979c, Potter and others 1979), and C. A. S. Hall and Day (1977), simulation is a tool that has much to offer resource management and land management planning. Models and simulators help us handle, manipulate, and use vast quantities of information in the context of our current environmental understanding. There are three main advantages to the manager of using simulation:

1) The ability to *compare* management strategies and alternatives before making a decision
2) The ability to *consider* a decision without either the cost or the consequences of implementing that decision
3) The ability to *predict* probable effects of a decision at future times.

Should simulators then tell the manager how to manage, or what decision to make? Certainly not. Simulators are designed (or at least should be designed) to display the probable consequences of various actions: If you do this, then you can expect the following to happen. They are not designed, and should never be designed, to replace the manager or to make the decision for him or her. Their purpose is to provide the best possible information to the manager so that a better, more enlightened decision can be made.

Each of the models, simulators, and data bases described earlier in this chapter, therefore, is simply a tool designed to aid the manager. For example, consider a manager seeking to develop a fire management policy. This manager must be concerned with fire behavior when a fire does occur, and with the short- and long-term ecologic impacts of alternative fire management policies. Such a manager can thus benefit from recent research on modeling fuels and fire behavior, the latest methods for stratifying plant communities and animal communities, and newly developed succession simulation methods. Note that I have said "benefit from" rather than "use"—*simply the fact that the research has been done does not mean the manager can use the results!* It is a naive researcher who thinks that publishing his or her data and simulators automatically makes them available to the land manager.

Let us return to our manager who is trying to determine fire policy. Fire weather and danger information is available from AFFIRMS, basic stylized fuel models can be found in NFDRS, fuel inference models are in the literature, and some ground truth samples can be collected in the forest. The manager also knows of several available fire behavior models. Which fire model should he or she use? Which fuel model should be input to that fire model? The data have been stratified by community types; should they also have been stratified by stand age, or should gradients have been used? Can the method of stratifying the fuels be changed now? Should it be changed? Can it also be used to stratify the vegetation? The animal communities? Should the manager predict postfire successions using a Clements approach or try the multiple-pathway model? For which fire model do the available fuels, weather, and terrain data provide appropriate resolution? What will limit the accuracy of the manager's predictions? Will they be reliable? How reliable?

One begins to see the scope of problems raised briefly in Chapter 1. Land management problems are complex, multifaceted problems; the manager must weigh alternative strategies that impact whole ecosystems. No single model or data base, no single family of models or data bases, can answer all the questions. Furthermore, even if all the models and data bases noted above are immediately available from his or her own computer terminal, the manager faces two additional problems:

1) The selection of each simulation tool (fuel model, fire model, succession model, etc.) from an ever-increasing group of models
2) The operation of different models with different input requirements, job control requirements, access procedures, and programming experience, all located on different computers.

The conclusion: Resource management modeling and simulation requires more than a multitude of separate models and data bases. Integrated systems are needed for resource management and land management planning, systems that provide broad capabilities under a single umbrella, systems that can use the output from one simulator or module as input to the next. Let us now look at some of these systems.

Integrated Systems

The construction of integrated systems for resource management is really nothing more than the development of appropriate linkages among the individual simulators and data bases. For example, suppose one has a set of stylized fuel models in tabular form and a computer program of the basic fire behavior model. Construction of a new program that includes the fire model, the fuel models, and the user option to input any fuel model to the fire model is a step closer to an integrated system. Inclusion of historical weather data, with a user option to choose weather station and cumulative percentile level, takes the process a step further. One might next add a fire hazard model, or a postfire succession model, or site-specific fuels data selected by geographic coordinates. And so the process continues.

Not quite. The process of building integrated systems is something more than stringing together a whole bunch of models and data bases, as too many modelers have learned the hard way. There are several reasons that such "add on" approaches prove unsatisfactory:

1) As noted in Chapter 2, management models and information systems should be accurate, comprehensive, flexible, synthetic, problem oriented, accessible, and economical. However, as shown in Fig. 3.4, meeting these objectives involves tradeoffs. As systems are expanded to make them more comprehensive, more accurate, and more flexible, they often become more expensive and less accessible; given fixed cost constraints, accuracy and comprehensiveness are often achieved at the expense of flexibility. A

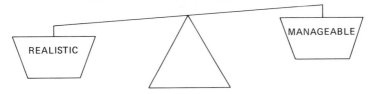

"MODEL": A Substitute Representation of Reality

To Meet Managerial Needs, a Modeling System Should Be:

1. Accurate
2. Comprehensive
3. Flexible
4. Synthetic
5. Problem Oriented
6. Accessible
7. Economical

Figure 3.4. The process of constructing integrated systems that meet the seven needs of resource management (described in Chap. 2) is a continual struggle to make models realistic while keeping them manageable.

modeler must understand and consider these tradeoffs before he or she constructs a large system.

2) As noted earlier in this chapter, resolution levels must be considered when models and data bases are being matched. It is not reasonable to take very crude, low-resolution input data and then run them through the largest, fanciest, most expensive simulator. Alternatively, the value of high-resolution input data should not be limited by use of a crude simulator. For example, Table 3.4 shows four levels of planning, their associated resolution levels, and the resolution required of fuels data, weather data, and fire behavior models.

3) The biggest, most detailed, most expensive system is not always the best system for a given management problem; in some terrain, an old Jeep or a trail bike is more valuable than a new Rolls-Royce. Too often, managers call out the Marines when they could do the job themselves with a slingshot! To put it more directly, as larger and more sophisticated systems become available, there is a greater risk of misuse by managers unfamiliar with their use and of overselling by exuberant researchers.

4) Environmental statistics often include large standard deviations and large errors of the estimate. What happens when we combine them, for example, by the input of noisy fuels, weather, and terrain data to a fire model? Are the errors compounded, or do they tend to cancel one another? The modeler must consider this problem before the numerous components are linked in a single system.

Given all of these problems, is it reasonable to construct large, integrated resource management information systems? Yes, recent successes

Table 3.4. Four Levels of Fire Management Planning and their Associated Resolution Levels Related to Required Fuel and Weather Inputs to Various Fire Behavior Models.[a]

Planning level	Resolution level	Fuels data	Weather data	Fire model
Area	Low	NFDRS fuel models	Weather percentile from FIREDAT	Noncomputer nomograms (Albini) of fire behavior
Unit	Moderate	NFDRS fuel models stratified by stand ages	NFDRS–AFFIRMS	Computer version of Rothermel fire model with unique inputs
Drainage	High	Locally developed fuel inference models	NFDRS–AFFIRMS refined by conditional forecasting and/or microclimatic extrapolation	Multistrata version of the fire behavior model
Project	State of the art	Site-specific fuels sampling	Site-specific, real-time measurements	Multistrata version of the fire behavior model further refined to consider discontinuous fuels

[a]Subsequent chapters give further details on fuels (Chaps. 9 and 17), weather (Chap. 14), and elaborations of the fire behavior model (Chap. 14). Linkage of the components under variable resolution levels is discussed in Chapter 19.

demonstrate that it certainly is reasonable and useful to do so. Then how does one build such systems and avoid these pitfalls? By very carefully considering the potential problems raised at each stage of the modeling process.

In developing large, integrated systems, one must very carefully define the objectives to be met as well as the constraints imposed by funding, available (or attainable) data, and state of the art understanding and capabilities. One must further carefully match the accuracy and resolution level of each data base, model, and simulator, realizing that some factor is always limiting but that no factor should significantly detract from the capabilities of the whole system. One must further tailor the system to meet the manager's objectives, recognizing that other approaches may be more exciting or lead to more interesting theoretical developments but be less desirable to the manager. Finally one must carefully study the behavior of the system, determine the composite effect of input errors, and evaluate the sensitivity of the system to those errors. This is a vital part of model validation and frequently leads to the identification of new research that should be conducted.

In line with the above considerations, we faced additional, specific problems in the development of the gradient modeling method in Glacier National Park. First, it was to meet certain basic resource management goals, which are outlined in Table 3.5. Next, we had certain cost and resolution constraints, namely that cost to develop, implement, and test the system would not exceed $1.50 per hectare and that resolution would exceed 20 m. The next constraint was that managers without automatic data processing training could be taught to use the system with no more

Table 3.5. Basic Resource Management Purposes of the Glacier National Park Gradient Modeling System *(BURN)*.

---

1. To provide a synecologic model of Glacier's tree species, important herb and shrub species, flammable ground fuels, diversity relationships, and successional patterns; this included:
   a) A quantitative description of the distributional, structural, and dynamic ecologic relationships of the natural communities that have developed in response to an extremely diverse habitat complex
   b) The basis for understanding the distribution of the component species, the dynamic ecologic processes that permitted the development and maintenance of these communities, and the manner in which they relate to one another and to habitat gradients
   c) The framework for a resource information system based on these community models
2. To develop the additional components necessary to produce a resource information system and real-time fire management system; this included:
   a) A resource inventory system that describes each stand's location on environmental gradients and within classification units such that a stand's phytosociologic composition may be accurately predicted without field sampling each individual stand
   b) Fuel moisture and microclimate models that extend the capabilities of the NFDRS and AFFIRMS to provide absolute meterologic inputs to fire behavior models
   c) Refined fire behavior models that permit real-time fire behavior simulation; this included modification of the Rothermel (1972) fire behavior model and the development of a refined ground fire model
   d) The necessary methods, algorithms, and programs to link these components into an efficient, easy to use system that would neither intimidate managers nor require the user to receive extensive computer training
3. To evaluate this new method of gradient modeling as a resource management tool, to compare its costs and benefits with other systems, and to make recommendations on its continued development, refinement, and application to other areas

---

than 8 hr of training. Further constraints included economic development and day-to-day use of the system on the National Park Service's computer facilities (an IBM 370/168 VS system) and implementation of the system on a 12-month contract (perhaps half of the basic research had been completed prior to that contract).

The development of *BURN* (the acronym for the Glacier National Park gradient modeling system) therefore required that we develop detailed, high-resolution community models within the constraints of compatability with existing or developable inventory systems, state of the art weather and fire modeling capabilities, and very rigid resolution, cost, manpower, and time constraints. *BURN* met or exceeded each requirement, cost slightly less than estimated, was completed on schedule, and demonstrated the utility of gradient modeling.

Because the Glacier National Park system was the original gradient model, and because components of this system have been used (with modification) in several subsequent systems, including *FORPLAN,* over half of this book is devoted to describing that work. Section II (Chaps. 4–12) details all of the field, analytic, and synthetic work that developed the basic gradient models for Glacier National Park. Section III (Chaps. 13–16) documents how this basic research was used to construct the management system, with emphasis on the construction and linkage of the various other components and the development of the computer software. Section IV (Chaps. 17–20) then highlights some subsequent applications of these basic methods.

SECTION II

THE GLACIER NATIONAL PARK STUDY

4

Introduction to Glacier National Park

An Act of the United States Congress in 1910 established 401 000 ha of wilderness astride the continental divide in northwestern Montana as Glacier National Park. It is a land of dense forests on the lower mountain slopes, with scattered prairie remnants and intrusions, giving way to subalpine mosaics and alpine tundra at the higher elevations. Alpine lakes, rock outcrops, talus slides, luxurious meadows, sparkling streams, permanent snowfields, and over 50 active mountain glaciers dot the high country. Lower forests give way to occasional meadows and marshes and exhibit a diverse mosaic of secondary successional patterns. Known as "The Crown of the Continent," Glacier is noted for its rugged majesty and wilderness solitude and includes some of North America's most spectacular scenery. Over a thousand species of vascular plants occur within the park (Kessell 1974); common mammals include the mule and white-tailed deer, elk, moose, mountain goat, bighorn sheep, black bear, grizzly bear, lynx, mountain lion, and marmot (Bailey 1918). Small populations of the timber wolf, wolverine, and fisher still exist within the park (Singer 1975a). Quite remarkably, notwithstanding extensive management and mismanagement since 1910, much of Glacier's native biota has been maintained, despite the obvious alterations to its pristine ecosystems that have been caused by human activity.

Ecology

Ecologic investigations, which reflect various degrees of thoroughness, sophistication, and imagination, have been carried out over the past 80 years. The first descriptive studies of Glacier's vegetation were made in 1898–1899 by the U.S. Geological Survey under the direction of H. B. Ayres (1900). He placed particular emphasis on mapping the distribution of major tree species and made numerous comments on the role of fire in shaping Glacier's communities.

The next three decades saw considerable effort devoted to the development of checklists of the park's flora and fauna and to compiling natural history notes on various aspects of the park's biota. Included in this work is that of Standley (1921, 1926), in preparing a flora of the vascular plants, as is Bailey's treatise (1918) on Glacier's fauna. Both Standley and Bailey extensively used the "life-zone" concepts of Merriam (1898). Merriam's system divided Glacier into four major zones: (1) a "transition zone," including the grassland and savanna communities found at the lower elevations; (2) a "Canadian zone," including the bulk of the continuous coniferous forest in the park; (3) a "Hudsonian zone," transition from the "Canadian" to the "Arctic–Alpine" zone; and (4) the "Arctic–Alpine zone" including alpine meadows and shrubfields, local patches of Krummholz, talus and rock outcrop areas, permanent snowfields, and active glaciers. A more detailed historical treatment of this early work and the life-zone concept is given by Habeck (1970a).

The next historical stage in the study of Glacier's vegetation includes the work of Rexford F. Daubenmire (1943, 1952, 1966, 1969). Daubenmire maintained that a science of vegetation cannot exist without a system of classification. He proceeded to develop such a classification scheme of climax associations; 14 climax associations were proposed for the four life zones that occur in Glacier National Park (Daubenmire 1952). In 1969, Daubenmire proposed a new classification system, which reflected his opinion that only one tree species can achieve dominance on a given site (Daubenmire 1969). Of his 36 recognized "habitat types," 28 potentially occur in Glacier. Although Dr. Daubenmire conducted little field work in Montana, his classification system was the best available for Glacier National Park until the late 1960s. His habitat type classification system for Montana has been modified and refined, following extensive field work by Pfister and others (1977).

During the early 1950s, Leroy H. Harvey (1954a,b) conducted fairly extensive taxonomic work in Glacier National Park and updated the park's checklist of vascular plants. In 1974, I completed still another, more complete checklist (and gratefully acknowledge Dr. Harvey's assistance with its preparation).

The third historical stage of Glacier's phytosociologic investigations includes the work of James R. Habeck during the late 1960s. Habeck approached Glacier with the concepts of the continuity of vegetation in response to environmental gradients. He searched for both description and an explanation of Glacier's plant communities in terms of spatial and temporal environmental and habitat gradients. His publications on Glacier's vegetation as a whole (Habeck 1970a,b) clearly reflect his belief that the complicated mosaic found in Glacier can best be described and understood on the basis of the underlying environmental variation. He attempted the construction of a generalized two-dimensional mosaic of mature communities (using elevation and moisture gradients) based on 70 stands sampled in northwestern Montana (Habeck 1970a). He also conducted more extensive studies of specific communities found in Glacier; these include work on the alpine mosaic at Logan Pass (Habeck and Choate 1963, 1967; Habeck 1969), the mesophytic cedar–hemlock *(Thuja plicata–Tsuga heterophylla)* forests near Lake McDonald (Habeck 1968), the hybrid origin of Glacier's spruce *(Picea)* populations (Habeck and Weaver 1969), and

Stephen F. Arno's doctoral work on alpine larch *(Larix lyallii)* populations (Arno 1971; Habeck and Arno 1972).

Most of Habeck's phytosociologic analysis relied almost entirely on indirect ordination performed using the Bray–Curtis method (Bray and Curtis 1957) and synthetic importance values. His results demonstrated that plants responded to gradients that undoubtedly reflected continuous variation in the environment. Since these were indirect gradients, however, and no real effort was made to determine which directly measurable environmental gradients corresponded to them, it was unfortunately not possible to determine a stand's location on the gradients prior to field sampling of the stand. As a result, many managers in Glacier decided that gradient analysis had nothing to offer resource management.

The current stage of vegetation work in Glacier National Park includes my work since 1972 (Kessell 1973, 1974, 1976a, b, 1977). The philosophy of continuous temporal and spatial variation in the environment influencing plant communities is maintained in this work, but the major effort has been to describe the vegetation in terms of readily identifiable gradients that can be elucidated for unsampled stands using such remote methods as aerial photographs and maps.

Geology, Geomorphology, and Climate

Except for a few areas of volcanic intrusions, the geologic strata of Glacier National Park are limestones, shales, and sandstones of sedimentary origin belonging to the Belt series. These were laid down during the Precambrian era when the Belt Sea filled a trough extending from Alaska to Mexico. The Belt Sea existed three times (the last occurred in the Mesozoic era); each existence allowed the deposition of Cretaceous sediments over the Belt series (Dyson 1962).

The advent of the Rocky Mountain Revolution near the close of the Cretaceous period brought about the gradual uplifting of the Belt geosyncline. Tectonic forces produced a fold in these uplifted beds now known as the Lewis Overthrust. This resulted in the older Belt sediments of the eastern limb of the overthrust now covering the younger Cretaceous sediments, whereas the sediments on its western limb remain in their original depositional sequence.

Erosion and weathering during the close of the Tertiary were accentuated by the continuing uplift and eliminated the Cretaceous beds on the western limb. Continued erosion, weathering, and glaciation have formed the rugged terrain now seen in Glacier National Park. The Lewis and Livingston Ranges, which provide the backbone of the park, were formed in this manner.

Until the Pleistocene, the valleys of the park were eroded by stream action alone. The advent of the Pleistocene brought large mountain glaciers, which originated at the higher elevations; these glaciers advanced and retreated at least twice (Dyson 1962).

The 50–60 mountain glaciers which remain in the park today are believed to have formed during the Recent period's cooler and wetter climate (Dyson 1962). These glaciers advanced until the late 1800s; rising temperatures and decreasing annual precipitation led to their retreat until about 1945. They continued to retreat until 1973 and have grown slightly in the past few years.

The action of the park's glaciers have produced the morains, U-shaped valleys, beautifully carved cirques and cirque lakes, aretes, and hanging valley waterfalls that give the park its character. Concurrently, Glacier's streams and free-flowing rivers and its steep slopes, precipitating talus slides, mud slides, and frequent avalanches, continue to dynamically shape Glacier's landscape.

The climate of Glacier National Park may be distinctly divided into two regimes, which occur east and west of the Continental Divide. The west portion of the park is influenced primarily by coastal systems, which follow a west to east track. The wet Pacific air masses that ascend the Coastal and Cascade Mountains of Washington State and British Columbia still retain considerable moisture when they cross the lower Purcell Mountains of extreme northwestern Montana prior to reaching Glacier Park. West Glacier, at the west central edge of the park, receives about 65 cm of precipitation annually at an elevation of about 1000 m. Higher elevations (over 2000 m) receive up to 250 cm of precipitation, mostly in the form of snow. Snow accumulations of 500 cm are not uncommon, and much deeper snow drifts occur in some areas.

North or south of West Glacier, on the west slope of the Continental Divide, precipitation decreases. The Bitterroot Mountains (230 km southwest of West Glacier) receive less precipitation; Missoula (about 200 km south-southwest of West Glacier) receives about 40 cm of precipitation annually (Habeck 1970a). The Polebridge area (about 45 km north of West Glacier on the park's west border) also receives less precipitation than West Glacier (Table 4.1), and the vegetation of the upper Kintla and Bowman drainages northeast of Polebridge reflect slightly higher precipitation than Polebridge. The Camas drainage, situated north of the McDonald drainage (West Glacier lies at the mouth of the McDonald drainage), reflects conditions intermediate to those found at West Glacier and Polebridge.

The eastern portion of Glacier National Park is subjected to lower temperatures and precipitation because of the influences of continental air masses and winter storms that travel in a south-southeast track along the east slopes of the continental divide. As a result, most of the park's eastern

Table 4.1 A Simple Climatic Summary for Three Stations In or Near Glacier National Park. (Based on Habeck 1970a)

	West Glacier	Babb	Polebridge
Temperature (°C)			
January average	−6	− 7	− 7
July average	18	16	16
Maximum temperature	38	39	39
Minimum temperature	−41	−42	−42
Precipitation (cm)			
January	7.8	2.3	
February	5.3	2.0	
March	4.3	2.8	
April	3.8	4.1	
May	5.1	6.1	
June	6.1	8.4	
July	3.0	4.9	
August	3.6	4.6	
September	5.1	5.3	
October	6.6	3.0	
November	6.9	2.5	
December	8.6	2.5	
Total for Year	66.5	49.3	55.1

border, originally vegetated with grassland communities and occasional aspen groves *(Populus tremuloides)* and limber pine *(Pinus flexilis)* savanna (Habeck 1970a), is unforested today.

A detailed analysis of the microclimate of the McDonald drainage was conducted by Mason and Kessell (1975) and is summarized in Chapter 14.

Plate 1. Aerial false-color infrared photo of the mosaic subalpine and alpine communities near Logan Pass. (Photo by the author.)

Plate 2. Aerial false-color infrared photo of the upper coniferous forests and alpine communities near Ruger Lake and Lake Evangeline in the upper Camas drainage. (Photo by the author.)

Plate 3. Aerial false-color infrared photo of the middle Camas drainage near Rogers Lake and Trout Lake. (Photo by the author.)

Plate 4. Aerial false-color infrared photo of the bottomland meadow and forest communities along the Middle Fork of the Flathead River between Apgar Village and West Glacier. (Photo by the author.)

5

Overview of Glacier's Terrestrial Communities

If one were to fly over Glacier National Park in a light aircraft, one would be immediately overwhelmed by row after row of high, snow-capped peaks (several of which reach over 3000 m) exhibiting bare rock and alpine tundra. Although about two-thirds of Glacier's land area is forested, it is this extravagant display of the high country that catches the eye. Following the slopes down to lower elevations, one would observe the alpine meadows give way to larger and larger patches of Krummholz vegetation and, by 1800-m elevation, one would observe primarily continuous, upright forest of *Abies lasiocarpa* and *Picea*.[1] At still lower elevations, the fir and spruce are replaced by *Pseudotsuga* and by either *Tsuga* and *Thuja*, or hybrid *Picea* in the mature stands, and by a complex successional mosaic including *Pinus, Larix, Populus,* and *Betula,* interspersed with patches of meadow, shrubfields, and the ever-present avalanche chutes. At the park's lowest elevations (about 1000 m on the west side and 1350 m on the eastern border), more extensive meadows and prairie remnants and intrusions are encountered, as are more frequently and more intensely burned forest communities.

Indeed, this mosaic is best observed using a light plane and the extended spectral range of false-color infrared film (Kodak Ektachrome Infrared), as shown in Plates 1–4. All were taken by me from a Cessna 182 flying from 600 to 1500 m above the terrain, using a 35 mm Nikon camera, 55 mm lens, and no. 12 Wrattan filter; more details on the use of infrared photographs to interpret vegetation are given in Chapter 14 and in Kessell (1976e). In such photographs coniferous vegetation shows up in various shades of purple, deciduous vegetation appears as shades of pink and red (usually shrubfields are scarlet and meadows are a lighter pink), talus and rock outcrop areas show as shades of grey and greenish, and snow remains white.

Plate 1 shows a good example of the alpine mosaic found above 2000-m elevation. The landscape is a complicated pattern of rock outcrops and talus slides, supporting a very sparse vegetation cover (perhaps 1%–3%), interspersed with areas of alpine meadows, shrubfields, and Krummholz patches. The most important factor contributing to the development of this mosaic is the general progression of primary succession, which occurs as large boulders weather, crack, and break up, forming the talus areas; these in turn support limited vegetation cover, which further hastens the weathering process. Sufficient weathering, soil development, and moisture permit

---

[1]Appendix 1 includes scientific and common names of all plant species mentioned in the text.

development of the meadow communities; these, in turn, give way, in sufficiently sheltered areas, to the shrubfields and patches of Krummholz.

Plate 2 shows the Ruger Lake area at the Camas Creek headwaters. Here we observe the effects of slightly lower elevation (about 1800 m) and topographic sheltering (the lake's outlet is to the north) on the vegetation's development; the Krummholz patches become larger and larger, giving way to *Pinus albicaulis* and almost continuous Krummholz *Abies lasiocarpa* and *Picea* with only scattered patches of meadow and rock outcrop. These are rapidly replaced by typical forests, again primarily composed of *Abies lasiocarpa,* with some *Picea* on the wetter sites and *Pinus albicaulis* in the drier areas and around the outcrops. Large talus slide areas, extending from the solid rock to the meadow–Krummholz interface, are also seen on the west side of the lake.

Plate 3 shows Rogers and Trout Lakes, which are located about 10 km down the drainage from Ruger Lake (the view is facing the northeast). Trout Lake, the upper of the two large lakes, has an elevation of about 1200 m; by now the *Abies–Picea* communities have given way to a dense forest of *Pseudotsuga, Picea,* and some *Abies.* In mature stands, *Picea* predominates on the wetter sites, whereas *Pseudotsuga* is dominant on the more xerophytic slopes; *Abies* and *Betula* are often present in smaller quantities. Late successional stands (which burned 130–200 years ago) also contain some *Pinus monticola* and *Larix occidentalis.* The photograph clearly shows a burn perimeter along the northwest side and outlet of Trout Lake and extending southwest (downstream) to Rogers Lake. This area burned around 1910; the brighter color is caused by the abundance of *Pinus contorta.* The remaining area around the lakes burned not fewer than 175 years ago. Numerous avalanche chutes may be seen on either side of Trout Lake.

Plate 4 was taken about 22 km south of Rogers Lake; it shows the outlet of Lake McDonald (elevation 920 m) and the Apgar area in the foreground, and West Glacier village and the Middle Fork of the Flathead River in the background. (The view is looking south; the stippled areas near the lake's outlet is the Apgar Campground development.) Most of the area between the lake and the river was burned in an intense fire in 1929; its vegetation is predominantly *Pinus contorta* with an understory of *Pseudotsuga* and some *Picea.* On the left side of the photograph are the Belton Hills (elevation about 2000 m); here the 1929 fire gave way to mosaic, patchy burns. Mature forest is found near the hills' summits. The steep, intensely burned, open southwest slopes are experiencing a slow successional recovery; the dominant formation type is shrub (*Acer* and *Ceanothus*), with scattered *Pinus contorta, Pseduotsuga,* and *Betula* stands.

Somewhat different communities occur about 45 km north of West Glacier, near Polebridge and the park's west border. Here, at the lower elevations along the North Fork of the Flathead River, another vegetation mosaic is observed. The area's lower precipitation, fire history, and river action combine to form numerous open mesophytic meadows and occasional prairie remnants. Both open *Pinus ponderosa* savanna and closed *Pinus ponderosa–P. contorta–Pseudotsuga* forests are found. The more mesophytic sites exhibit extensive *Picea–Populus trichocarpa* forest bottomland development, whereas *Populus tremuloides* grovelands are found adjacent to the meadows, prairie intrusions, and some of the ponderosa pine savannas. A complex interaction of climate, soil deposition, and natural fire has established and maintained these communities.

A very different community complex is observed on the east side of the Continental Divide in Glacier National Park. Here, the eastern prairie abruptly reaches Glacier's peaks and the Continental Divide. A complex pattern of xerophytic prairie, mesic and submesic meadows, bottomland successional forests, shrub communities, and the mature *Abies–Picea–Pseudotsuga* forest occur adjacent to *Populus tremuloides* groves, *Pinus contorta–Pseudotsuga* low-elevation successional forests, *Pinus albicaulis–Abies* high-elevation successions, numerous rock outcrop areas, and some *Pinus flexilis* savanna.

Factors Important in Terrestrial Community Development

Even this brief survey of Glacier's major plant communities demonstrates the wide range of processes and conditions that affects community development and species composition. In the high country, the principal effect of primary succession is augmented by sheltering from high wind; local topography; aspect and soil moisture; hydric effects (both near temporary ponds and in the numerous ravines); the common mud slide, rock slide, and avalanche areas; and to a lesser extent by elevation and the infrequent high-elevation fires. At the lower elevations, most areas are sufficiently sheltered and exhibit sufficient soil development to permit the development of forest communities. Here, soil moisture, elevation, and especially fire shape the communities. At the lowest west-side elevations, fire continues to be a determining factor, as does the action of water along the lake bottomlands, the streams, and especially the larger river systems. At low elevations east of the divide, a complex interplay of soil moisture, fire, and exposure to the area's high winds and severe winter storms determine community development.

Glacier National Park also shows the effects of wide phytogeographic influences. An extension of the Pacific coastal climatic peninsula into northwestern Montana (Kirkwood 1922, Habeck 1970a) makes Glacier the eastern limit of the range for several tree species, including *Tsuga heterophylla, Thuja plicata, Abies grandis, Pinus monticola,* and *Taxus brevifolia,* and for numerous herb and shrub species. These species are common only in the mesophytic McDonald valley but do occasionally occur in other drainages. A major portion of the *Abies–Picea* forests in Glacier shows elements indigenous to the Rocky Mountains south of the park; *Picea engelmannii* reaches the northern limit of its range just north of Glacier. The prairie margins and *Populus* grovelands on the park's eastern border contain numerous species common in the grasslands of eastern Montana and southeastern Alberta, whereas alpine elements of Glacier's flora appear to derive from areas of the Rockies north of the park (Habeck 1970a); Glacier lies near the southern edge of the ranges of *Picea glauca* and several of the *Juniperis* and alpine *Salix* species (Habeck and Weaver 1969, Habeck 1970a). Grasslands in the Polebridge area exhibit species common to the Palouse grasslands centered in the states of Washington, Oregon, and Idaho; common species include *Agropyron spicatum, Festuca idahoensis, F. scabrella,* and *Stipa richardsonnii* (Koterba 1968). A final example is the *Larix lyallii* communities found in the northwestern part of the park; the species is generally restricted to alpine communities in the north Cascades and Inland Empire region (Arno 1971).

Glacier's vegetation mosaic is further complicated by the fact that it has all developed during the past 10 000 years, since the last ice retreat. As a result, primary succession is still the single most important determinant of plant communities at the higher elevations, and numerous species exhibit hybridization, wide ecotypic differentiation, and considerable genetic variability, as discussed in Chapters 9, 10, and 11.

Detailed pollen analysis was conducted in two small lakes in the McDonald drainage (Johns Lake and Fish Lake) by Hansen (1946). Results indicate that freshly exposed areas on the west slope were first invaded by *Pinus contorta,* with lesser quantities of *P. monticola, P. ponderosa, Pseudotsuga, Picea,* and *Abies.* As the exposed substrate was developed, *Pinus contorta* dropped from peak pollen percentages of about 80% and gave way to these other species. About 5000 years ago the climate apparently became warmer and drier, as the percentages significantly increased for *P. ponderosa* and *Pseudotsuga.* More recent layers show increased quantities of *Abies, Picea, Thuja,* and *Tsuga* pollen; no *Larix* pollen was found, although the species undoubtedly occurred in the area during this time (Hansen 1946, Habeck 1970a). (Today, the forest around these lakes is *Tsuga* and *Thuja,* with some *Pinus monticola* and *Larix occidentalis; Pinus contorta* is abundant for the first century after a fire. Higher elevation forests around the lakes are predominantly *Abies* and *Picea.*)

The Special Role of Fire

I should like to reemphasize that natural fires have been an extremely important determinant of Glacier's vegetation, especially on low-elevation and xerophytic sites, ever since the Pleistocene. Results of this study (Chap. 9) and of Singer's (1975a) suggest that over 95% of Glacier's forests below 1500 m have burned within the past 400 years. Fire periodicity in

Glacier National Park ranges from a few years for xerophytic grasslands, to a stretch of 10–15 years for open xerophytic savanna, to perhaps 25–100 years for other low-elevation subxeric forest communities (Singer 1975a). Even most of the mesophytic *Tsuga–Thuja* forests have burned in the past 250 years, and the periodicity for high-elevation mesophytic communities, and for much of the alpine, is perhaps 200–600 years.

Most of Glacier's fires are natural ignitions by dry lightning during July and August; simultaneous starts of several dozen fires from a single storm are not uncommon (O'Brien 1969). These fires were allowed to run their course prior to the park's establishment in 1910; for all practical purposes, effective fire control was not possible before 1930. Since that time, the park has operated on a policy of total fire suppression; as a result, most of Glacier's fires are now stopped while still of class A size (less than 0.1 ha). Occasionally, such as in 1929, 1936, and 1967, extremely dry conditions and high winds permit large conflagrations; there is little humans can do to control such a fire in a remote wilderness area.

Ayres (1900) devoted considerable time to describing the effects of natural fire on Glacier's vegetation; he noted it as one of the most important determinants of its plant communities. The park's policy of total fire suppression has undoubtedly significantly altered the pristine ecosystem the National Park Service is charged with protecting by threatening several fire-dependent plant species and thus decreasing plant species diversity and detrimentally affecting some animal populations; it has also permitted unnatural fuel buildups and expended many million tax dollars. The Service has been severely criticized by Habeck (1970b), by myself, (Kessell 1976c, 1978), and by others for this policy of total fire suppression; more enlightened fire management planning over the past 2 years indicates that some natural fires may be permitted to burn beginning in 1979.

Whereas the Service exhibits less cautious safeguards on such projects as sewer construction, bear management, and back-country planning, the park's extremely conservative posture on fire has been a disappointment to many. As a result, a major goal of the gradient modeling system was the development of an ecologically sound and administratively feasible plan for restoring fire to its natural role in Glacier National Park.

6

Field Methodology

In conducting the field work to construct the gradient model, Glacier National Park was divided into three geographic units (Fig. 6.1). Area A is the old West Lakes district; it includes that portion of the park west of the Continental Divide, from the Canadian border south to and including the Lincoln drainage (the next major drainage south of the McDonald drainage). Its total area is about 160 000 ha. This was the original study area in Glacier; vegetation sampling was conducted in this area from 1972 through 1974. Fuel sampling was conducted in this area during 1974 and 1975. The results, analyses, conclusions, and incorporation of these data into the gradient model are the subject of Chapters 7–12.

Area B includes the Saint Mary and Many Glacier drainages which lie east of the continental divide. Their total area is about 90 000 ha. Extensive vegetation sampling was conducted in this area during the summer of 1975; fuel sampling was conducted in this area during the summers of 1974 and 1975. The fuel data collected during the summer of 1974 were used to develop the initial gradient fuel models (Kessell 1976a, b); however, the fuel analyses presented here are based on data collected west of the continental divide.

Section C includes the Belly River and Waterton drainages at the northeast corner of the park, and the Middle Fork of the Flathead River, Cut Bank, Two Medicine, and East Glacier areas at the south portion of the park. The total area is about 150 000 ha. Except for limited fuel sampling, no field work has yet been conducted in these areas.

In terms of how it contributes to the gradient model system, the field work may be grouped into six convenient categories: vegetation sampling, fuel sampling, shrub dimension analysis, fuel moisture and microclimate study, aerial infrared photography, and fire behavior monitoring. Each is described below.

Vegetation Sampling

Field vegetation sampling was conducted to obtain detailed site descriptions, overstory composition (both density and basal area) by species, understory composition (cover and frequency) by species, and any other unusual or botanically interesting information. A total of 696 stands was sampled in the West Lakes area; an additional 413 stands were sampled east of the Continental Divide during 1975.

As described in Chapter 3, Whittaker (1973d) notes that vegetation

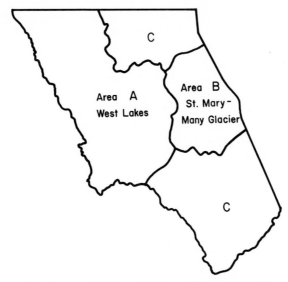

Figure 6.1. Sketch map of Glacier National Park showing the study areas.

samples should be sufficiently large to represent the community, homogeneous, efficient, and appropriate to the purposes of the study. A large, remote, terribly inaccessible, and frequently inhospitable wilderness such as Glacier National Park requires a sampling method tailored to collect the necessary and most interesting data during short summer seasons for a wide range of plant communities.

Stands were selected to obtain representative samples from virtually all possible combinations of environmental conditions found in the study area. During 1972 and 1973, they were selected by hiking all the major trails and taking samples whenever a new vegetation complex was noted. A significant attempt was made to obtain replicate samples for each environmental complex. By the summer of 1974, virtually all common habitats had been sampled, and the emphasis shifted to obtaining the unusual combinations of site factors that are not often encountered; these included burns in mesophytic sites, low-intensity burns in xerophytic sites, burns during years of low fire occurrence, sheltered areas at the higher elevations, and the like.

No attempt was made to select stands at random (stands were defined as areas as large as, or larger than, the normal sample size that exhibited fairly homogeneous environmental and vegetative composition). In an area as large and as inaccessible as Glacier, a truly random sampling technique would be an extremely inefficient way to obtain the necessary representative samples. However, once a stand was selected, the areas selected for canopy data collection and the locations of all herb and shrub quadrats were selected at random (from a table of random distances and directions, by throwing a rock, or some other appropriate method).

When the area for the canopy sample was established, a detailed site description was recorded. This included a sequential plot number, date, a notation on the type of sampling method (discussed below), the 7.5 minute topographic map code number, and the Universal Transverse Mercator (UTM) coordinates to the nearest 100 m. We then recorded the elevation, topographic type(s) (among bottomland, ravine, draw, sheltered slope, open slope, undifferentiated slope, peak, ridge, or xeric flat), and aspect (on an eight-point magnetic scale). Next a notation was made of whether the stand had burned during the past 200 years; if the stand was less than 200 years old, the burn intensity was estimated on a scale of full canopy, mosaic canopy, partial canopy, understory only, or available fuel (the latter category indicates an area without a developed canopy). The number of years since the last fire was also estimated; this was usually done by taking increment cores from several trees but occasionally was taken from fire records if the stand had been burned by a large fire since 1910. Cover type was recorded along with an estimate of the stands location on the primary succession gradient. Any indications of other disturbances (slides, hydric, etc.) were also noted. Finally, drainage area and a location code were recorded. In addition to providing a detailed site description, these data are sufficient to approximately locate the stand on all the gradients and within the categories of variation used by the model (Chap. 8).

Eight different sampling combinations were used. Type A samples included the regular overstory sampling procedures. In addition, for an area of about 15 × 30 m, the location, species, and diameter of every tree were recorded and mapped. Two increment cores were taken from each tree for later correlation to macroclimatic variation. Field notes were recorded as appropriate. Type B samples included the regular overstory sampling procedures and field notes as appropriate. In type C samples, extensive field notes on species composition were recorded *in lieu* of the regular overstory sampling procedure. Type B-with cores samples were identical to type B samples except they also included increment cores taken from several trees for later correlation to macroclimatic data; however, unlike the type A sample process, the locations of individual trees were not mapped (this technique was used primarily in the alpine areas, whereas type A samples were taken in the *Thuja–Tsuga* forests). Types AH, BH, CH, and BH-with cores samples were similar to types A, B, C and B-with cores samples, except herb and shrub quadrat data were also recorded. In addition, in type AH samples, detailed soil moisture readings and an herb quadrat were taken in each 5 × 5 m portion of the sampling area.

Next, overstory species composition was recorded. (All species nomenclature follows Hitchcock and Cronquist 1973 and Kessell 1974; see Appendix 1). Tree species, here taken as species with over 2.54 cm dbh (diameter at breast height) were recorded by species and size class (2.5–7.6 cm, 7.6–15.2 cm, and by 15.2 cm increments above that value). The method does not use a sample of fixed area, nor is the exact size of the land area sampled recorded. I have been criticized for this technique, but I defend it and use it for the following reasons.

1) No single canopy sampling area is efficient throughout the study area in Glacier National Park. The standard tenth-hectare sample is too small and would include only a few trees in an open *Pinus ponderosa* savanna, but it is much too large and would include over 4000 trees in an early seral *P. contorta* community.

2) No single number of individuals to be included in a sample is efficient throughout the study area. For example, a technique that required sampling 100 trees in every stand would be inadequate for some late seral low-elevation forests and alpine Krummholz communities but unnecessarily large for a *Tsuga* ravine.

3) Because of the refinement of the model (six gradients and four additional categories of disturbance), it is often necessary to take a very small sample in order not to include portions of the community with different locations on the gradients (such as a different aspect, topography, cover, or age). This is especially true for some small draws and ravines that occur on slopes and may include only a handful of individual trees; it is also true for small pockets of low burn intensity located in a primarily high-intensity fire site.

4) Although the vegetation sampling technique provides relative density and relative basal area, it does not provide absolute density and basal area. However, these latter parameters are determined from the fuel sampling (described below). The total absolute density and basal areas (all species) determined from the 0.01-ha canopy fuel samples may be multiplied by the species' relativized values to obtain absolute density and basal area by species if desired.

Under normal circumstances, about 75–125 trees were included in each sample, but this was sometimes modified for the reasons noted above.

Understory species composition sampling was added to the sampling procedure during the summer of 1973. To sample herb and shrub species composition, quadrats were located at random within the area sampled for overstory species composition. Once again, the number of herb quadrats and their sizes varied considerably, depending on the diversity of the understory, the homogeniety of the understory, the distribution of individuals (random, uniform, or contagious) within a homogeneous area, and the size of the understory plants. When plants were small and the area was fairly homogeneous, two 2 × 2 m quadrats were used. When plants were larger or the area was rather heterogeneous, larger quadrats, such as 5 × 5 m or 5 × 10 m values, were used. Herbs and shrubs were recorded by absolute cover for each quadrat on an eight-point scale:

0	Not present
T	Trace (less than 1%)
1	1–5%
2	5–25%
3	25–50%
4	50–75%
5	75–95%
6	95–100%

The numbers and sizes of the quadrats were recorded, and a notation was made if the different quadrats exhibited obviously different soil moisture regimes.

Finally, field notes were taken to indicate any common species in the stand that did not fall in the tree or understory sampling quadrats; they included any other descriptive information that the field sampler believed might be of value in the data reduction and analysis.

Fuel Sampling

The purpose of field fuel sampling is to collect data that allow determination of:

1) Fuel loadings (dry weight per unit area) for various size classes of dead and down branchwood fuel, litter, standing grass and forbs, and woody shrubs (separated by foliage and branchwood)
2) Parameters that describe the vertical and horizontal spatial distribution of these fuels
3) Parameters that describe the canopy dimensions and species composition

Fuel sampling was initiated during the summer of 1974 using a method very similar to the planar intercept technique of Brown (1971a, 1974) but modified to use uniform metric (rather than English) measurements. Approximately 150 stations were sampled by Glacier's forestry technicians using this method under the supervision of the Fire Management Specialist. Concurrently, I applied and modified a method he devised for efficiently describing the horizontal spatial distribution patterns for various fuel categories and classes; another 55 stations were sampled using both these techniques and the planar intercept method. Experience gained during this period permitted the development of a new set of methods designed to meet the purposes stated above. Over 600 two-station samples were recorded during the summer of 1975 using this new method; approximately 75% of these samples were taken west of the continental divide in the West Lakes area.

The general layout of a standard two-station fuel sample is shown in Fig. 6.2a. At each station, parameters were recorded to permit derivation of dry weight per unit area for each ground fuel category and size class; fuel depth (differentiated by litter, duff, grass and forbs, shrubs, highest dead fuels, and average dead fuel depth); and cover (differentiated by litter, grass

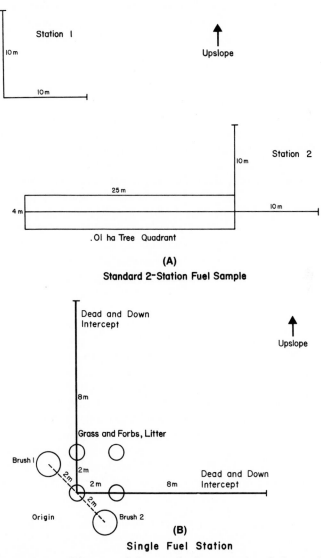

Figure 6.2a and b. a Diagram of the standard two-station fuel sample plot. b Diagram detailing the layout of a single fuel sample station.

and forbs, and shrubs). A 0.01-ha canopy sample was also taken; recorded data included species and diameter (to the nearest centimeter) for all trees, canopy cover, and paired height to top of ground fuel–height to lowest living branch at six points along the transect. In addition, while three crew members were collecting these data, the fourth member recorded variance/ mean contagion estimates for six fuel classes on appropriate grid sizes. The entire procedure required about 45–60 min per sample with a crew of four trained individuals.

When the sampling crew members reached a fuel sampling site, they recorded site parameters using the same method described for vegetation sampling. Once this was completed, the first fuel station was located at random within the larger stand.

The layout for a single fuel station is shown in Fig. 6.2b. After the origin was pinned, aspect and dip were recorded to the nearest $10°$ and $1°$, respectively. A 2-m intercept line was then placed along the slope ($90°$ counterclockwise from the aspect direction), and duff depth was recorded (in millimeters) at points 0.3, 0.6, and 0.9 m from the origin. Heights of the highest dead and down branchwood fuel intercepting the line were also recorded (in centimeters) along the 0.0–0.3 m, 0.3–0.6 m, and 0.6–0.9 m portions of the intercept. Dead and down intercepts for three size classes (1-, 10-, and 100-hr time lag)[1] were then counted along the entire 2-m line. Next, the line was extended an additional 8 m (to a total of 10 m), and greater than 100-hour intercepts were recorded by diameter (and classified as sound or rotten) along this 10-m line.

Once this was completed, an identical transect was taken in the upslope direction ($180°$ from the aspect direction). Note that the original Brown method takes a single intercept in a random direction. However, the conversion of intercept counts to weight per unit area uses a method that assumes a random orientation of branchwood (see Chap. 7). This assumption is probably unrealistic on steep slopes; therefore, intercepts were taken in both the across-slope and upslope directions to minimize this bias.

After the intercepts were completed, four 0.1-m^2 area circular plots were established (as shown in Fig. 6.2) for grass and forbs and litter sampling. The circular plot with the estimated greatest volume (and so the greatest dry weight) of standing grass and forbs was selected and declared the "grass and forbs base plot." The average height of standing grass and forbs in this plot was recorded; the percent cover (estimated) and percent green (also estimated) were also recorded. For each of the other three circular plots, the estimated relative volume (compared to the base plot), percent

---

[1]Fuels are classified by moisture time lag, an estimate of the delay time before they respond to a humidity change. One-hour fuels are less than 0.64 cm in diameter; 10-hr fuels range from 0.64 to 2.54 cm in diameter; hundred-hour fuels range from 2.54 to 7.62 cm in diameter; greater than hundred-hour fuels are greater than 7.62 cm in diameter.

green, and average height were then recorded. When this was completed, all standing grass and forbs in the base plot were clipped and returned to the lab for oven drying and weighing. (The method assumes that the dry weight for the other three plots may be closely estimated by multiplying the dry weight of the base plot grass and forbs by the relative volume recorded for the other circular plots).

Litter loadings were determined in a similar fashion. The same circular plots were used; the one with the estimated greatest volume of litter was declared the "litter base plot" (it was not necessarily the same plot declared as the grass and forbs base plot). The depth of the litter (in centimeters) and percent cover were recorded for the base plot. Next, the relative volume (compared to the base plot) and average litter depth were recorded for the three remaining circular plots. When this was completed, all litter in the base plot was bagged and returned to the lab for oven drying and weighing. Litter loading estimates for the three other plots were determined by the same method described for grass and forbs. When the litter sampling is complete, "average fuel depths" (the depth that includes the combined lower 80% of the dead litter, 1-hr, and 10-hr branchwood) were estimated for the circular plot at the origin and for the circular plot diagonally opposite the origin.

Next, two circular plots with areas of 1.0 m² were established, as shown in Fig. 6.2b, to record standing (woody) shrub fuel. Within each circular plot, the average height of and percent cover by standing woody shrubs were recorded. For each shrub, a six-digit species code was recorded, and each shrub was classified by diameter (classes of 0–5 mm, 5–10 mm, 10–15 mm, 15–20 mm, 20–30 mm, and 30–50 mm; all shrubs with a diameter greater than 50 mm were arbitrarily considered as trees for fuel sampling purposes). The analysis to determine fuel loadings from species and diameter use the results of the shrub dimension analysis described below.

This completed the single fuel station; the second station was located by consulting a table of random distances and directions and pacing its origin from the first station's origin. Sampling procedure for this station was exactly like the procedure for the first station. (Note that in unusual circumstances, samples might include as few as one or as many as four stations; however, the vast majority consisted of two stations.)

When the second station was completed, a 0.01 ha canopy sample was conducted. The origin for this sample was the origin of the last station. From this point, a 25-m line is placed across the slope (90° from the aspect direction); the sample quadrat extends 2 m on both sides of this line. Every tree located within this quadrat (trees were defined for fuel sampling purposes as woody vascular plants over 50 mm in diameter) was recorded by species and diameter to the nearest centimeter. Average percent canopy cover over the quadrat was estimated. At six points along the quadrat centerline (beginning at the origin, and then at 5.0-m intervals) the height of the highest dead ground fuel and the height of the lowest living canopy branch, within a circle of 1-m radius, were recorded.

While the above work was being conducted, one individual determined an estimate of the spatial heterogeniety of ground fuels. This was done by locating a "contagion origin" a random distance and direction from the first station's origin. Starting at this origin and working in the upslope and across-slope directions, various 5×5 element grid systems were superimposed on the ground fuel. The grid dimensions depended on the fuel type being measured; grids were squares 0.1 m on a side for litter, 0.5 m on a side for grass and forbs, 1.0 m on a side for shrubs, 0.2 m on a side for 1-hr branchwood, 1.0 m on a side for 10-hr branchwood, and 5.0 m on a side for 100-hr and larger branchwood. (Thus a total of five different grid systems was used.) For each fuel type, the importance value of the fuel type in each of the 25 grid elements was recorded. For litter, grass and forbs, shrubs, and 1-hr fuels, the importance value was percent cover; for 10-hr and larger fuels, the importance value was relative volume. The mean, variance, and variance/mean ratio were then calculated (using electronic calculators in the field). The variance/mean ratios recorded using these synthetic importance values were later transformed to variance/mean ratios that corresponded to an importance value for loading (metric tons per hectare). Although the fuel sampling method seems rather complicated, it efficiently recorded all the necessary parameters on ground fuels and no more than 1 week was required to train a proficient crew.

Experience with the above method during the summer of 1975 has suggested several modifications and improvements to the method. These are currently being tested by Gradient Modeling, Inc. personnel in the United States, Canada, and Australia.

Shrub Dimension Analysis

The shrub fuel sampling procedures described above require estimating shrub foliage and branchwood dry weight per unit area from species and diameter alone. Although it is feasible to bag and return to the lab litter and grass and forbs samples, bulk limitations (and an attempt to minimize destructive sampling) require the use of an indirect method for estimating shrub mass. Because of this, a shrub dimension analysis was conducted during the summer of 1975. Glacier's 20 common shrub species (and three tree seedling species) were collected, including 20–40 individuals of each species. The purpose was to estimate dry weight of foliage and branchwood and to divide branchwood into the three time-lag categories, from species and diameter at ground height (dgh) alone using log–log regressions. The results are summarized in Chapter 9.

Fuel Moisture and Microclimate Study

The efficient use of a real-time fire behavior model for a wilderness area requires the capability to extrapolate fuel moisture and microclimate parameters from base stations to remote wilderness sites. The required site parameters are recorded as part of the field vegetation samples and resource inventory. The key weather parameters therefore must be recorded under a variety of site conditions to permit the extrapolation of both diurnal and seasonal variations. To accomplish this, 10 new remote weather stations were established in the McDonald drainage during the summer of 1975. Two contained recording thermometers and barographs and five contained recording hygrometers. Each station was visited every day, between 1200 and 1600 hours MDT; 1-hr and 10-hr time-lag moisture sticks[1] were recorded at this time, along with temperature, humidity, wind speed, and wind direction. Every few days, samples of litter and grass and forbs were collected, oven dried, and weighed to determine moisture content of these fuels.

Glacier's four manned fire lookouts were also equipped with portable weather equipment and recorded temperature, humidity, wind speed, and wind direction four times a day; the 1400 hours MDT reading was radioed to the fire laboratory for use in fire control planning. The Swiftcurrent lookout recorded these parameters, along with 1- and 10-hr time-lag fuel sticks, every hour from 0700 to 2100 hours MDT for the entire summer.

In addition, most fuel samples taken by Gradient Modeling, Inc. crews were preceded by a weather observation (temperature, humidity, wind speed, and wind direction) using standard belt weather kits. Weather personnel also took numerous remote readings with portable equipment throughout the summer and established a number of wind vertical profiles. Preliminary results of this work were reported by Mason and Kessell (1975); more detailed information is included in Chapter 14.

False-Color Infrared Aerial Photography

The resource inventory procedure described in Chapter 14 requires coding detailed site information from aerial photographs. Although the park financed a set of 1:16 000 black and white vertical aerial photographs in 1968, no regular color or false-color aerial photos were available when this project was initiated. As a result, and because of numerous phytosociologic

---

[1]Moisture sticks are physical fuel moisture analogs, manufactured from *Pinus ponderosa* and trimmed to a dry weight of 100 g.

and management uses for false-color infrared aerial photos, I took a complete set of aerial obliques (approximately 1500 35-mm transparencies) during the 1972–1975 period. These cover the entire West Lakes, Saint Mary, and Many Glacier drainages (250 000 ha) and are archived with Gradient Modeling, Inc. Examples are included in Plates 1–4 in Chapter 5.

Fire Behavior Monitoring

Any fire behavior model's predictions must be checked against observed fire behavior to test the model's accuracy. Because of this need for detailed fire behavior information, all significant fires that have occurred in Glacier National Park (beginning with the 1974 fire season) have been manned (and photographed in false-color infrared from the air) by Gradient Modeling personnel. These include the 1974 Curley fire, the 1974 Snyder fire, the 1975 Redhorn fire, and the 1976 Larsen fire (Kessell 1976b). Although fuel loadings are predicted by the gradient model, field fuel samples were taken either during or just after the fire to check fuel inference and prediction accuracy. During the 1974 Curley fire, I established fuel samples that burned a few hours later, before the fire was controlled. In addition, project personnel have provided fire intelligence data to the Fire Management Officer for each of these fires. The results of this work are included in Chapter 15.

7

Analytical Methodology

The field data collected using the methods described in the preceding chapter were used to construct the gradient vegetation and fuel models and the resource and fire management model. The environmental gradients, and hence the gradient models, were derived solely from the vegetation data. Once the gradient framework was established, the distribution of flammable fuels on these gradients was determined. This chapter describes the analytical procedures used to tabulate and reduce these data and to construct the gradient models. The reduction and tabulation of the results of other field work required for the management and fire model, including the shrub dimension analysis, fuel moisture and microclimate study, infrared photography, and fire behavior monitoring, are discussed in Section III.

Vegetation and Fuels Data Tabulation

Before the field data can be used for gradient analysis, modeling, or other synecologic purposes, it is necessary to reduce them to a meaningful form and to prepare tabulations or stand summaries. For vegetation samples, such tabulation should include a résumé of sampling procedures, a detailed site description, quantitative information on the overstory and understory species composition, and at least a notation of, or an abridged version of, appropriate field notes. Because diversity relationships were of major interest in the development of this model, rather detailed overstory and understory diversity calculations were also included in the summaries.

A computer program was developed to prepare stand summaries from the keypunched stand sample records. Summary information includes a full site description, overstory species by density and basal area, understory species by absolute cover and frequency, and abridged field notes. In addition, diversity was calculated separately for overstory density, overstory basal area, and understory cover, using six different diversity measures: number of species present, the Shannon information index (log base e), e raised to the Shannon index value, the Simpson index, the reciprocal of the Simpson index, and Whittaker's E_c (the inverse of a straight line fit to the dominance–diversity curve).[1] A sample printout for two stands is shown in Table 7.1; the full set of stand summaries for all samples is archived in Glacier National Park and with Gradient Modeling, Inc.

---

[1]These diversity measures are described in detail in Chap. 12; *cf.* Peet 1974.

Table 7.1. Sample Printouts for Two Stands from the Vegetation Samples Tabulation Program.[a]

PLOT 523 KINTLA WATERSHED

USGS 7.5 MIN. TOPO. MAP - MOUNT CARTER

TYPE OF SAMPLE - B H WITH CORES   DATE SAMPLED - 080673

LOCATION - BOULDER PASS

    UTM - 5426900 NORTH 0712200 EAST

COVER TYPES - MEADOW - FOREST

ELEVATION - 2287. M (7500. FEET)

TOPO. TYPE - OPEN SLOPE
EXPOSURE NORTH

PRIMARY SUCCESSION GRADIENT - 50

ALPINE (WIND - SNOW) GRADIENT - PASS

DISTURBANCES - NONE

TREE DATA -

SPECIES	NUMBER OF STEMS	REL. NUMBER OF STEMS (0/0)	BASAL AREA (SQ. M)	REL. BASAL AREA (0/0)
ABIES LASIOCARPA	5	11.4	0.01900	7.6
PICEA	3	6.8	0.01500	6.0
LARIX LYALLII	36	81.8	0.21600	86.4
TOTAL	44	100.0	0.25000	100.0

    3 SPECIES PRESENT

DIVERSITY MEASURES (RELATIVE DENSITY) -
    HPRIME = 0.594 E TO THE HPRIME= 1.812 C= 0.687
    1/C = 1.456 EC = 2.780

DIVERSITY MEASURES (RELATIVE BASAL AREA) -
    HPRIME = 0.491 E TO THE HPRIME= 1.634 C = 0.756
    1/C = 1.323 EC = 2.590

HERB AND SHRUB DATA -

QUADRATS USED -
    3 5×5 M DIFFERENT MOISTURE CONDITIONS

Table 7.1. Sample Printouts for Two Stands from the Vegetation Samples Tabulation Program[a] (*Continued.*)

	SPECIES	MEAN COVER	FREQUENCY
150108	JUNCUS DRUMMONDII	1	1.00
150200	LUZULA SPECIES	2	1.00
160100	CAREX SPECIES	0	0.33
160100	CAREX SPECIES	2	1.00
160100	CAREX SPECIES	2	1.00
160115	CARIX DIOICA	2	1.00
160137	CAREX NIGRICANS	2	1.00
160158	CAREX SPECTABILIS	1	0.33
172800	POA SPECIES	0	0.33
172804	POA CUSICKII	1	0.67
221301	VERATRUM VIRIDE	0	0.33
250202	SALIX ARCTICA	0	0.33
250217	SALIX NIVALIS	3	0.67
300201	OXYRIA DIGYNA	2	1.00
300312	POLYGONUM VIVIPARUM	0	0.33
360204	ANEMONE OCCIDENTALIS	1	0.67
360708	RANUNCULUS IN AMOENUS	1	1.00
360709	RANUNCULUS MACOUNII	0	0.33
400207	ARABIS LYALLII	0	0.33
440501	PARNASSIA FIMBRIATA	1	0.33
440608	SAXIFRAGA LYALLII	2	1.00
470804	POTENTILLA DIVERSIFOLIA	1	1.00
620401	EPILOBIUM ALPINUM	1	1.00
681003	PHYLLODOCE INTERMEDIA	1	0.33
760304	PHACELIA SERICEA	1	0.67
800205	CASTILLEJA OCCIDENTALIS	1	0.67
800206	CASTILLEJA RHEXIFOLIA	0	0.33
800904	MIMULUS LEWISII	0	0.33
801505	VERONICA WORMSKJOLDII	0	0.67
860104	VALERIANA STICHENSIS	1	0.67
880101	ACHILLEA MILLEFOLIUM	1	0.67
880501	ANAPHALIS MARGARITACEA	1	1.00
880709	ARNICA MOLLIS	2	1.00
881909	ERIGERON PEREGRINUM	2	1.00
882604	HIERACIUM GRACILE	0	0.67
883709	SENECIO RESEDIFOLIUS	1	1.00
883711	SENECIO TRIANGULARIS	2	1.00
990000	MOSS	2	1.00

TOTAL COVER = 6   38 SPECIES PRESENT

DIVERSITY MEASURES (COVER) -
$\quad$ HPRIME = 3.050 E TO THE HPRIME = 21.109 C = 0.063
$\quad$ 1/C = 15.926 EC = 20.266

Table 7.1. Sample Printouts for Two Stands from the Vegetation Samples Tabulation Program[a] (*Continued.*)

FIELD NOTES -

BOULDER PASS TRANSECT
DETAILS ON ORIGINAL FIELD SHEETS.

– –

PLOT 557 MCDONALD WATERSHED

USGS 7.5 MIN. TOPO. MAP - LAKE MCDONALD EAST

TYPE OF SAMPLE - B H   DATE SAMPLED - 082373

LOCATION - TRAIL LAKE MCDONALD - SPERRY CHALET

    UTM - 5388000 NORTH 0289250 EAST

COVER TYPE - FOREST

ELEVATION - 1159. M (3800. FEET)

TOPO. TYPES - OPEN SLOPE - SHELTERED SLOPE
EXPOSURE SOUTHWEST

PRIMARY SUCCESSION GRADIENT - N/A

ALPINE (WIND-SNOW) GRADIENT - N/A

DISTURBANCE - BURN
    TYPE OF BURN - UNDERSTORY
    APPROX. YEARS SINCE BURN 70

TREE DATA -

SPECIES	NUMBER OF STEMS	REL. NUMBER OF STEMS (0/0)	BASAL AREA (SQ. M)	REL. BASAL AREA (0/0)
PICEA	7	5.9	0.22000	12.1
THUJA PLICATA	4	3.4	0.03500	1.9
PINUS MONTICOLA	14	11.9	0.26100	14.4
PINUS CONTORTA	35	29.7	0.25300	13.9
PINUS PONDEROSA	1	0.8	0.38500	21.2
LARIX OCCIDENTALIS	1	0.8	0.00200	0.1
PSEUDOTSUGA MENZIESII	56	47.5	0.66200	36.4
TOTAL	118	100.0	1.81800	100.0

    7 SPECIES PRESENT

Table 7.1. Sample Printouts for Two Stands from the Vegetation Samples Tabulation Program[a] (*Continued.*)

DIVERSITY MEASURES (RELATIVE DENSITY) -
   HPRIME = 1.330 E TO THE HPRIME = 3.782 C = 0.332
   1/C = 3.011 EC = 4.004

DIVERSITY MEASURES (RELATIVE BASAL AREA) -
   HPRIME = 1.589 E TO THE HPRIME = 4.898 C = 0.232
   1/C = 4.302 EC = 2.778

HERB AND SHRUB DATA -

QUADRATS USED -
  3 1×1 M

SPECIES		MEAN COVER	FREQUENCY
10102	LYCOPODIUM ANNOTINUM	0	0.33
61201	PTERIDIUM AQUILINUM	1	0.33
170000	GRAMINEAE	0	0.33
172800	POA SPECIES	1	0.67
221401	XEROPHYLLUM TENAX	2	1.00
471001	ROSA ACICULARIS	1	0.67
471400	SPIRAEA SPECIES	2	1.00
680101	ARCTOSTAPHYLOS UVA–URSI	2	1.00
680301	CHIMAPHILA UMBELLATA	1	0.67
850101	LINNAEA BOREALIS	2	1.00
861301	VACCINIUM CAESPITOSUM	2	0.67
882602	HIERACIUM ALBIFLORUM	1	1.00
990000	MOSS	2	1.00

TOTAL COVER = 5   13 SPECIES PRESENT

DIVERSITY MEASURES (COVER) –
   HPRIME = 2.215 E TO THE HPRIME = 9.163 C = 0.124
   1/C = 8.052 EC = 8.801

FIELD NOTES -

PARTIAL CANOPY BURN OF MUCH OLDER BURN. P. PONEROSA +
LARIX SURVIVING BOTH BURNS. P. CONTORTA RESULTING FROM
LAST BURN.
CANOPY COVER 3 - 4. SEEDLING - SHRUB COVER 3-4.

[a]Herb and shrub species importance values are described in Chapter 6; on the printout, a mean cover of 0 indicates trace (less than 1% absolute cover).

A computer program was also developed to reduce fuel sample data and prepare stand summaries. This program both tabulates and performs simple statistics on fuel properties measured directly in the field (such as fuel depths and percent cover) and computes the dead and down branchwood loadings from the intercept data. This is accomplished by converting the intercept counts to branchwood volume per unit area of ground surface and then converting branchwood volume to branchwood dry weight. The procedure is documented in Brown (1971, 1974). The fuel packing ratios are also calculated. The program both lists the measured variance/mean contagion values for each fuel type using the field synthetic importance values and computes the transformed variance/mean ratios when fuel loading (metric tons per hectare) is the importance value. The program also prepares a table of overstory species importance by species and size class.

Output for a single sample stand is shown in Table 7.2. The first page of output gives a detailed site description and fuel summary for all fuel stations. This is followed by a one-page summary for each individual fuel station recorded within the stand (usually two). The last section gives the overstory and contagion data. A complete set of printouts for over 600 fuel samples is archived in Glacier National Park.

These stand by stand vegetation and fuel summaries were used to construct the gradient model of Glacier National Park's vegetation and fuel.

Gradient Analysis

In studying a complex vegetation pattern, plant ecologists seek a description and understanding on three levels—environmental factors, species populations, and community characteristics. As reviewed in Chapter 3, studies that seek to arrange samples in relation to one or more environmental gradients or axes are termed gradient analysis (or ordination). Studies that seek to group samples together on the basis of shared characteristics into abstract units are termed classification (Whittaker 1973a). Although often viewed as alternative, competitive, or mutually exclusive approaches, most vegetation studies that hope to achieve an understanding of environmental factors, species populations, and community characteristics require the integrated application of both approaches (Whittaker 1967, 1973a, Kessell 1976b, 1979b).

Despite being termed a gradient model and placing very great reliance on gradient analysis, the Glacier National Park study used both ordination and classification of the vegetation. However, the philosophical approach was to assume a response of the vegetation to environmental gradients until otherwise demonstrated. Therefore, classification was used only (1) when gradient analysis demonstrated that classification was possible without a

Table 7.2. A Sample Printout for One Stand from the Fuel Samples Tabulation Program, Including Site Description, Fuel Loadings, Packing Ratios (Betas), Contagion Values, Shrub Stratum Summary, and Canopy Summary.

PLOT 2232 BOWMAN DRAINAGE.

USGS 7.5 MIN TOPO MAP - MOUNT CARTER

UTM - 5424.0 NORTH 713.9 EAST

NUMBER OF STATIONS - 2   DATE 750821 TIME 1200

LOCATION - TRAIL : BROWN PASS - BOWMAN LAKE

COVER TYPE - TYPICAL (UPRIGHT) FOREST

ELEVATION - 1329. M (4360. FEET) MSL

TOPO-MOISTURE - OPEN SLOPE

NORTHEAST ASPECT

PRIMARY SUCCESSION - 99

ALPINE (WIND SNOW) GRADIENT - NONE

FIRE HISTORY -

    BURNED ABOUT 200 YEARS AGO

    INTENSITY - FULL CANOPY

OTHER DISTURBANCES - NONE

STAND SUMMARY

LIVE AND STANDING DEAD -	
GRASS AND FORBS	3.18 METRIC TONS / HECTARE
SHRUB FOLIAGE	0.20 METRIC TONS / HECTARE
SHRUB BRANCHWOOD	0.15 METRIC TONS / HECTARE
DEAD AND DOWN -	
LITTER	1.47 METRIC TONS / HECTARE
1 HOUR D AND D	2.57 METRIC TONS / HECTARE
10 HOUR D AND D	4.78 METRIC TONS / HECTARE
100 HOUR D AND D	2.55 METRIC TONS / HECTARE
>100 HOUR D AND D	157.96 METRIC TONS / HECTARE

AVERAGE FUEL DEPTH = 0.8 CM

Table 7.2. A Sample Printout for One Stand from the Fuel Samples Tabulation Program, Including Site Description, Fuel Loadings, Packing Ratios (Betas), Contagion Values, Shrub Stratum Summary, and Canopy Summary (*Continued.*)

BETAS—
    COMPLEX BETA 1 (AV FU DEP) = 0.1903
    COMPLEX BETA 2 (G + F DEP) = 1.5266
    MEAN COMPLEX BETA = 0.8584

    STRATUM BETA (LITTER) = 0.0383
    STRATUM BETA (G + F DEP) = 1.3337
    STRATUM BETA (SHRUBS) = 0.5000

		---CONTAGION---	
FUEL	MEAN	VAR/MEAN	TRANSFORMED VAR/MEAN
GRASS+FORBS	55	18.34	1.06
SHRUBS	14	21.71	0.55
LITTER	70	11.79	0.25
1 HR D + D	2	0.96	1.24
10 HR D + D	9	1.42	0.75
100+ > D + D	12	2.72	36.38

(PLOT 2232) STATION 1 (ASPECT = 0 DEG, DIP = 25 DEG)
LIVE AND STANDING DEAD--
    GRASS AND FORBS   2.63 METRIC TONS/HECTARE
    SHRUB FOLIAGE   0.41 METRIC TONS/HECTARE
    SHRUB BRANCHWOOD   0.30 METRIC TONS/HECTARE
DEAD AND DOWN--
LITTER   2.90 METRIC TONS/HECTARE
1 HOUR D AND D   2.51 METRIC TONS/HECTARE
10 HOUR D AND D   6.41 METRIC TONS/HECTARE
100 HOUR D AND D   5.10 METRIC TONS/HECTARE
>100 HOUR D AND D   266.33 METRIC TONS/HECTARE

AVERAGE FUEL DEPTH = 1.5 CM

BETAS---
    COMPLEX BETA 1 (AV FU DEP) = 0.1536
    COMPLEX BETA 2 (G + F DEP) = 1.5266
    MEAN COMPLEX BETA = 0.8584

    STRATUM BETA (LITTER) = .0383
    STRATUM BETA (G + F DEP) = 1.3337
    STRATUM BETA (SHRUBS) = 0.5000

                      ---OTHER CHARACTERISTICS---

DUFF DEPTH = 2.0 CM   DEAD AND DOWN HEIGHT = 8.3 CM
GRASS AND FORBS HEIGHT = 2.3 CM   LITTER HEIGHT = 0.8 CM
GRASS AND FORBS PERCENT GREEN = 98.0%

Table 7.2. A Sample Printout for One Stand from the Fuel Samples Tabulation Program, Including Site Description, Fuel Loadings, Packing Ratios (Betas), Contagion Values, Shrub Stratum Summary, and Canopy Summary (*Continued.*)

---SHRUBS---

PERCENT COVER = 20. %   NUMBER OF SPECIES = 1
SHRUB HEIGHT = 22.50 CM

(PLOT 2232) STATION 2 (ASPECT = 0 DEG, DIP = 25 DEG)
LIVE AND STANDING DEAD --
  GRASS AND FORBS   3.72 METRIC TONS/HECTARE
  SHRUB FOLIAGE   0.0 METRIC TONS/HECTARE
  SHRUB BRANCHWOOD   0.0 METRIC TONS/HECTARE
DEAD AND DOWN --
  LITTER   0.03 METRIC TONS/HECTARE
  1 HOUR D AND D   2.64 METRIC TONS/HECTARE
  10 HOUR D AND D   3.16 METRIC TONS/HECTARE
  100 HOUR D AND D   0.0 METRIC TONS/HECTARE
  >100 HOUR D AND D   49.59 METRIC TONS/HECTARE
AVERAGE FUEL DEPTH = 0.0 CM
BETAS --
    COMPLEX BETA 1 (AV FU DEP) = 0.2270
    COMPLEX BETA 2 (G + F DEP) = 0.5764
    MEAN COMPLEX BETA = 0.4017

    STRATUM BETA (LITTER) = 0.0011
    STRATUM BETA (G + F DEP) = 0.4609

--- OTHER CHARACTERISTICS ---
DUFF DEPTH = 7.8 CM   DEAD AND DOWN HEIGHT = 6.0 CM
GRASS AND FORBS HEIGHT = 2.0 CM LITTER HEIGHT = 0.0 CM
GRASS AND FORBS PERCENT GREEN = 98.0 %

--- SHRUBS ---
PERCENT COVER = 0. % NUMBER OF SPECIES = 0
  SHRUB HEIGHT = 0.0 CM

(PLOT 2232)   CANOPY
  --- FORMATION PARAMETERS ---
CANOPY COVER = 25 % MEAN GROUND FUEL HEIGHT = 1.8 M
MEAN LOW BRANCH HEIGHT = 7.2 M
(NOTE: 16% OVER 20 M EXCLUDED.)
MEAN DIFFERENCE = 5.4 M
INDIVIDUAL DIFFERENCES (M): 10.0 5.6 2.0 5.5 4.5

--- MEAN CANOPY COMPOSITION ---
ABIES LASIOCARPA   ABUNDANT
PICEA   ABUNDANT
PSEUDOTSUGA MENZIESII   COMMON

Table 7.2. A Sample Printout for One Stand from the Fuel Samples Tabulation Program, Including Site Description, Fuel Loadings, Packing Ratios (Betas), Contagion Values, Shrub Stratum Summary, and Canopy Summary *(Continued.)*

--- .01 HECTARE PLOT ---

DIA	CM	AL	PIC	PTM
0-	4	0	0	0
4-	8	0	2	2
8-	12	2	0	0
12-	16	4	0	0
16-	20	0	0	0
20-	24	0	0	0
24-	28	0	1	0
28-	32	0	0	0
32-	36	0	1	0

1200 TREES/HECTARE

SPECIES	NUMBER OF STEMS	REL. NUMBER OF STEMS (0/0)	BASAL AREA (SQ. M)	REL. BASAL AREA (0/0)
ABIES LASIOCARPA	6	50.0	0.0794	33.2
PICEA	4	33.3	0.1538	64.4
PSEUDOTSUGA MENZIESII	2	16.7	0.0057	2.4

loss of resolution, precision, or understanding; or (2) for certain environmental continua or vegetation sequences in which insufficient data were available for the proper use of environmental gradients.

Direct gradient analysis seeks understanding of the vegetation in response to direct, measurable, spatial or temporal environmental gradients. It uses either a direct environmental index (such as elevation, stand age, or soil pH) or a species weighted-average method (in which species are assigned to various classes—such as xerophytic or mesophytic—and sample indices are derived from the weights of its included species) to arrange samples along one or more gradients or axes (Whittaker 1956, 1973d, Whittaker and Gauch 1973). Ordination or indirect gradient analysis techniques are comparative in that they arrange samples in a sequence (along one or more axes) by comparing their compositional similarity. Thus, with direct gradient analysis, the environmental gradients are already established, and samples are arranged along these gradients. The user may then study species relationships and often will attempt various curve-fitting techniques to abstract the response of communities, species, and populations to these gradients. However, the user of ordination methods must rely on the samples' similarity to abstract the ordination axis. This axis may or may not correspond to an observable, measurable, or meaningful environmental gradient. If the axis does correspond to such a gradient, the methods of direct gradient analysis may be employed to further reduce and understand the data. If the axis does not correspond to such a gradient, different samples or different techniques must be tried or an imperfect understanding of the basis for these relationships must be accepted. The determination of an axis of variation that does not correspond to a direct environmental gradient may be of some usefulness by itself to a phytosociologist and, more importantly, may allow him or her to then determine more direct and more meaningful gradients from these intermediate results. Unfortunately, by itself, an indirect ordination axis that does not correspond to a measurable environmental gradient is of very limited or no value to the land manager, because statements cannot be made concerning the composition or dynamic processes on a specified piece of land if it cannot be determined where that land lies on various ordination axes. This is undoubtedly one reason many land managers believe that gradient analysis as a whole has little to offer management.

This discussion does not imply that direct gradient analysis and ordination are alternative or mutually exclusive techniques. Just as the development of a vegetation model will undoubtedly require both the broad application of both gradient analysis and classification, the development of a gradient model requires the application of both direct gradient analysis and ordination. Neither approach alone is adequate. Although the usual approach in a discussion of this sort is to describe direct gradient analysis and ordination separately, I shall not do so here because of the continuous and complex interplay of both approaches in developing the gradient model. Even the most direct of the direct environmental gradients were either verified or verified and modified by the application of ordination techniques; conversely, continuous variation along what are now recognized as three distinct direct gradients was first recognized from the results of numerous ordinations. Furthermore, several sets of directed variation that were originally described along either direct gradients or indirect ordination axes may now be treated in terms of discrete classification units. Detailed examples are given in Chapter 8.

As noted above, direct gradient analysis using environmental indices requires the plotting of samples along one or more gradients—the limitations of two dimensional paper usually require plotting not more than two gradients at a time. If more than two gradients are being considered, this is accomplished by holding the gradient indices along the other gradients constant. For example, assume one is performing a direct gradient analysis and is dealing with three direct gradients—elevation, topographic-moisture, and stand age. Either one could plot the samples on the elevation and topographic-moisture gradients and hold the other index constant by plotting only samples of similar ages, or one could plot the samples on the elevation and stand age gradients while holding topography and aspect constant, etc. It immediately becomes apparent that a large number of samples, a large amount of paper, and large amounts of expensive computer time are required as more and more gradients are added.

Once the samples are plotted, one usually wishes to abstract certain kinds of information (species distributions, diversity relations, and the like). This can be accomplished by various regression and/or curve-fitting techniques. If one had uniformly spaced, noiseless samples (no random variation) recorded with no sampling error, and if one were using an ideal gradient scheme that accounted for all directed variation in natural communities, then a curvilinear fitting technique, such as orthogonal polynomial curve fitting (cf. Kessell 1972, Orloci 1975), could produce final species and community nomograms. Unfortunately, real sample data are beset with sample noise, random variations, sampling error, and other stochastic effects; also, the researcher's gradient scheme undoubtedly does not include all the macro- and microenvironmental gradients that influence the distribution of species populations and community patterns. As a result, samples representing stands with identical gradient indices are by no means identical in terms of sample similarity, and the internal association (IA)[1] is less than 100% (Whittaker 1967, 1973c, Kessell 1976b). Although this does not cause a serious problem for many informal curve-fitting techniques, it does cause considerable distortion with the orthogonal polynomial curve-fitting techniques, even when a particular level of smoothing is specified. The problem is complicated by the fact that real samples are almost never uniformly spaced within the rectangular coordinate framework provided by the direct gradients, and so polynomial extrapolations through areas of few or no samples often cause considerable additional distortion. As a result, after 18 months' use of such curve-fitting programs, I have abandoned them as causing more distortion and problems than they are worth (Kessell 1972, 1979a).

The alternative is to seek simpler curve-fitting techniques for use in direct gradient analysis. These could include polynomials of lower degrees and various regression procedures. However, the use of such techniques requires an assumption as to the shape of the resulting distribution (linear, Gaussian, polymodal, etc.), whereas the whole purpose of their application is to elucidate the shape of the distribution. A trial and error approach may produce reasonable results, but once again the user expends considerable time and energy in correcting the "automatic" results.

My solution has been to contour curves by hand and eye *after* stands with similar environmental indices have been averaged. This procedure requires an understanding of the normal internal association values so that the curve fitter realizes whether variation is trivial (no more than is expected from the internal association value) or meaningful. A very helpful check on this procedure is the use of relativized importance values; when relative importance values are used for all species in an area or strata, the sum of the importance values for all species should equal 100% for any

---

[1] Usually defined as the percent similarity of replicate samples. Identical replicates have an IA of 100%.

combination of gradient indices. If they do not sum to 100%, an error has been made in contouring the data and the results should be revised until the "100% check" condition is met.

These procedures were used to construct the nomograms of vegetation species and fuel distributions in response to the direct environmental gradients. However, such a method assumes the researcher has quantitatively defined all gradients that influence the vegetation or fuel. This is virtually impossible to accomplish by simple inspection of the stand summaries from several hundred samples. One must therefore use other techniques to first determine the important environmental gradients.

It would be most fortunate if some theoretical phytosociologist *cum* mathematician were to develop an ordination technique that allowed the determination of several (or all) major environmental gradients from a single ordination of many hundred samples. Unfortunately, even the best techniques can seldom extract more than two (often only one) meaningful axes from a subset of the stand samples (Gauch and Whittaker 1972a, Whittaker and Gauch 1973, Gauch and others 1974, 1977, Kessell and Whittaker 1976); and, of course, there is no guarantee that the extracted axes will correspond to direct gradients, offer ecologic insight, or even represent anything except obtuse mathematical artifacts.

It is therefore necessary to choose the sample set for ordination critically and carefully. One must broadly know, or at least suspect, which trends of variation, which kinds of unexplained variation, or which environmental effects need to be distributed along new ordination axes. This is difficult to grasp in general terms, but it is easy to visualize using examples. Suppose an ecologist has performed a direct gradient analysis using gradients that correspond to elevation, topographic moisture, and stand age but still observes considerable variation among "replicate" samples (replicate in that their elevation, topographic-moisture, and stand age indices are similar). Instead of subjecting the several hundred samples to the ordination, one will gain greater understanding by ordinating the subsets of "replicate" samples to observe directed variation among them. Such ordinations (performed for each group of replicates) may well show that at the lower elevations, the variation corresponds to fire intensities on xeric sites, intensity and/or periodicity of flooding on bottomlands, different drainage (watershed) systems, or the like, whereas the ordinations of high-elevation samples may show directed variation is response to degree of sheltering from wind on the slopes or in response to the intensity and/or periodicity of slides in the draws and ravines. This does not imply that every ordination will offer such insights, but it does demonstrate the procedure by which the ecologist advances from unexplained variation, or "noise," to a possible trend of variation, to an ordination axis, to a direct gradient with measurable environmental indices. The development of the Glacier models required the completion of several hundred such ordinations to identify and verify the direct gradients and variation categories used in the final model.

The next decision faced by the researcher is the choice of ordination techniques. A wide variety of techniques has proliferated which offer various degrees of accuracy, flexibility, efficiency, and lucidity. These techniques have been highlighted in Chapter 3 and are extensively reviewed in Whittaker and Gauch (1973), Gauch and Whittaker (1972a), Gauch and others (1977), and Kessell and Whittaker (1976). A brief summary indicating the selection for the Glacier work, and reasons for this selection, follows.

As noted in Chapter 3, the two most widely used ordination techniques are Bray–Curtis (Wisconsin comparative) polar ordination (Bray and Curtis 1957, cf. Cottam and others 1973, Gauch 1973b) and principal components analysis (PCA) (reviewed in Orloci 1966, 1973, 1975). Bray–Curtis ordination arranges samples by their similarity to two endpoint samples using a Pythagorean algorithm; the user may select the choice of similarity measurement, endpoints, and relativization method. Principal components analysis uses a sample matrix to extract axes of variation; although it objectively selects the endpoints, it assumes a linear relationship between species importance and the axes and therefore performs very poorly with the typical curvilinear and polytonic species distributions observed in the field (Gauch and Whittaker 1972a, Whittaker and Gauch 1973, Kessell and Whittaker 1976). PCA was hesitantly attempted with some of the overstory data from Glacier; it gave very poor, inconsistent, and misleading results, contradicted the results of several other techniques, and was finally abandoned as totally inappropriate and hopeless for noisy polytonic data exhibiting moderate beta diversity.

Two new techniques that offer promise include Gaussian ordination (Gauch and others 1974) and reciprocal averaging (RA) (Gauch and others 1977). Gaussian ordination assumes Gaussian species distributions rather than linear relationships. Both formal testing with simulated data (Gauch and others 1974) and application in Glacier indicated it to be an effective technique if the majority of the species distributions were approximately Gaussian. It was successfully used for many understory ordinations, especially along the primary succession gradient (see Chap. 8); however, it proved of marginal value, and it was on occasion misleading or even ridiculous, for overstory species that exhibited bimodal or polymodal distributions. Reciprocal averaging also shows promise from testing using field and simulated data (Hill 1973a, Gauch and others 1977); unfortunately, it did not become available until virtually all the ordination work had been completed for Glacier. It is, however, being used in the construction of the gradient model for the Saint Mary and Many Glacier drainages and will be discussed in detail in a paper comparing the major techniques using the Glacier data (in preparation).

As indicated in tests using simulated data (Gauch and Whittaker 1972a, Gauch 1973a, b, Whittaker and Gauch 1973, Kessell and Whittaker 1976), Bray–Curtis ordination consistently gave the most illuminating and most meaningful results of the techniques used. Because Bray–Curtis ordination is affected by the selected similarity measures, choice of these parameters merits discussion.

Three similarity measures are ordinarily associated with Bray–Curtis ordination—coefficient of community (CC), percent similarity (PS) and euclidean distance (ED). Coefficient of community is a measure of presence or absence of species; it does not consider their importance values:

$$CC = S_{ab}/(S_a + S_b - S_{ab}) \qquad (7.1)$$

where S_{ab} is the number of species in both samples being compared, S_a is the number of species in the first sample, and S_b is the number of species in the second sample. Percent similarity considers the importance values of species in both samples (importance value may be density, cover, frequency, basal area, etc.):

$$PS = 2\Sigma\min(x,y)/\Sigma(x + y) \qquad (7.2)$$

where x and y are the importance values of each species in the first and second samples, respectively. For relativized data, this simplifies to:

$$PS = \Sigma\min(x,y) \qquad (7.3)$$

Distance measure are the complemented coefficient of community (CD = 1 − CC) and percent difference (PD = 1 − PS). The third measure, euclidean distance, squares the distance term:

$$ED = \sqrt{\Sigma(x - y)^2} \qquad (7.4)$$

where x and y are the importance values for each species in the first and second samples, respectively. Euclidean distance has been found to offer greater distortion (under all circumstances using single standardization) than either CD or PD (Gauch and Whittaker 1972a, Whittaker and Gauch 1973, Kessell and Whittaker 1976); therefore, it was not used in Glacier.

Both CD and PD, used properly, are valuable similarity measurements for use in Bray–Curtis ordination. Percent difference is most useful when several samples exhibit the same component species but different importance values, as CD would show no compositional differences among such samples. Percent difference is most valuable, therefore, when low beta diversity (between-habitat diversity) and relatively few species are present. Because it considers importance values, however, PD is much more sensitive to sample noise (IA less than 100%) than is CD and may also be predominated by one or two dominant species unless doubly relativized data are used (Gauch 1973a, Whittaker and Gauch 1973). Given these precautions, Bray–Curtis ordination using PD was proved beyond question as the most valuable ordination technique for Glacier's overstory. [Note: All ordinations were performed by computer using modified versions of the Cornell Ecology Program (CEP) series (Gauch 1973c).]

However, percent difference caused numerous problems when applied to the understory. Its sensitivity to both noise and strong dominance caused distortions and inconsistencies in most cases of ordinating individual samples; the situation improved considerably when several samples were grouped into an average and these averages were then ordinated. For some of these cases, CD proved useful; it certainly solved the problems of dominance and most of the noise trouble by not considering importance values. However, it ignores importance values while placing all computational significance on presence or absence, and this is a double-edged sword; noise among species of very low cover (0–2%) will cause them to appear in one quadrat and disappear in the other (while the other species remain relatively constant). When absolute cover on a 0–100% scale was used as the importance value, therefore, neither the CD nor the PD measure proved especially satisfactory or reliable.

At this point, much of the herb and shrub ordination work was shifted over to the Gaussian technique. Gaussian ordination using the 0–100% cover importance values proved a bit more illuminating than the Bray–Curtis technique in most circumstances, as it seemed to smooth over the minor (and sometimes even major) irregularities caused by noise, microgradients, and the like. However, for perhaps one Gaussian ordination in five, an almost totally incomprehensible arrangement was generated; invariably, some fairly common species (or a few such species) with polymodal distributions were responsible (good arrangements were obtained when the culprit species were eliminated from the data set). The results were unsettling.

The problem was solved by returning to Bray–Curtis ordination using the PD measure, but using it on the original eight-point sampling scale (described in the Chap. 6 section on "Vegetation Sampling"), not the 0–100% cover scale. This is also suggested by Whittaker and Gauch (1973); the result is to diminish the exaggerated differences in importance values caused by noise and other effects. Use of Bray–Curtis ordination with this importance value calculation gave by far the most consistent and most easily interpretable results. (Note that use of this scheme with Gaussian ordination worsens the results, as the Gaussian algorithm is now approximating a continuous function by only eight nonlinear importance value classes.)

The results of this work, 18 months after it was initiated, indicated that (with very few exceptions) the best technique was Bray–Curtis ordination using PD (0–100% density importance values) for the overstory, with Bray–Curtis ordination using PD (0–6% cover scale) for the understory (with Gaussian ordination, importance values on a 0–100% cover scale) running a distant second. Because of this, all ordinations of all sample sets included these methods regardless of any methods also attempted.

Once again, the reader is reminded that the development of a gradient model involves a continuous interplay of direct gradient analysis and ordination. One does not conduct a limited direct gradient analysis, then perform several hundred ordinations, and then return to complete the direct gradient work. Instead, one evaluates the direct gradient framework, locates a source of unexplained variation, performs numerous ordinations in an attempt to fix this variation along an axis, then attempts to identify the environmental factors responsible for this axis, quite possibly performs new ordinations, reevaluates the new axis, finally incorporates it back into the model as a new gradient, and then starts tracking down the next source of unexplained variation. The procedure requires considerable patience, computer time, and graph paper.

This general procedure is undoubtedly required whenever a multidimensional gradient framework is being constructed when its author is initially unaware of all the important environmental gradients. It is recognized that for some phytosociologic purposes, it is not necessary to have every gradient correspond precisely to environmental features that can be determined from site characteristics (either in the field or by remote methods) and require no compositional sampling. However, recall from Chapter 3 that the purposes of the development of this gradient model were to (1) describe species distributions, (2) gain understanding of the environmental basis of these distributions, and (3) use the gradients as the framework for an information retrieval and management model. This latter purpose required the identification of gradients such that a site's indices on each gradient could be determined *solely by remote methods,* without visiting the site in the field. Therefore, considerable emphasis was devoted to identifying gradients the indices of which could be determined in this fashion. This purpose becomes even more obvious in the next chapter, which describes the actual framework of the gradient model.

8

The Gradient Model

To describe the distribution of the vegetation, fuel, and community properties, the Glacier National Park gradient model simultaneously uses six continuous environmental gradients, including:

1) Elevation
2) Topographic-moisture
3) Time since the last burn
4) Primary succession (soil and vegetative cover development)
5) Drainage (lake and geographic moisture) influence
6) Alpine wind–snow exposure

and four additional categories of variation and disturbance:

7) Intensity of the last burn
8) Slide disturbances (fellfields, avalanche fields, mud and rock slides)
9) Hydric disturbances (floodplains, bogs, marshes, lake edges)
10) Influences of heavy winter ungulate grazing

This chapter describes the development, identification, and application of these gradients and influences; the remainder of Section II (Chaps. 9–12) describes the distribution of populations, species, and communities along these gradients.

Elevation Gradient

A literature review (especially Habeck 1970a) prior to my initiating work in Glacier National Park in 1972 suggested that at least three direct gradients—elevation, topographic-moisture, and time since the last burn—would be significant in explaining the distribution of the vegetation. Previous work in Glacier and elsewhere suggested that both elevation and time since the last burn could be handled directly with little reliance on ordination techniques, but that determination of a reasonable topographic-moisture sequence would require some ordination.

The elevation gradient is one of the most readily visible in Glacier National Park; it represents a quantitative application of Merrian's (1898) life-zone approach. A stand's elevation index may be simply determined from a topographic map. As the elevation index increases, higher precipitation, lower temperatures, a shorter growing season, higher (in a very nonlinear fashion) winds, greater snow depth, greater solar radiation, and a decrease, followed by an increase, in fuel moisture are all encountered (Habeck 1970a, Kessell 1976a, b, Mason and Kessel 1975).

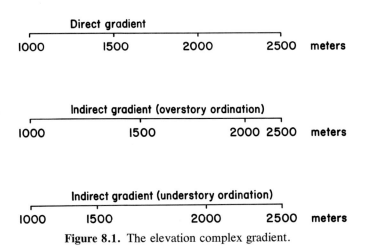

Figure 8.1. The elevation complex gradient.

Although elevation is normally viewed and scaled as a linear gradient (equal ecologic distances between 1000 and 1500 m versus 2000 and 2500 m), ordination of Glacier's overstory and understory indicated a nonlinear relationship, especially for the overstory. This is demonstrated in Fig. 8.1, which shows the average placement of the overstory and understory by ordination methods, compared to the linear sequence. The high-elevation end of the gradient was compressed for the overstory, as very little species compositional change (in response to elevation) was observed over 2000 m (the species distribution curves tend to plateau, as shown in Chapter 9). This effect was not demonstrated by the understory; instead, the gradient was slightly compressed at the lower elevations because of the lower beta diversity observed among these samples. Beta diversity remained rather constant above about 1700 m (Chap. 12). Although the gradient nomograms that appear in Appendix 2 indicate a linear elevation gradient, the reader should realize the actual relationships are those shown in Fig. 8.1.

Time Since Burn Gradient

The time since burn gradient quantifies the length of time since the community was last subjected to fire. Like the elevation gradient, it (theoretically) offers a quantified sequence that begins at zero and approaches infinity. Unfortunately, unlike the elevation gradient, it is much more difficult to accurately determine a stand's index along this gradient.

Fire history maps are useful for large fires that have occurred since 1910 and for smaller fires that have occurred during the past 30 years. For older stands, age must be estimated, usually by taking increment cores from a number of individual trees. The estimated values are usually correct to within 10% for stands not over 100 years old, and within 20% for stands not over 200 years old. Age estimation is very difficult for older stands but causes minimal problems, as shown in Fig. 8.2.

Figure 8.2 shows average ordinations of the overstory and understory along the time since burn gradient. As shown in detail for individual species in the next chapter, the distribution curves plateau at about 150–250 years after the burn, with minimal compositional change after that time. The time since burn gradient is therefore strongly compressed at the older end. This effect is even more apparent when understory samples are ordinated; the older end is extremely compressed, whereas the younger end is rather stretched. This latter effect is caused by high beta diversity during the first 50 years after a burn, which includes both the high herb and shrub productivity and the within-stand (alpha) diversity immediately following the burn and the canopy closure and resulting sharp alpha diversity drop (and change in species composition) that occurs 15–30 years after the burn.

As with the elevation gradient, the nomograms simplify it as a linear relationship; the plateau effect at the older extreme is readily apparent for virtually all species.

Topographic-Moisture Gradient

The topographic-moisture gradient represents an attempt to arrange soil moisture and evapotranspiration regimes that are functions of topographic location and aspect, as opposed to those resulting from elevation, geographic location, lake effects, and the like. It has been successfully used in numerous other gradient analyses, including the Siskiyou Mountains of Oregon and California (Whittaker 1960), the Santa Catalina Mountains of Arizona (Whittaker and Niering 1965), the Great Smokey Mountains of Tennessee (Whittaker 1956; Kessell 1979a), and the Linville Gorge Wilderness of North Carolina (Kessell 1972). It is the first of the Glacier gradients that required ordination work to establish the sequence and relative widths of gradient categories.

The topographic-moisture gradient arranges stands that are otherwise similar (i.e., same elevation, age, cover) along a gradient defined by both topography and aspect (Fig. 8.3). It uses nine topographic definitions, including: bottomlands, ravines, draws, sheltered slopes, undifferentiated slopes, open slopes, peaks, ridges, and xeric flats; these are arranged in six groupings: bottomlands; ravines; draws; sheltered slopes; open slopes; and peaks, ridges, and xeric flats. (Undifferentiated slopes are given an intermediate location between sheltered slopes and open slopes.) Each of these six

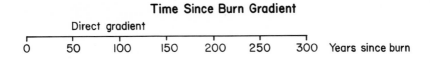

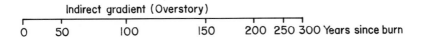

Figure 8.2. The years since burn (stand age) complex gradient.

categories or groupings is divided into eight compass points, arranged from the wettest to the driest aspects: NE, N, E, NW, SE, W, S, and SW. Together, topography and aspect provide 48 possible topographic-moisture categories.

The topographic-moisture gradient at the bottom of Fig. 8.3 shows an abstracted, idealized arrangement. Such an arrangement is an average sequence determined from the ordination of sample sets located in different elevation, age, cover, and drainage area regimes. Of these different regimes, elevation exerts the greatest influence upon the topographic-moisture sequence, as shown in the top three arrangements in Fig. 8.3.

In each arrangement, considerable overlay among the topographic categories is observed. In addition, at 1000 m, the xeric end of the gradient is compressed, whereas the more mesic end is stretched. This is because of both the greater absolute moisture differences and greater beta diversity at the mesic end, and the lesser topographic and vegetational differentiation among open slopes, ridges, and xeric flats at the xeric end. At 1500 m, the sequence approximates the idealized scheme (except for the recognized overlap). By 2000 m, the mesic end of the gradient is compressed, primarily because influences other than topography and aspect (primarily sheltering from wind) are determining the moisture regime. Conversely, the xeric end of the gradient is stretched because of the greater differentiation among slopes, ridges, and peaks.

Numerous examples of species distributions along the elevation, time since burn, and topographic-moisture gradients are given in the following chapters. To simplify their drafting, the nomograms showing the topographic-moisture gradient do not show the overlap among topographic categories; it must be realized that both the overlap and the high- and low-elevation gradient compression exist.

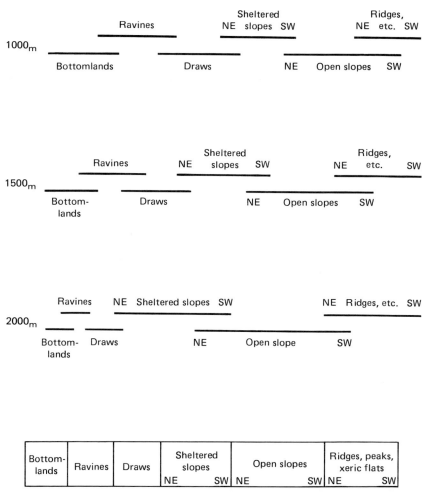

Figure 8.3. The topographic-moisture complex gradient, showing overlap of topographic types and compression of the relative widths at various elevations. Within each topographic type, communities are arranged on an eight-point aspect sequence of NE–N–E–NW–SE–W–S–SW.

Drainage Area (Lake Influence) Gradient

The previous three gradients—elevation, time since burn, and topographic-moisture—all influence the vegetation's moisture regime either directly or indirectly. Another important determinant of the moisture regime is a community's location among the various drainages that comprise the West Lakes area.

The Lake McDonald area represents the most mesophytic extreme of a geographic moisture gradient that extends (at least) as far north as Pole-bridge. As a result, the overstory vegetation at the lower elevations in the McDonald drainage is radically different from the vegetation at similar sites among the other drainages. It has been argued that this effect results from: (1) the McDonald area's local storm patterns; (2) the topographic sheltering of the entire drainage; (3) the presence of a large body of water (Lake McDonald); or (4) some combination of the above (cf. Habeck 1968, 1970a). Whatever the reason, the influence must be recognized and dealt with accordingly.

The Camas drainage (just north of the McDonald drainage, separated from it by the 1500-m elevation Howe Ridge) exhibits compositional elements similar to both the McDonald and the Polebridge regimes (but considerably more like the Polebridge vegetation—see Chap. 9). One therefore may view these drainages as a continuum that may be arranged along a geographic moisture, drainage area, or "lake influence" gradient.

This was attempted by ordinating stands that were similar except for their drainage system. Three major watershed areas were recognized: the McDonald, the Camas, and the North Fork of the Flathead River (Pole-bridge area). The McDonald was further divided into both the Lake McDonald area proper and the Avalanche Creek drainage; Avalanche exhibits the most mesophytic communities in the park, apparently because of its extreme topographic sheltering in the midst of Mount Brown, Mount Cannon, Edwards Mountain, the Little Matterhorn, and Gunsight Mountain. The North Fork area was further divided into the Bowman and Kintla drainages.

Figure 8.4 shows a Bray–Curtis ordination (percent similarity,[1] or PS, using relative density of the overstory) of 46 individual stands (mature, forested ravines and draws between 1000 and 1300 m) located among the five drainage areas (top of ordination axis); it also shows the results of ordinating (same technique) the average compositional values of the five drainage areas (bottom of ordination axis). The results indicate a strong similarity between the Bowman and the Kintla drainages of the North Fork watershed; a strong (but lesser) similarity between the Avalanche and the Lake McDonald proper drainages of the McDonald watershed; and an intermediate location for the Camas drainage (but one more similar to the North Fork than to the McDonald). The distribution of major overstory species is shown in the bottom portion of Fig. 8.4.

Figure 8.5 shows a similar ordination of 52 stands (mature, forested sheltered slopes between 1000 and 1300 m) and the five drainage areas. This ordination shows slightly less similarity between the Bowman and Kintla drainages, slightly greater similarity between the Avalanche and the Lake McDonald proper drainages, and slightly greater similarity between the Camas and the North Fork watersheds. On the whole, the ordinations are

---

[1]As defined in Equation 7.2, Chapter 7.

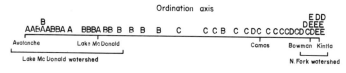

Drainage Gradient

Mature, Forested Ravines and Draws, 1000–1300 m

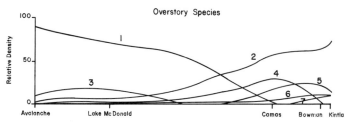

Stands (n = 46)

Drainages (n = 5)

Figure 8.4. Sketch and typical overstory species distributions along the lake influence–drainage area complex gradient for ravines and draws (1000–1300 m). Drainage areas include: A, Avalanche; B, Lake McDonald; C, Camas; D, Bowman; and E, Kintla. Species include: 1, *Tsuga heterophylla;* 2, *Picea;* 3, *Thuja plicata;* 4, *Abies lasiocarpa;* 5, *Pseudotsuga menziesii;* 6, *Larix occidentalis;* and 7, *Betula papyrifera.*

strikingly similar. The distribution of major overstory species is shown at the bottom of Fig. 8.5.

Other ordinations were performed using low-elevation, mature forested bottomlands and open slopes; early and midseral, forested, low-elevation slopes; and low-elevation meadows; they produced similar results. At higher elevations, the differences among the drainages diminish; by 1800-m elevation, other differences greatly overshadow the effect of drainage area on the overstory. However, the understory apparently continues to reflect some minor differences among a few of the less common alpine herb species; furthermore, the North Fork area is the only West Lakes drainage with abundant *Larix lyallii* populations (Chap. 9).

These ordinations give the modeler two choices: He or she can recognize that the drainages do form a continuous sequence and treat them as direct gradients. When this is done, there is a large uncertainty involved in locating an individual stand at the proper gradient index. Alternatively, the modeler may view this drainage sequence as creating three rather natural drainage units and classify samples as belonging to one of these three units. (Note that the Quartz, Logging, Anaconda, and Dutch drainages, which fall between the Bowman and the Camas drainages geographically, fall

Drainage Gradient

Mature, Forested Sheltered Slopes, 1000‑1300 m

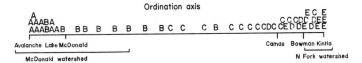

Ordination axis

Stands (N= 52)

Drainages (N= 5)

Overstory species

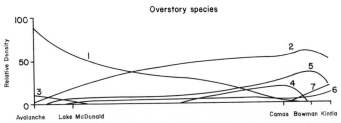

Figure 8.5. Sketch and typical overstory species distributions along the lake influence–drainage area complex gradient for sheltered slopes (1000–1300 m). Drainage areas include: A, Avalanche; B, Lake McDonald; C, Camas; D, Bowman; and E, Kintla. Species include 1, *Tsuga heterophylla;* 2, *Picea;* 3, *Thuja plicata;* 4, *Abies lasiocarpa;* 5, *Pseudotsuga menziesii;* 6, *Larix occidentalis;* and 7, *Pinus contorta.*

between the Camas and the North Fork watersheds on the gradient.) I chose the latter alternative for purposes of simplification; the uncertainty of locating a stand on the continuous gradient roughly equals (IA[1] of about 75%–80% for the overstory by either method) the lost resolution from lumping the stands into three gradient units. The gradient nomograms of overstory species shown in the following section, therefore, are divided among the McDonald, Camas, and North Fork drainages; the reader should remember that these units represent three modes on a nearly continuous gradient.

Primary Succession Gradient

For stands with a typical forest formation type, the four gradients discussed above explain almost all overstory species distributions (i.e., the similarity between the observed and the predicted composition is only slightly below the standard IA of 75%–80%); exceptions are discussed below. However, this four-dimensional gradient scheme offers no information about forma-

---

[1]Internal association as described in Chapter 7.

tion development reflecting soil development, sheltering, and the general
primary succession development.

Considerable effort was devoted to determining a scheme permitting
description of formation development and species composition which did
not require field sampling of every stand. The multiple factors that determine these relationships—rock weathering, sheltering, soil development,
etc.—are not amenable to wide area, direct observation in the field, much
less by remote methods.Various attempts ended in failure.

Finally a sequence was devised that simply arranged stands from no
vegetation cover (glaciers and rock) through maximum cover of herbs,
shrubs, and trees. This arrangement is shown in Fig. 8.6. The arrangement
recognizes six major formation groups (glacier, rock, talus, meadow,
Krummholz and shrub, and typical forest), with as many intermediate categories as one cares to resolve. The gradient was determined to present
greatest continuity of herb and shrub cover and overstory cover, not species composition.

A single problem with the scheme is its combining both shrub and
Krummholz communities into a single category. This poses no problems at
the lower elevations, which have no Krummholz development, but it does
create difficulties at the higher elevations where both formation types
occur. The problem is that although cover forms a smooth continuum
across this category, species composition often does not (different species
comprise both the overstory and the understory). The gradient, therefore,

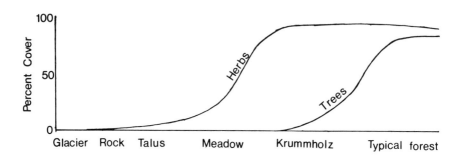

Figure 8.6. The formation types and vegetation cover along the primary succession
complex gradient. (Kessell 1976a)

should probably be visualized as two gradients, a major primary succession gradient and a minor Krummholz—shrub gradient, as shown in Fig. 8.7. In practice, however, the problems created by this second minor gradient have been minimal, and therefore it is not shown in subsequent nomograms using the primary succession gradient.

Once this gradient was determined from formation type and overstory–understory percent cover, numerous ordinations were performed on the understory (because overstory species occur only at one extreme of the gradient); both Bray–Curtis (PS using the eight-point herb cover scale) and Gaussian ordination were used extensively. Ordinations included stands that were similar except for their formation development. The ordination results were compared to the results of direct gradient analysis by using the stand's formation type as its gradient index.

The results were, in most cases, very similar; Bray–Curtis ordination, Gaussian ordination, and direct gradient analysis gave the same general locations for stands. On occasion, the direct gradient arrangement contradicted the results of one of the ordinations (usually the Gaussian ordination); the other ordination method seldom offered such a contradiction. (As noted earlier, this contradiction was partially resolved by eliminating the polymodal species from the Gaussian ordination—all three methods were plagued by high sample noise and the unquestionable occurrence of microgradients within a stand.) The final analysis demonstrated that the direct gradient method was as good as, and much simpler than, the indirect methods; the formation arrangement was therefore accepted as a direct gradient. The (smoothed) results of locating sample stands on this primary succession gradient by direct gradient analysis are demonstrated in Fig. 8.8.

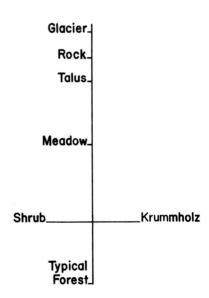

Figure 8.7. The two-axis primary succession complex gradient.

Mature Draws and Ravines 1400 m

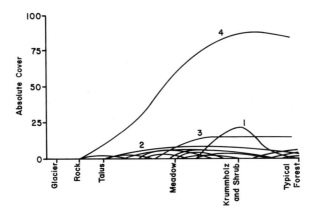

a

Mature Open Slopes 1400 m

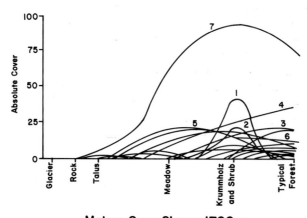

b

Mature Open Slopes 1700 m

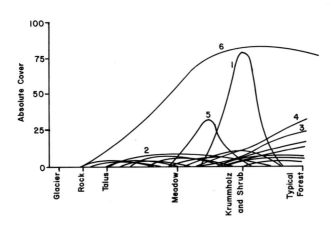

c

Mature Open Slopes 2000 m

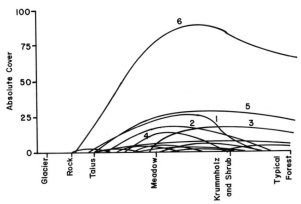

d

Figure 8.8 a–d. a Distribution of herb and shrub species along the primary succession gradient for ravines and draws at 1400-m elevation. Common species include 1, *Alnus sinuata;* 2, *Achillea millefolium;* and 3, *Rubus parviflorus;* 4 is total cover by all species. **b** Distribution of herb and shrub species along the primary succession gradient for open slopes at 1400-m elevation. Common species include: 1, *Alnus sinuata;* 2, *Veratrum viride;* 3, *Spirea* spp.; 4, *Vaccinium* spp.; 5, Graminae; and 6, *Thalictrum occidentale;* 7 is total cover by all species. **c** Distribution of herb and shrub species along the primary succession gradient for open slopes at 1700-m elevation. Common species include: 1, *Alnus sinuata;* 2, *Fragaria* spp.; 3, *Xerophyllum tenax;* 4, *Vaccinium* spp.; and 5, *Thalictrum occidentale;* 6 is total cover by all species. **d** Distribution of herb and shrub species along the primary succession gradient for open slopes at 2000-m elevation. Common species include: 1, *Epilobium angustifolium;* 2, *Erigeron* spp.; 3, *Vaccinium* spp.; 4, Graminae; and 5, *Xerophyllum tenax;* 6 is total cover by all species.

Alpine Wind–Snow Exposure Gradient

The last gradient, and the most difficult to apply using remote sensing methods, is the alpine wind–snow exposure gradient. The gradient was identified almost accidentally when unexplained variation was noted under two major circumstances in the high country—both along and near the passes and in the ravines and draws.

As discussed in detail in the next chapter, both *Abies lasiocarpa* and *Pinus albicaulis* are common tree species in subalpine habitats. They show a marked distributional difference in response to the topographic-moisture gradient; *Abies* predominates on the mesophytic and submesic sites, whereas *Pinus* becomes common (over 20% relative density) only on the more xeric sites. However, two major exceptions were observed.

On and very near the passes, virtually no *Pinus* was found, despite the fact that such areas were undoubtedly drier (because of wind) than the adjacent slopes: The topographic-moisture nomogram would predict more, not less, *Pinus* on such sites. Moreover, within the mesic—often hydric— ravines and draws, virtually no *Abies* was found, despite the opposite expectation.

The solution to this apparent contradiction was found in the species' growth forms. The pass areas are subjected to high, admittedly drying winds; however, a major effect of such winds is to remove almost all of the snow during the winter months. *Abies'* Krummholz growth form at such altitudes allows it to stay buried in the remaining snow, whereas *Pinus* (which seldom exhibits the Krummholz form) sticks out of the snow and is frozen, and the next high wind causes severe mortality. This was checked

Boulder Pass
Alpine wind–snow gradient

Bray–Curtis ordination (PD) of :

Species distribution along direct wind–snow gradient :

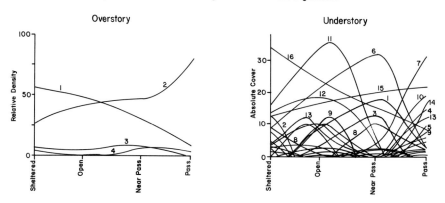

Figure 8.9. Distribution of overstory and understory species and results of Bray– Curtis (PD) ordination and direct gradient analysis for the alpine wind–snow gradient at Boulder Pass. (L is length of ordination axis in PD units.) Overstory species include: 1, *Abies lasiocarpa;* 2, *Larix lyallii;* 3, *Pinus albicaulis;* and 4, *Picea engelmannii.* Understory species include: 1, *Juncus drummondii;* 2, *Luzula hitchcockii;* 3, *Luzula piperi;* 4, *Carex dioica;* 5, *Carex nigricans;* 6, *Carex specta- bilis;* 7, *Salix nivalis;* 8, *Parnassia fimbriata;* 9, *Oxyria digyna;* 10, *Saxifraga lyallii;* 11, *Hypericum formosum;* 12, *Epilobium alpinum;* 13, *Arnica mollis;* 14, *Aster laevis;* 15, *Erigeron peregrines;* and 16, *Senecio triangularis.*

both from the ground and during low-altitude aerial reconnaissance during the winter of 1973.

In the ravines and draws, however, it is *Pinus'* flexible branches that allow it to outcompete *Abies*. These high-altitude ravines and draws are subjected to frequent and severe snow, mud and rock slides (discussed in the section "Categories of Disturbance and Variation"); the more brittle *Abies* branches are destroyed by such slides, and high mortality results, whereas *Pinus'* mortality is comparatively low. This was also later confirmed in the field following spring slides.

Analysis of the distribution of *Abies* and *Pinus* along Swiftcurrent, Ahern, and Hidden Lake Passes suggested a gradient that, at one extreme, merges with normal slopes and then increases its index as a pass is approached. Ordinations of the overstory were not terribly satisfactory, as there are usually only three species to ordinate (*Abies, Pinus,* and *Picea*). Subsequent work at Gunsight, Lincoln, Brown, and Boulder Passes, however, and in other areas subjected to high wind, confirmed this effect. (Boulder Pass was especially useful as it included an additional species, *Larix lyallii,* the distribution of which also corresponded to this gradient, as shown in Fig. 8.9.)

When overstory and understory ordinations were compared, it was also found that herb species generally corresponded to the same gradient sequence. The gradient was therefore adopted and is shown for Boulder Pass in Fig. 8.9. The single problem in using the gradient as part of an information model is the frequent difficulty in locating stands along it prior to field sampling. The matter is made fairly simple along the passes and in other areas of extremely high wind by using aerial photographs. There are numerous other stands, however, for which the effect is more subtle, although nonetheless significant, and these cannot always be identified from the photographs. Fortunately, just walking through one of these stands on a windy day (all days are windy at high elevations in Glacier) allows one to approximately locate it on the gradient without compositional sampling—fairly efficient ground mapping therefore is possible.

Categories of Disturbance and Variation

The alpine wind–snow exposure gradient provided the sixth and last continuous environmental gradient used by the model. Yet other influences also affect the distribution of the vegetation. As noted earlier, these are treated as classification categories because it is possible to lump effects without significantly decreasing resolution, or because we lack sufficient data to treat them as continuous gradients, or because of a combination of these two factors.

Burn Intensity

Natural fires may range in intensity from a creeping ground fire that destroys only part of the litter and almost none of the duff or living plants, to a raging inferno that destroys all ground fuels, the entire duff down to mineral soil, and all living vegetation (herbs and trees alike); all intermediate intensities also are possible.

Until the recent development of a predictive fire behavior model (Rothermel 1972), burn intensity was estimated only qualitatively based on the portion of the canopy and/or ground fuel and/or duff destroyed. It is now possible to calculate fire intensity in terms of either total energy release per unit area or energy release per unit area per unit time. Unfortunately, this cannot be applied to past fires because of insufficient information on the fuel loadings and fuel moisture when the stand burned. As a result, we use the five intensity categories noted earlier (Chap. 7) to approximate the intensity continuum. Since there is a fair correlation between the portion of the duff destroyed and the portion of the crown destroyed by a fire (other things being equal), for a given habitat there is undoubtedly a reasonable correlation between measured (or predicted) intensity and these two parameters. However, for purposes of ecologic modeling and succession prediction, an even cruder system is adequate.

As shown in the next chapter, the critical value for overstory destruction required for fire succession predictions is approximately 50% of the crown destroyed. If more than 50% of the crown in destroyed, sufficient mineral soil and light usually will be available to permit the normal "full canopy fire" successions of overstory species (such as *Pinus contorta* at low, subxeric sites). However, if much less than 50% of the canopy is destroyed, only a moderate portion of the duff will be destroyed; this creates conditions less than favorable for *P. contorta,* and other species (such as *Pseudotsuga,* with lesser *Larix occidentalis* and *Picea*) will predominate.

The system is admittedly crude, but it seems adequate for most purposes; it is quite possible that other deterministic and stochastic elements act as greater controlling factors on species successions than does an expanded fire intensity scale, as discussed in Chapter 10.

Slide Disturbances

Fellfields, avalanche fields, mud slides, and rock slides comprise the second major additional disturbance influence. Slides are very numerous in the draws and ravines of the high country; individual sites undoubtedly exhibit a wide range of slide periodicities and intensities. One effect of slides (on the *Abies–Pinus* balance) has been noted above; they have numerous other effects on virtually all plants that occur in such habitats (on occasion, they will destroy trees as large as 60 cm dbh). Slides also exhibit a major effect on diversity, usually by significantly increasing it (the most diverse stand yet found in Glacier is a ravine–slide area with meadow cover which burned in 1967—see Chap. 12).

Undoubtedly, if sufficient data were available, much of the vegetation development in such habitats could be attributed to periodicity, intensity, and type (snow, mud, rock, etc.) of slide. Unfortunately, because slide areas exhibit a wide range for these parameters and occur at various elevations, with various formation types, at various aspects, and in various stages of recovery from fire, the data collected in this study were in no way adequate to construct such a scheme. We are forced by lack of data, therefore, to lump slides into a single special disturbance effect category superimposed on the gradient scheme.

Hydric Disturbances

Hydric disturbances (river floodplains, lake edges, bogs, and marshes) comprise the third disturbance category. Two major rivers—the North and Middle Forks of the Flathead River—and numerous smaller streams occur in the West Lakes area. All are subject to periodic flooding of various intensity. Such a situation may be viewed in terms of the traditional riparian successions from gravel bars, through herbs, and then invading shrubs, and finally a "mature riverine ecosystem" to be washed away by the next large flood (Brower and others 1972).

Singer (1975a and personal communication) recognizes a seven-step fluvial succession gradient along the North Fork of the Flathead; a stand's location on this gradient may be determined indirectly through two of our gradients (topography of bottomland, plus the appropriate formation, or "primary succession" type, and the hydric disturbance notation). Such riverine succession patterns are discussed in Chapters 10 and 11.

Other hydric disturbances include the *Carex–Juncus–Salix* communities along the edge of many lakes and ponds. These occur because of high water in the spring; they are simply noted as a disturbance type, and no attempt has been made to refine the system. Still others include wet meadows, bogs, and marshes, which are discussed in Chapter 11.

Grazing

Heavy winter ungulate grazing is the final other disturbance category. It is included because, in a few areas, the significant presence of large ungulates during the winter suggests that they may affect the stand's vegetational composition and successional patterns. In the West Lakes area, this is suspected on the Belton Hills near West Glacier. This area was burned by high-intensity fires in 1910 and 1929 which destroyed the entire canopy (Cattelino and others 1979); the area is steep south and southwest open slopes between 1000- and 1900-m elevation. Despite the recognized nonanimal effects of such a burn (producing a slow succession pattern), the very low conifer seedling density and predominance of shrubs (especially *Acer* and *Ceanothus*) is hypothesized to result from the combined effects of short interfire periods and the presence of a large wintering elk population. It is suggested that these elk maintain this composition by both devouring virtually all the conifer seedlings and rapidly recycling the nutrients needed by the shrubs to maintain high productivity. If this hypothesis is correct, the elk may well maintain a stable or semistable shrub community much like the one created by Niering and Goodwin (1973) through selective herbicide application.

This completes the development of the gradient framework of the model—the next chapters describe, and attempt to explain, the distribution of the West Lakes area vegetation, fuel, and community relationships within this complex of spatial and temporal environmental gradients.

9

Forest Communities

As discussed in the last chapter, ordinations of stands that were environmentally similar except for the drainages in which they occurred produced the lake influence–drainage area complex gradient. Although the distribution of stands along this gradient formed a near continuum, three distinct modes or groupings were recognized that corresponded to the McDonald, Camas, and North Fork drainage units. In the following discussions, communities and species are treated by using this tripartate division of the study area. Comparisons between and among drainages are made, and similarities as well as differences in these drainages' communities and species composition are noted.

Distribution of Tree Species

The six major gradients that were determined to have a major influence on the West Lakes vegetation and community development are described in Chapter 8. Because the forested stands occur at only one extreme of the primary succession gradient, one need not deal with it as a continuum when describing the distribution of trees. If one ignores those alpine stands that show a strong influence of the wind–snow effects, one may also ignore the alpine wind–snow gradient. Furthermore, if one treats each of the three drainage modes as a separate unit, one may dispense with the drainage gradient. Thus, within a single drainage unit, if the wind–snow influences are ignored the distributions of tree species may be described on three major gradients: elevation, topographic-moisture, and stand age (time since burn). Superimposed on this three-dimensional gradient structure will be the other disturbances (fellfields, burn intensity, etc.).

Appendix 2, Figs. 1–28, includes relative density nomograms for all tree species that occur in the McDonald drainage of Glacier National Park; these include all tree species that occur within the park, except *Pinus flexilis,* which is restricted to areas east of the continental divide. For each species (except *Larix lyallii* and *Betula occidentalis,* both very rare in the area), nomogram sets are paired; the first set plots elevation versus topographic-moisture for four age classes (20–50 years, 50–90 years, 90–150 years, and more than 150 years since the last canopy fire); the second set plots stand age (time since the last canopy fire) versus topographic-moisture for up to five elevation classes of 305 m (1000 ft) elevation each (if the species occurs in each elevation class). In this fashion, a three-dimensional habitat space framework can be expressed in nomograms plotted two-dimensionally.

Figures 29–43 in Appendix 2 are relative density nomograms for all tree species that occur in the Camas drainage. The nomograms plot elevation versus topographic-moisture for the four stand age classes of each species. Nomograms plotting stand age versus topographic-moisture are not included, but the continuous changes along the stand age gradient can be seen by visualizing the four nomograms "stacked" to form the third axis.

Figures 44–54 of Appendix 2 are relative density nomograms for all tree species that occur in the North Fork drainage area north of the Camas drainage. For each species, nomograms plot elevation versus topographic-moisture for the four stand age classes.

McDonald Drainage

Inspection of the nomograms in Figs. 1–28 (Appendix 2) immediately demonstrates the individualistic distributions of tree species' and their importance along the three gradients. One may speak of stand or community types, but one must often do so in an abstract or subjective sense only—clear-cut boundaries between species' distributions or community types are not readily observed. The following discussion covers both the species changes observed as a function of the environmental variation expressed by the gradients and the environmental and competitive influences that interact with species' adaptations to define their natural ranges within the gradient framework.

Low-elevation bottomlands are found along the floodplains of the larger streams and rivers in the area (primarily the Middle Fork of the Flathead River and lower McDonald Creek) and along smaller streams and lakes. The former are usually subjected to fairly intensive flooding at a periodicity from a few to perhaps 50 years; species composition therefore reflects such influences upon stands already very mesophytic or even hydrophytic. Predominant tree species include *Picea, Thuja plicata, Pseudotsuga menziesii, Populus trichocarpa,* and *Betula papyrifera. Abies lasiocarpa* is often present at low density. These species distributions as shown on the nomograms reflect stands with minimal flooding or else a low flood periodicity. Let us now superimpose a "hydric disturbance" effect, axis, or gradient upon these low-elevation bottomland stands. Such a construct shows the stands subjected to frequent flooding exhibit a much higher density of *Populus trichocarpa* and almost no *Thuja plicata;* alternatively, the other extreme of the axis (those bottomlands along lakes with no real flooding) show a higher density of *Thuja* and much lower (often zero) density of *Populus trichocarpa.* The densities of other species do not change significantly.

We may now consider seral stands on the low-elevation bottomlands. Most successional development observed here is caused by stand retrogression (or destruction) following heavy flooding, whereas a few such stands are recovering from the infrequent fires that occur on these wet sites. The seral nomograms (age classes of less than 150 years) for the species noted above indicate that *Abies lasiocarpa* is more common in the seral communities than in the mature stands (reaching a peak relative density of about 30% at about 100 years after a disturbance). *Picea,* in contrast, exhibits the opposite effect; *Picea* exhibits a bimodal distribution along the stand age gradient for these stands, with minimum density at a stand age of about 75 years. *Larix occidentalis* is common in the early seral stands (20- to 50-year-old stands show a relative density of 20%), drops somewhat in midseral stands (from 15% down to 1% relative density in the 50–150 year range), and is finally absent or very rare in the mature stands. *Pinus monticola* is also present in the seral low-elevation bottomlands at the 5% level up to about 75 years after the disturbance. Although *Pinus contorta* is rare in the early seral stands (being outcompeted by *Picea* on such mesic sites), it does enter at low densities and holds a 10% relative density until about 75 years after the disturbance. Presumably its occurrence in nonburned flooded stands is a result of the flooding's exposure (or deposition) of mineral soil and eradication of a light-blocking canopy. *Pinus ponderosa* is also a rare element of the canopy in the early seral bottomland stands, and *Pseudotsuga* does not reach a 10% density level until nearly 100 years after the disturbance. *Tsuga* is usually absent from such stands until they approach maturity (at about 175 years), whereas *Thuja* exhibits a strong increase in density at a stand age of about 130 years. *Populus trichocarpa* exhibits a bimodal distribution along the stand age gradient similar to that of *Picea* but with lower densities. The more subtle influences of the exposure of each stand to flooding potential, combined with the periodicity and intensity of past floods, doubtlessly explain some of the variations in seral species composition noted above. However, in most cases, the stands converge to a *Picea–Thuja–Populus* overstory.

As one moves from the bottomlands to the ravines and draws along the topographic-moisture gradient, strong species turnover (beta diversity) is observed. The steep-walled, spatially narrow topographies usually confine flooding to the lower reaches of the ravines (which almost always have a meadow, talus, or gravel cover instead of forest) but still provide a cool, mesophytic environment to the adjacent canopy species. Within the McDonald drainage, ravines are strongly dominated by *Tsuga hetero-phylla;* relative density of this species ranges from 45% to 98%, with a mode of about 75–80%. Its predominance in mature ravine and draw forests is sufficient to split the distribution of *Thuja plicata* into two distinct modes along the topographic-moisture gradient; one occurs in the draws and north aspect sheltered slopes, and the other occurs on the bottomlands with low-flood frequency. This problem of Thuja's distribution was addressed by Habeck (1968), who maintained that *Thuja* preferred the more mesic sites and *Tsuga* dominated the less mesic stands, and by Daubenmire (1966), who reached the opposite conclusion (*Thuja* preferring more xeric sites than does *Tsuga*). The distributions observed in this study suggest that both researchers were correct but that each observed only a part of the total picture. As yet uninvestigated allelochemical effects may also be involved.

Other species that are often present in the mature ravines are *Picea, Pseudotsuga,* and *Betula papyrifera*. In the somewhat less mesic draws, *Abies grandis* is localized but may reach relative densities as high as 30%. The draws often include some residual *Larix occidentalis* (long lived from earlier seral stands) up to a 5% density, and somewhat more *Pseudotsuga* than in the ravines (10% versus 5% averages). *Betula* holds at a rather constant 5%, whereas *Picea* holds at about 10%.

Early seral communities in the forested draws and especially the ravines are difficult to locate in abundance. This is because the forested portions lack major flooding yet preserve sufficiently mesic conditions to preclude fire (fuel moisture is above the extinction value—see Chap. 14) under all except extreme drought conditions. Early seral communities that were found and sampled were strongly dominated by *Picea,* with relative densities approaching 75%. *Picea* density drops very rapidly when stands reach 50–60 years of age and then moderately increases after about 150 years of age. Both *Tsuga* and *Thuja* enter these stands soon after a fire, but neither becomes very important until at least 75 years after the burn. *Thuja* shows a slow increase to 5% density by 70 years after the burn and then gradually increases to about 10%–15% by stand maturity. *Tsuga* gains predominance in the draws sooner than it does in the ravines (60 years to reach 50% density in the draws, but 150 years to do so in the ravines), although it usually achieves a greater dominance by maturity in the ravines. Apparently in each situation *Thuja* and *Tsuga* are replacing the early *Picea,* and *Picea* manages better to postpone replacement in the more mesophytic ravines. In both topographies, some *Pinus contorta* is present in early seral stages, but it seldom achieves a density greater than 15% and it cannot compete well with *Picea* as an early seral component under these mesophytic conditions.

In the ravines, draws, and north- and east-aspect sheltered slopes, the seral species composition problem is also complicated by fire intensity. Fires are infrequent in such communities, and when they do occur they are often only ground fires (frequently mild ones that do not destroy the duff) or else only occasionally destroy a tree here and there but do no serious canopy damage. In such low-intensity fires, many stands' recoveries do not exhibit the usual succession patterns; instead, destroyed individuals are replaced with those of the same species—usually *Tsuga* and *Thuja*. Furthermore, in the most mesic sites, most of *Thuja's* reproduction occurs vegetatively by layering. Some of these forest communities are shown in Figs. 9.1–9.3.

Figure 9.1. Montane forests of *Picea* and *Abies lasiocarpa* give way to alpine communities along Going to the Sun Highway in Glacier National Park. (Photograph by Mel Ruder for *The Hungry Horse News*)

As one moves from low-elevation mature mesophytic sites to more xeric communities (sheltered and open slopes, and then ridges and xeric flats), one encounters a smooth transition of species along the topographic-moisture gradient. *Abies lasiocarpa* and *A. grandis* are virtually absent from such stands, whereas *Larix occidentalis* maintains a 5%–10% density on all but the most mature low-elevation slopes. *Picea* is still present at the

Figure 9.2. Low-elevation forests of *Tsuga heterophylla, Thuja plicata,* and *Picea* are the predominant feature of the McDonald drainage. This view shows the Garden Wall (Continental Divide) in the background and a black bear on Going to the Sun Highway. (Photograph by Mel Ruder for *The Hungry Horse News*)

5%–10% level on sheltered slopes, but it too drops out as more xeric conditions are approached. Some residual *Pinus monticola* can be found in all but the most mature open slope stands, but only at the 1%–2% level. *Pseudotsuga's* density increases on the lower slopes and reaches a peak of 30%–40% on the more exposed sheltered slopes and the open slopes. Both

Figure 9.3. *Picea* forest, shrubs, and nonwoody vegetation predominate low-elevation hydric sites in the McDonald drainage. Moose are common in these habitats. (Photograph by Mel Ruder for *The Hungry Horse News*)

Thuja and *Tsuga* show declining densities on the sheltered slopes and are no longer present in stands more xerophytic than the sheltered slopes. Finally, at the extreme xerophytic end of the topographic-moisture gradient, mature stands are very rare, as such xerophytic sites have a high fire frequency and severity, and perhaps less than 1% of such stands have seen no fires in the past 125 years. In those rare mature, xerophytic low-elevation stands, *Betula papyrifera* is often strongly dominant.

Seral development on the slopes, ridges, and xeric flats also strongly reflects the influence of the topographic-moisture gradient. Not until the sheltered slopes with south and southwest aspects are reached does *Pinus contorta* become a major seral species, but its influence rapidly increases as more xeric conditions are met; on the most xeric sites, its density in early seral stands will exceed 75% for the first 50 years after a fire. The understory will usually include a mixture of *Picea* and *Pseudotsuga*, with *Picea's* density increasing as conditions become more mesophytic. Other common seral species are *Larix occidentalis*, which reaches a peak density of about 25% on open slopes from 100 to 125 years after a burn; *Pinus monticola*, which reaches lower densities and achieves peak densities (10%) on sheltered slopes about 70 years after a fire; and *Betula papyrifera*. *Betula* shows a bimodal distribution along the stand age gradient for the open slopes, with a minimum at about 100 years, and a similar but even more pronounced bimodal distribution for sheltered slopes with a minimum years of age. Examples of seral forests are shown in Figs. 9.4–9.5.

Not only do species composition and community characteristics change from the low elevation (below 1200 m) forests to high-elevation stands, but the very nature of the topographies, formation characteristics, and the

Figure 9.4. A mosaic of meadows and conifers (*Pinus contorta, P. flexilis,* and *Pseudotsuga menziesii*) found near Rising Sun, east of the Continental Divide. Numerous wildlife species, including the mule deer shown here, utilize these communities. (Photograph by Mel Ruder for *The Hungry Horse News*)

Figure 9.5. Seral forests result from the frequent canopy fires ignited by lightning. The standing snags are *Larix occidentalis* and serve as vital habitats for cavity dwelling animals. (Photograph by Mel Ruder for *The Hungry Horse News*)

intensity and types of other effects—especially disturbances—also vary. Bottomlands may be found above 1200 m, but they are not the broad, gentle bottomlands of the low elevations: They are often steep; are subjected to frequent snow, mud, and rock slides; and often support meadow or talus cover (rather than forest). Ravines and draws are more difficult to differentiate from one another and from the bottomlands and seldom support forest cover. Slopes are still forested but exhibit a much greater variation in vegetation because of exposure to or sheltering from the elements (open versus sheltered slopes) and the more pronounced effects of differing aspects. Therefore, beta diversity is low on the mesophytic end of the topographic-moisture gradient but rapidly increases under more xeric conditions.

Midelevation forests (1200–1600 m) on the mesophytic sites (bottomlands, ravines, and draws) are a mixture of *Abies lasiocarpa* and *Picea*. On the wettest sites, especially at the higher elevations, *Abies* predominates, whereas *Picea* becomes more common on the draws and sheltered slopes of the midelevations. *Abies* also shows peak densities (up to 80%) in the late seral communities (90–150 years old), rather than on the most mature stands (peak densities here of 55%). *Abies* also maintains its high densities until rather xerophytic conditions are encountered on the open slopes with southeast, west, south, and southwest aspects. *Pinus albicaulis* × *monticola* is present in all but the most mature stands within this elevation range on most slopes but reaches a moderate density only in the 60- to 130-year-old stands—it becomes neither common nor clearly *P. albicaulis* until higher elevations are reached. *Pseudotsuga* rapidly replaces both *Picea* and *Abies* on the open slopes; here one finds communities strongly dominated by *Pseudotsuga* (up to 75% density). *Pinus contorta* is the most important seral species on the drier sites and is often dominant (in terms of density) until 100 years after a fire. (In terms of basal area, *P. contorta* is not replaced by *Abies, Picea,* or *Pseudotsuga* as fast as the density nomograms suggest, as *Pinus* invariably has much larger stems and canopy coverage than the three competitors during this first 100 year period; however, this fact is ignored when relative density alone is considered.) *Populus tremuloides* is also an important seral component during the first 50 years after a burn on the 1300–1600 m open slopes—it is often clumped, or in savanna-like stands, surrounded by meadow and talus cover, especially on the steepest slopes.

High-elevation forests (over 1600 m) develop in the McDonald drainage whenever the stands exhibit sufficient soil development and sheltering from high winds and slides. These stands are invariably composed of *Abies lasiocarpa, Picea,* and *Pinus albicaulis,* with extremely rare *Larix lyallii* on xeric sites over 2000 m, frequent *Pseudotsuga* extending up to 1850 m on the open slopes, and *Larix occidentalis, Pinus contorta,* and *Populus tremuloides* extending in the seral stands up to 1700–1900 m. Above 1900–2000 m, seral development involving new (strictly seral) species does not occur; at most, the importance values of the mature species will change through succession. Above about 2300 m, even the importance values will not change during fire succession, primarily because of the mild fire intensity and plants resprouting from a surviving root structure.

These high-elevation forests show peak *Abies* density on the most mesophytic sites, peak *Pinus albicaulis* densities on the most xeric sites, and *Picea* covering the transition with low (5%–10%) densities. As exposure to the wind increases, winter snow cover is blown away, exposing and freezing branches for the next wind to break off; therefore the forests rapidly are transformed to a low Krummholz mat. Species composition does not change much from the composition of the typical forests at the same elevations—*Abies* and *P. albicaulis,* with some *Picea,* and extremely rare Krummholz *Pseudotsuga* and typical *Larix lyallii.* Here the alpine wind–snow gradient rapidly becomes an important determinant of both formation characteristics and species composition. Extreme wind conditions preclude the development of even a patchy Krummholz forest. More moderate conditions permit Krummholz development, with solid Krummholz mats of *Abies* common under all but the very xerophytic conditions. *Pinus albicaulis* is either adjacent to or interspersed with the Krummholz *Abies* under xerophytic conditions.

As discussed in Chapter 8, here occurs the phenomenon of *Abies* replacing *Pinus* under the most extreme conditions along the passes. The distribution of the two species on the topographic-moisture gradient suggests that *Pinus* should be more common under such xeric conditions but, as noted earlier, only the Krummholz growth form of *Abies* allows it to exist here. Virtually all the ravines and draws at these elevations are also fellfields, however, subjected to frequent slides. Under these conditions, *Abies'* rigid structure is a liability, and the flexible stems of *P. albicaulis* permit it to compete successfully under conditions too mesophytic for it under ordinary circumstances.

At still higher elevations, even Krummholz is rare—meadows, talus, low shrubfields, and bare rock–glaciers predominate and are discussed in Chapter 11.

Camas Drainage

Nomograms of tree species that occur in the Camas drainage are shown in Figs. 29–43 of Appendix 2. Comparison of these nomograms to those for the McDonald drainage reveals several major differences:

> *Tsuga* and *Thuja* have very limited ranges and localized distributions when compared to the McDonald drainage.
> *Picea* exhibits major distributional differences on all three gradients.
> *Pinus albicaulis* is more important at the higher elevations.
> *Pinus contorta* remains an important seral component for a longer period of time.
> *Pseudotsuga* is a more important seral species, especially on the open slopes.
> *Betula papyrifera* does not achieve strong dominance on the mature, xeric low-elevation stands.

Except for these compositional differences, successional processes and species relationships are similar to those in the McDonald drainage.

Extreme low-elevation stands (below 1100 m) are quite rare in the drainage. The absence in low, mesophytic stands of *Tsuga* and *Thuja* leads to a much higher density of *Picea*—usually between 55% and 75% in the ravines, draws, sheltered slopes, and north-aspect open slopes. As is undoubtedly the case in the McDonald drainage, the *Picea* in the Camas drainage is of hybrid origin, and most of it should be considered *Picea glauca* × *engelmannii* below about 1400-m elevation (Habeck and Weaver 1969). However, although it shows a distinct bimodal distribution (with modes in lowest, wettest sites and higher, submesic environments) in the McDonald drainage's mature stands, this is not observed in the Camas drainage except in the seral stands. Even in the seral communities, the bimodality is primarily along the topographic-moisture gradient, not along the elevation gradient, whereas in the McDonald drainage it is observed along both gradients. In the McDonald drainage, the bimodality of *Thuja's* distribution corresponds with *Tsuga's* peak density. This is not the case for *Picea* in the Camas drainage: its distribution appears caused both by genetic and by competitive influences; that is, populations on the mesophytic extreme are most like *P. glauca* (than like *P. engelmannii*), whereas those on the xeric end of the gradient are more like *P. engelmannii* (cf. Habeck and Weaver 1969). Combined with this effect is competitive pressure in the seral communities from *Abies lasiocarpa* on the mesic side of *Picea's* range, and from *Pseudotsuga* (and possibly even *Larix occidentalis*) on the xeric side of its range.

Inspection of the nomograms shows that several species are displaced toward (topographically defined) more mesophytic conditions, undoubtedly because of the geographically drier climate of this drainage. This is especially pronounced for *Pinus albicaulis* and *Abies lasiocarpa* (and possibly for *Picea*) at the higher elevations.

Another major seral difference between the McDonald and Camas drainages is the more important role of *Pseudotsuga* in the latter drainage. *Pseudotsuga* is a very strong seral species on the drier sites at low to moderate elevations; together with *Pinus contorta,* it almost totally dominates such midseral stands. Apparently the different distributions of *Picea,* including its displacement toward the mesic end of the topographic-moisture gradient, permit this strong dominance by *Pseudotsuga* and *Pinus.* Instead of *Picea's* being very common on the low, dry, mature slopes, as it is in the McDonald drainage, much stronger dominance by *Pseudotsuga* is observed, and greater longevity of *P. contorta* (along with early *Populus tremuloides*); *Betula's* lack of success here is discussed in Chapter 10.

Tsuga, Thuja, and *Abies grandis* do achieve fair density (20%) in a few localized, isolated stands, apparently because of a combination of locally favorable (mesophytic) conditions and unknown stochastic elements. Except for the differences noted above, middle- and high-elevation stands are very similar to those found in the McDonald drainage.

North Fork Drainage

Nomograms of tree species that occur in the North Fork of the Flathead River drainage are shown in Figs. 44–54 of Appendix 2. Comparison of these nomograms to those for the Camas drainage reveals these differences:

> *Larix lyallii* is a strong dominant in many high-elevation stands in the upper Kintla and Bowman drainages and therefore strongly affects the distributions of *Abies lasiocarpa, Picea* and *Pinus albicaulis.*
> *Thuja, Tsuga, Abies grandis,* and *Betula occidentalis* are absent.

> *Populus tremuloides* is much more common, occurs on (topographically defined) more mesophytic sites, and forms moderately large groves.
>
> *Populus trichocarpa* extends to higher elevations on the bottomlands.
>
> *Pinus ponderosa* is much more common.
>
> Distributional differences for *Pseudotsuga* and *Larix* are noted.
>
> *Pinus contorta* occurs on (topographically defined) more mesophytic sites, is more common, and exhibits greater longevity.
>
> Evidence for intergradation between *Pinus monticola* and *P. albicaulis* is absent.

The North Fork area therefore shows some major differences from the Camas drainage and is very different from the McDonald drainage. Much of this difference results from the interfacing of forests, savannas, and meadows–prairies at the lower elevations and from a more severe alpine environment (with more area in earlier primary succession stages). Therefore, these communities are discussed further in Chapter 11.

Mature bottomlands are often not forested; those that are forested are, like the ravines and many of the draws, almost pure *Picea*. When these stands are influenced by hydric disturbances, *Populus trichocarpa* is usually present, albeit at densities below 5%. Recently burned bottomlands and ravines are rare; such stands are usually dominated by *Picea* and *Larix occidentalis,* with some *Pinus contorta* and *Pseudotsuga* and rare but persistent *Abies lasiocarpa. Abies* survives to maturity better than do the other seral species. On the low-elevation mature slopes, a smooth transition from *Picea* to *Pseudotsuga* is encountered as more and more xeric conditions are reached. *Pinus ponderosa* forests, often mixed with *P. contorta,* and almost pure *P. ponderosa* savannas (under a natural fire regime of frequent ground fires but rare canopy fires) are encountered on the south aspects of open slopes with gravelly soils. These two pines, as well as *P. monticola,* are displaced to the (topographically defined) more mesic sites, as are the seral stands of *Larix occidentalis* and *Pseudotsuga.* Low-elevation, xeric mature stands are very rare; under a natural fire regime, most of the low, open slopes and more xeric sites should exhibit a *Pinus contorta–Pseudotsuga–Larix occidentalis* disclimax, with stands interspersed by *P. ponderosa* savanna, *Populus tremuloides* groves, meadows, and prairie remnants.

In the midelevation forests, mature stands are dominated by *Abies lasiocarpa, Picea,* and *Pseudotsuga.* Strong dominance by *Picea,* probably supported by a strong gene pool resulting from its great abundance at the lower elevations, appears to have displaced *Abies* toward the xerophytic end of the topographic-moisture gradient and to have (competitively) split Pseudotsuga's distribution into two modes. Midelevation successional communities show a decline in *Abies* and *Picea,* dominance by *P. contorta* in all but the most mesophytic sites, and *P. contorta* gradually replaced by

Pseudotsuga on the xerophytic sites. *Larix occidentalis* is not a strong component of such seral communities.

High-elevation stands show major changes from both the Camas and the McDonald drainage, primarily because of strong dominance by *Larix lyallii* on many high-elevation slopes. *Larix lyallii's* dominance apparently results from its ability to successfully invade high, cold, rocky (mixed rock–talus–meadow fringes) unsheltered sites that are too severe for its major competitors (*A. lasiocarpa* and *P. albicaulis*) (Habeck and Arno 1972). By this ability it achieves a foothold at the point in the environmental space where these two competitors are replacing one another, a point, therefore, for which neither is especially well adapted. From this foothold, it is able to achieve dominance and strongly displace both species.

Larix lyallii displaces *Abies* to lower, drier sites and subsequently lowers *Abies's* density; it also affects *P. albicaulis* by displacing it toward lower elevations. The presence of *Larix* undoubtedly combines with the lower precipitation of this area to force some displacement of *Pinus* along the topographic-moisture gradient. (This effect is also shown in Fig. 8.9, Chap. 8, on the Boulder Pass alpine wind–snow gradient). It is also possible that: (1) the strong displacing and density lowering effects that *Larix* has on *Abies;* (2) the much stronger effect *Larix* has on *Abies* than on *Pinus;* and (3) the displacement of both *Abies* and *Pinus* by geographical moisture differences, all combine to actually permit *Pinus* to achieve higher densities here than it can in the other drainages (which in fact *Pinus* does). Conversely, *Pinus'* ability to tolerate xerophytic conditions might permit it to achieve even higher densities were *Larix* not present. Neither this study nor the work by Arno (1971) satisfactorily answers the question.

Distribution of Important Herb and Shrub Species

Figures 55–87 of Appendix 2 show the nomograms of herb and shrub cover, herb and shrub diversity, and the importance values (absolute cover) for the Gramineae family, 12 genera and 18 individual species, on the elevation and topographic-moisture gradients for five age classes (0–20 years, 20–50 years, 50–90 years, 90–150 years and more than 150 years since the last canopy fire) for stands with a forest overstory. (The distribution of herb and shrub species in stands without a forest overstory are shown in Appendix 2, Figs. 88–112. These nomograms include samples from all three drainage units (McDonald, Camas, and North Fork) for three reasons: (1) There were insufficient field samples to construct them individually for the Camas and North Fork drainages; (2) samples from Camas and North Fork drainages were very similar to predictions using preliminary nomograms constructed from McDonald drainage samples; (3) sample noise was obviously much higher for understory species than for the canopy and subtle differences among the three drainages are obscured by such noise. Because of this last problem, all nomograms use the 0–6% importance value scale defined earlier (see Chap. 6, the section on "Vegetation Sampling") and do not attempt to predict species' cover variations within a stand; instead, they predict the average abundance within a stand as a function of the macrogradients. The entire process of constructing nomograms for understory species is greatly complicated by the microgradients—especially moisture—which affect the understory in a much more obvious fashion than they affect the overstory. Again, as a result, a prediction of cover "2" (5%–25%) for *Xerophyllum* indicates a predicted average of 5%–25% cover within the stand; some quadrats may show 50% cover and others may have no cover by *Xerophyllum.*

The total herb and shrub cover nomograms show that the greatest total cover is reached at the lower elevations, with cover dropping rapidly above 2000 m. Mesophytic sites reach 95% cover within 25 years after a fire, whereas low-elevation xerophytic sites may require over 150 years to reach 95% cover.

As discussed in Chapter 12, attempts to construct nomograms of various diversity measures have been unsuccessful; therefore diversity is represented here by the rather crude method of contouring average number of species in two random 2 × 2 m plots within the stand. The results show that the lowest, wettest sites exhibit the greatest herb and shrub diversity regardless of stand age, and that diversity drops as elevations become higher or more xerophytic conditions are encountered. For a given site, the youngest communities are usually the most diverse, whereas the late seral and mature stands are the least diverse.

Species nomograms show the whole spectrum of individualistic responses. Some species, such as *Athyrium filix-femina* and *Sorbus scopulina,* are virtually absent until a stand reaches late seral conditions. *Athyrium* shows less than 1% cover until 150 years of stand age and then jumps to 5%–25% cover in the low, mesophytic forests. *Sorbus* shows less than 1% cover until 90 years of stand age and then increases to 5%–25% cover in the midelevation xeric and subxeric forests.

Several species are most common in the mature forests but enter these communities well before maturity. *Xerophyllum tenax* enters a stand soon after burning but does not exceed 5% cover until 50 years of stand age. *Clintonia uniflora* enters low, submesic forests about 20 years after a fire but does not reach peak density or maximum habitat range until stand maturity. *Disporum* also enters low, mesic stands soon after a fire but expands its cover and habitat range only in the older stands.

Tiarella trifoliata is absent until 90–100 years after a burn, reaches a good habitat range soon after entering the midseral community, but does not reach peak cover until stand maturity. *Amelanchier alnifolia* enters a stand soon after the fire and slowly increases both its cover and its habitat range as the stand matures. *Pachistima* behaves much like *Amelanchier* until stand maturity and then decreases its habitat range somewhat. *Aralia nudicaulis* enters a very narrow habitat range within 20 years after a fire but does not expand its range or abundance for another 75 years. Conversely, *Osmorhiza* rapidly reaches its maximum habitat range soon after entering the midseral stands. *Chimaphila umbellata* exhibits a bimodal distribution on the moisture gradient in midseral forests but then drops the wet mode and expands both its habitat range and density as the stand matures. Although *Linnaea borealis* achieves a bimodal distribution only in mature forests, it exhibits a broader habitat range in the seral forests.

Other species, including *Calochortus elegans, Thalictrum occidentalis,* and *Adenocaulon bicolor,* reach peak densities and/or habitat range in the midseral or late seral forests. *Calochortus* is present in trace but consistent quantities in the low-elevation xerophytic forests from 50 to 90 years after a burn. *Thalictrum* is common in a narrow habitat range soon after a fire, drops in both habitat range and cover by 25 years after the burn, and then reaches peak cover and habitat range in the 50–90 years after burn period. *Adenocaulon* shows a complex bimodal distribution in the mature and late seral stands and reaches peak density in these latter communities.

Two common species, *Epilobium angustifolium* and *Galium boreale,* exhibit greatest density and habitat ranges in the early seral communities but are also present in older stands. *Epilobium* is very common on the drier sites (below 1800 m) soon after a burn and shows a declining habitat range and density in older stands. *Galium boreale* reaches peak cover in the early seral stands (0–20 years after burn) and exhibits a similar habitat range but lower cover in the mature stands. [On Fig. 84 (Appendix 2) the curves in the 20–50 years after burn part and in the wet mode of the 50–90 years after burn part are *Galium triflorum.*]

No understory species of types that are common soon after a fire and then absent from late seral stands were found in the study area. However, several species and groups, including *Fragaria, Rosa, Rubus parviflorus, Spiraea betulifolia,* and *Heracleum lanatum,* seem to be virtually unaffected by stand age. *Fragaria* and *Rosa* are common on all low, xerophytic sites regardless of stand age. *Rubus parviflorus* is common in all but the driest low-elevation forests but does (barely) expand its habitat range as the stand matures. *Spiraea betulifolia* is common in all but the most mesic forests below 1600 m and slowly shifts to slightly dries sites as a stand matures. *Heracleum lanatum* is common in all low-elevation mesic and submesic forests, shifts slightly toward the mesic end of the topographic-moisture gradient as the stand matures, and, surprisingly, shows an unexplained bimodal distribution in the mature forests. A discussion of the role of disturbances, especially fire, in maintaining these species and their distributions is included in Chapter 10.

Distribution of Ground Fuels

Fuel data were reduced using standard techniques as described in Chapters 6 and 7. Shrub branchwood and foliage weights were determined using the results of the shrub biomass dimension analysis shown in Table 9.1; when regression constants were unavailable for a species, the constants for the species with the most similar growth form were used. Preliminary analysis of limited fuel data from Glacier (Kessell 1976a, b, 1977) indicated that the primary succession, stand age, elevation, and topographic-moisture gradients were all useful in explaining the distribution of ground fuels. However, from the extensive fuel data collected during 1975, and from the very large within-site variations of fuels, we learned that the first two gradients, plus fire intensity, were the only ones consistently significant for quantifying fuels distributions.

In addition, results of the basic data analysis showed that within each (gradient) stratification group, the frequency distributions for each fuel characteristic was skewed to the right; ratios of the median to the mean usually ranged from 0.4 to 0.8. Small fluctuations are also found in most of the curves in the far right-hand tail of the distribution; these small relative maxima are the values from a few samples taken in localized areas of abnormally heavy fuels. Such "jackpots" are the result of localized areas of high tree mortality, which may be caused by ice damage, windthrow, insect infestation, or avalanche activity (Kessell and others 1978). As a result, median rather than mean values are more indicative of the central tendency of the population.

Table 9.2 summarizes the response of fuels to postfire forest succession (time since burn gradient); all data are from stands that support a forest canopy. Plots sampled in the 1–10 years since burn category were obtained on sites where fire destroyed more than 90% of the overstory. Litter and 1-hr time-lag branchwood were almost totally consumed by these fires; these two fuels therefore exhibit very low loads. Loads of 10-, 100-, and 1000-hr time-lag branchwood are high (2.8, 6.0, and 27.8 tons/ha, respectively) as snags created by the fire fall to the forest floor, adding to the existing large fuels not totally consumed by the fire. Shrubs, grasses, and forbs flourish at this time because of increased sunlight and the enriched, exposed mineral soil seedbed.

As the stands succeed from the 10-year postburn level, the median fuel loads exhibit three types of behavior. Certain fuels decrease through time, others increase through time, and still others plateau at a particular level and then remain relatively constant through time. The live fuels consisting of herbaceous plants and shrubs decrease as the stand matures. When the canopy closes, grass and forb loads decrease 40% and continue this decline to 0.2 tons/ha by an age of 155 years. Total shrub load also shows a 40%

Table 9.1 Regression and Correlation Coefficients among Shrub Diameter at Ground Height (dgh), Foliage Weight, and Branchwood Weight.[a]

Species	N	Branch weight				Foliage weight			
		A	B	R	R^2	A	B	R	R^2
Tsuga heterophylla[b]	16	−1.28	2.67	0.98	0.97	−1.25	2.44	0.98	0.96
Pinus contorta[b]	14	−2.06	3.15	0.99	0.98	−1.23	2.30	0.96	0.93
Pseudotsuga menziesii[b]	17	−1.10	2.47	0.97	0.94	−0.74	1.91	0.94	0.88
Taxus brevifolia	20	−1.18	2.63	0.98	0.95	−0.94	2.20	0.93	0.86
Juniperis communis	25	0.09	1.50	0.51	0.26	0.40	1.10	0.47	0.23
Salix spp.	13	−1.74	2.90	0.98	0.97	−1.37	2.16	0.98	0.96
Alnus incana	36	−2.01	2.92	0.79	0.63	−1.60	2.38	0.93	0.86
Alnus sinuata	34	−1.89	3.02	0.98	0.97	−1.67	2.31	0.97	0.94
Ribes lacustre	33	−1.27	2.62	0.93	0.86	−0.51	1.27	0.68	0.46
Ribes viscosissimum	38	−2.28	3.58	0.97	0.95	−1.94	2.94	0.94	0.89
Amelanchier alnifolia	21	−1.34	2.70	0.94	0.88	−1.12	1.79	0.76	0.58
Rubus parviflorus	32	−1.22	2.34	0.93	0.86	−0.50	1.49	0.65	0.42
Sorbus scopulina	24	−2.06	3.21	0.99	0.98	−1.94	2.53	0.94	0.89
Spiraea betulifolia	28	−1.16	2.63	0.94	0.88	−0.80	1.78	0.82	0.68
Pachistima myrsinites	23	−1.45	2.69	0.87	0.75	−0.74	1.68	0.89	0.78
Acer glabrum	21	−1.91	3.21	0.96	0.93	−1.50	2.27	0.94	0.89
Shepherdia canadensis	27	−1.37	2.72	0.83	0.70	−1.44	2.15	0.80	0.64
Oplopanax horridum	38	−2.65	3.34	0.94	0.88	−1.48	2.11	0.82	0.67
Cornus stolonifera	20	−2.07	3.32	0.96	0.93	−1.53	2.22	0.87	0.76
Vaccinium membranaceum	21	−1.36	2.70	0.97	0.93	−0.78	1.36	0.90	0.81
Lonicera involucrata	26	−1.19	2.49	0.94	0.89	−0.94	1.68	0.87	0.77
Menziesia ferruginea	29	−1.64	2.93	0.98	0.96	−1.97	2.39	0.97	0.95
Symphoricarpos albus	29	−1.31	2.72	0.98	0.97	−1.24	1.96	0.81	0.65

[a]The form of the equation is: $\log_{10} Y = a + b \log_{10} X$, where Y is the dry weight in grams and X is the dgh in mm. The equations have not been tested for diameters greater than 60–70 mm.
[b]Tree seedlings.

Table 9.2 Median Fuel Loadings (by Fuel Types and Size Classes) for Typical Upright Forests Stratified by Stand Age (Time Since the Last Canopy Fire). [a]

	Stand age (years)								
	0–10	11–44	45–69	70–89	90–109	110–129	130–149	150–169	Over 169
Sample size (N)	12	7	21	58	15	29	6	18	48
Fuel loads (tons/ha) for:									
Litter	0.67	2.24	2.00	2.08	1.56	1.74	3.97	2.30	2.15
Grass and forbs	0.48	0.45	0.33	0.30	0.34	0.24	0.17	0.36	0.21
Branchwoods									
1 hr	0.42	1.05	1.06	0.72	0.88	0.96	0.86	1.02	1.02
10 hr	2.77	1.18	1.62	1.18	1.24	1.35	1.65	2.19	1.70
100 hr	6.02	3.37	3.38	1.69	3.51	3.43	5.98	1.86	1.70
1000 hr	27.80	32.51	8.93	15.51	9.65	15.00	29.86	29.49	46.71
Shrubs									
Foliage	1.13	0.74	0.40	0.41	0.42	0.44	0.31	0.44	0.20
Branchwood	3.76	5.39	1.35	1.75	1.43	1.60	1.08	1.24	0.68

[a]The text explains the temporal changes in the ground fuel array. (Kessell and others 1978.)

(a)

(b)   (c)

Figure 9.6 a–c. Vegetation and fuel changes during secondary succession following canopy fires in coniferous forests. **a.** Six years after a fire; **b.** 60 years after a fire; and **c.** 174 years after a fire. Photographs **b** and **c** provided by William Fischer, Northern Forest Fire Laboratory.

decrease at canopy closure, remains fairly constant from 70 to 150 years after the fire, and finally drops again by 155 years. Litter and 1-hr time-lag branchwood tend to increase and then plateau from canopy closure to maturity. Litter remains near 2.2 tons/ha except for an increase that occurs about 150 years after the burn; this increase probably results from the mortality of successional *Pinus contorta*, as discussed earlier in this chapter. One-hour time-lag branchwood reaches 1.1 tons/ha and shows no tendency to accumulate through maturity. Ten-hour time-lag branchwood drops from its earlier level, which resulted from snag deposition, and then remains fairly constant. Hundred-hour time-lag fuels show rather erratic behavior that may have unknown causes. Thousand-hour fuels exhibit a marked increase through time. Consistent with fuel loads, stratal packing ratios reflect the characteristics of their component fuel loads. Examples of forest floor fuels for three stand ages are shown in Fig. 9.6.

10

Forest Successions

Throughout the discussion of Glacier's forest communities in the last chapter, the words "seral" and "successional" kept appearing and reappearing. Although the discussion of succession has been held off until this chapter, it is simply not possible to understand or describe Glacier's communities without continual reference to successional development. Nearly all of the park shows community development that reflects varying stages and rates of recovery from natural disturbances—some are simply more obvious than others.

Modeling Succession in Glacier

As noted in Chapter 3, we used a Clements (1916) approach to modeling succession during the development of the Glacier Park gradient model. There were several reasons for this choice:

1) It is a simple approach; the time since burn gradient directly provides the deterministic, single-pathway predictions for plant and fuel recovery following a fire.
2) More sophisticated approaches, such as those of Botkin (Botkin and others 1972) and Horn (1974), require very high resolution, tree by tree data that were simply not available.
3) The Noble and Slatyer (1977) multiple-pathway model described in Chapter 3 had not yet been developed.

However, following the Noble and Slatyer work, we went back to the Glacier data base to see whether their model could improve the prediction of postfire succession in Glacier, especially for some of the "problem" communities, such as *Populus tremuloides* and *Betula papyrifera*. Highlights of that work are described by Cattelino and others (1979) and are reported below. (The reader unfamiliar with the Noble and Slatyer model should review the summary provided in subsections of the Chap. 3 section on "Succession Modeling").

Application of the Multiple-Pathway Model

The application of the Noble and Slatyer model to Glacier's overstory species revealed that many of the species were of the dispersal (DT or DI; see Chap. 3) type and would therefore establish themselves in a stand following a disturbance regardless of the interfire period. (D species never become extinct because new propagules are dispersed from surrounding communities.) For D species, therefore, results of the Noble and Slatyer model converge with the results of a Clements approach and do not provide new insights regarding replacement patterns following disturbance.

However, when there were problems predicting the successional patterns in Glacier Park, the Noble and Slatyer method provided considerably better results than the Clements method. For example, 9 *Tsuga heterophylla* forest just north of Lake McDonald on Glacier's west side burned approximately 75–80 years ago; it returned to a pure *Tsuga* community instead of being invaded by *Pinus contorta* as predicted by a Clements model. Here, *Tsuga* acted as a DT species and *Pinus* as a CI (see Chap. 3) species. The stand was in an old growth stage when it burned, and as a result, the intolerant *Pinus* was lost from the community prior to the burn (Fig. 10.1). With no seed source for *Pinus* from the surrounding communities, it could not establish in the postfire community. Furthermore, if this stand had burned in the early stages of development, then CI lodgepole pine also would have been lost from the stand, with *Larix occidentalis* and *Tsuga* invading the site. This stand, if left undisturbed, would return to a DT *Tsuga* community.

An even better illustration of the advantages of the Noble and Slatyer method is a *Populus tremuloides* community found in the North Fork of the

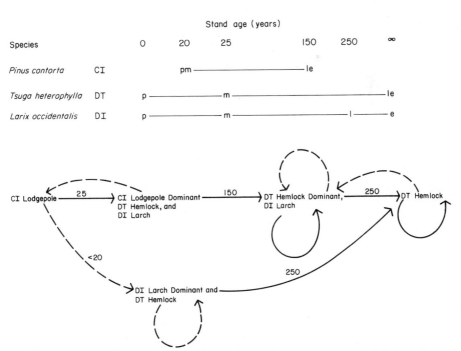

Figure 10.1. Life history characteristics and species replacement sequences for *Pinus contorta, Tsuga heterophylla*, and *Larix occidentalis* in northwestern Montana hemlock communities. Ages and transitions are indicated as in Fig. 3.3.

Flathead River drainage in western Montana. Typical species composition over time for these stands is given in Table 10.1 (also see Appendix 2). Normal fire periodicity is about 51 years (Singer 1975a); the stands are usually a mixture of *Populus tremuloides, Pinus contorta, Pinus ponderosa, Pseudotsuga menziesii, Picea glauca × engelmanii,* and *Larix occidentalis.* A fire in these stands under normal periodicities leads to a succession that includes all of these major species. However, obvious exceptions have been observed:

1) When there have been long interfire periods, some stands have lost *Populus* after the fire, with *Pinus contorta, Larix occidentalis, Pinus ponderosa,* and/or *Pseudotsuga* increasing their importance.
2) Also, when there have been long interfire periods, some stands have lost *Populus* and *Pinus contorta* after a fire, with a dramatic increase in *Larix occidentalis.*
3) It is suspected that when there have been very short interfire periods some stands have lost the *Pinus contorta,* with a dramatic increase in *Populus.*

As shown in Fig. 10.2, *Populus* is a VI type (Chap. 3) with a life span of about 130 years (therefore, extinction, e, also occurs at 130 years). *Pinus contorta* is a CI type with a life span (and therefore e) of about 150 years, whereas *Larix occidentalis* is a DI type with a much longer life span, 300–400 years (e occurs at infinity for D species). Now, if this *Populus* community is burned while both *Populus* and *Pinus contorta* are present (that is, if fewer than 130 years have passed since the last fire), both *Populus* and *Pinus contorta* (along with other DT and DI species) will be present in the postfire succession. If the stand burns with both *Pinus contorta* and *Populus* are extinct in the community, both species are lost in

Table 10.1. Changes in Relative Density (Stems over 7.62 cm dbh) as a Function of Stand Age for *Populus tremuloides* Community in Northwestern Montana.[a]

Species	Stand age (years)			
	21–50	51–90	91–150	150+
Populus tremuloides	25	25	3	0
Betula papyrifera	0	2	2	0
Pseudotsuga menziesii	20	20	25	38
Pinus ponderosa	15	15	15	0
Pinus contorta	20	18	8	0
Pinus monticola	2	2	0	0
Picea glauca × engelmannii	11	11	44	60
Larix occidentalis	5	5	3	2
Abies lasiocarpa	2	2	0	0

[a]These average values, from Appendix 2, are for interfire periods of less than 130 years.

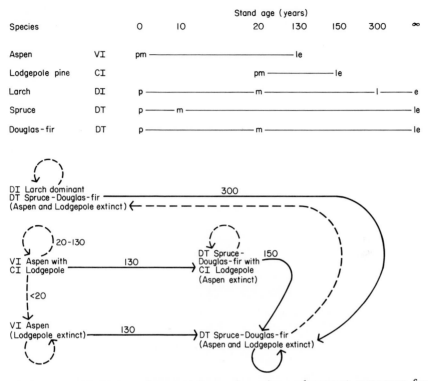

Figure 10.2 Life history characteristics and species replacement sequences for *Populus tremuloides, Pinus contorta,* and *Larix occidentalis* in northwestern Montana aspen communities. Ages and transitions are indicated as in Fig. 3.3.

the postfire succession. In this case, *Larix occidentalis,* with its much greater longevity than either *Pinus contorta* or *Populus* (and its D mechanism), will provide a seed source, and a stand where *Larix* is predominant will occur. This event is observed in portions of Glacier National Park. Finally, if the stand burns within 20 years after a fire, the model predicts the loss of *Pinus contorta* from the community, since *Pinus* does not reach the critical replacement stage of propagule, p, until an age of about 20 years.

Multiple Attributes for a Single Species

We found that the failure to explain certain successional patterns in Glacier National Park was not entirely a result of the modeling method applied; instead, it became apparent that some species could be assigned two different sets of vital attributes within a single community type. For example, as a CI species, *Pinus contorta* will establish dense stands (as a result of good reproduction) shortly after a fire; when it functions as a DI species,

Figure 10.3a. "Successional pockets" formed as a result of fire in a portion of the Belton Hills of Glacier National Park, Montana. Note the extensive pocket of *Pinus contorta* in the background, and smaller pockets of mixed conifer and *Betula papyrifera* in the foreground.

its reproductive success and resulting density will be much reduced. The *P. contorta* situation is probably further complicated by its cone serotiny. Unlike the classical CI situation, this characteristic delays extinction (e) beyond loss (l) of adults from the community.[1] (This is shown in Fig. 10.4.)

As an example of this situation, consider a section of the Belton Hills in Glacier National Park (Fig. 10.3). Portions of this area burned in 1910 and

---

[1]This delayed loss of a C type population is now recognized by Noble and Slatyer (1979) as a G type. Therefore, under their revised system, *Pinus contorta* is a GI species.

Figure 10.3b. *Betula* pocket with small amounts of mixed conifer and *Pinus contorta*.

again in 1929. Prior to the earlier burn, the area consisted of a mosaic of different ages of mixed conifer stands. As a result of the first (1910) fire, several "successional pockets" occurred, each following a somewhat different successional sequence. They included areas where:

1) All species present before the burn were reestablished in the postfire community.
2) *Pinus contorta* invaded poorly in the postfire community.
3) *Pinus contorta* was not retained in the postfire community.

These pockets can be explained by the species composition of the mosaic cells, which is a function of stand age. These pockets are illustrated in Fig. 10.3.

As shown in Fig. 10.4, in the first case all species were present in the stand at the time of the fire; the periodicity was such as to allow retention of all species in the postburn community. In the second case, lodgepole pine invaded poorly because it had become extinct from the community before the fire and was forced to act as a DI rather than as a CI species (Ie occurred before the fire for CI lodgepole pine) and so exhibited a lower density. In the third case, CI lodgepole pine became extinct in the prefire community in the same manner; however, owing to factors of geography and/or chance, there was no seed source available for it to function as a DI species. It therefore remained extinct in the postfire community.

In 1929, when portions of this area burned again, similar pockets were formed. The areas where all species were present before the second burn lost CI lodgepole pine because the stands (regenerating from the 1910 fire) had not existed long enough for lodgepole pine to reach maturity and replenish the propagule pool. This situation favors *Betula papyrifera* which acts as a VT and establishes by vegetative resprouting as shown in Fig. 10.4.

In areas where *Pinus contorta* was poorly established after the 1910 burn, it may or may not establish again, depending on the availability of a seed source and therefore its ability to act as a DI species. (CI *Pinus contorta* becomes extinct in the community if it cannot function in this fashion.) Establishment by the DI mechanism, in this case, depends on the mosaic's geography, wind, and other factors.

In this example, *Pinus contorta* must utilize the DI pathway, as an alternative to the more efficient CI mechanism when interfire periods are very long or very short. Likewise, *Betula* will regenerate as a DT species in the prefire community but will recover from fire disturbance via the VT pathway. These combined attributes of tolerance and vegetative reproduction allow birch to reestablish itself in a suitable stand regardless of the interfire period.

Several other overstory species in western Montana also exhibit this multiple-attributes trait:

> *Populus tremuloides* (VI or very poor DI)
> *Pinus monticola* (DI or CI)
> *Larix occidentalis* (DI or CI)
> *Thuja plicata* (invades as a DT but regenerates without disturbance primarily as a VT)

The pathway followed is probably determined by the size of the burn (distance from seed source), presence of the species in the preburn communities, and presence of the species in the surrounding undisturbed communities. Note that many of the key species involved in western Montana coniferous forest successions follow variable pathways in a distinct departure from traditional successional concepts. A revised scheme (first described in Noble and Slatyer 1978) incorporating multiple attributes has

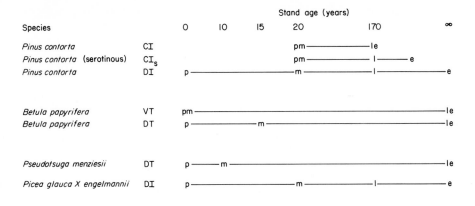

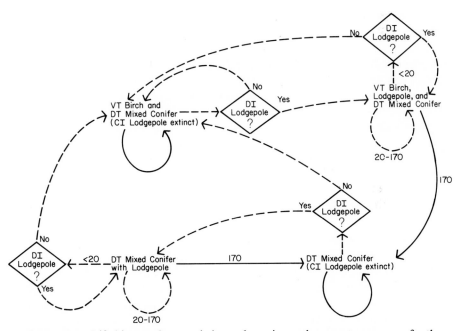

Figure 10.4. Life history characteristics and species replacement sequences for the Belton Hills in Glacier National Park, Montana. Species include *Pinus contorta, Betula papyrifera, Pseudotsuga menziesii,* and *Picea glauca* × *engelmannii*. Ages and transitions are indicated as in Fig. 3.3. Note the strong effect on community succession that the presence or absence of DI *Pinus contorta* has in communities where CI *Pinus contorta* is extinct.

been prepared by Noble and Slatyer (1979) to better describe these pathways.

Need for a Dispersal Model

The above discussion of single species possessing multiple attributes (with different reproductive success) demonstrates the need for a seed dispersal model. Although we recognize the possibility of some species invading the postfire community with a D attribute, we are as yet unable to establish the probability of this occurrence (Cattelino and others 1979). Recent studies of cone production (Franklin 1968, Franklin and others 1974) and seed dispersal (Franklin and Smith 1974a, b) in the Cascades, Olympic Mountains, and Coast Ranges (northwestern United States) suggest that these kinds of data, augmented by distance and wind parameters, may permit the construction of a stochastic model of seed dispersal. Such a model is currently (1979) under development by Gradient Modeling, Inc. personnel in cooperation with Franklin and other U.S. Forest Service researchers.

Walter (1977) developed a model of fire effects on vertebrates that describes species populations' tolerance in response to three fire gradients: fire frequency, fire intensity, and an area–terrain type. Substitution of a "dispersal probability" for the "area–terrain type" should provide a good framework for describing plant species' responses to fire in various communities. If a successful dispersal model can be developed, it should improve considerably the predictions of the multiple-pathway succession model.

Environmental Effects on Vital Attributes

In addition to some species' ability to function with different attributes within a single community, the attributes of several species change from one community type to another. For example, in Glacier National Park *Abies lasiocarpa* is an intolerant (I) seral species at low elevations but functions as a tolerant (T) climax species in subalpine communities; its reproduction is by dispersal at low altitudes but is often vegetative at higher elevations. Numerous other examples are available. The application of the multiple-pathway succession model therefore requires determination of vital attributes for each community type; alternatively, they may be expressed as functions of environmental gradients.

Application to the Understory Species

Although the multiple-pathway model provided considerable insight into successional patterns of coniferous forest overstories, we encountered problems when we applied it to individual understory species.

The first problem was difficulty assigning vital attributes to the individual understory species. Very little has been reported in the literature regarding how understory plants persist through a disturbance, or about the conditions necessary for their establishment (as defined by the Noble and Slatyer model). Information on critical events in their life histories is frequently lacking. Even when this information is known, the problem is often further complicated by microenvironmental differences within the site. Although the site may be "uniform" on the scale of the overstory trees, microenvironmental differences on the scale of herbaceous plants will produce a compositional mosaic. This, in turn, can cause a single herbaceous species to follow different successional pathways within the community mosaic. These preliminary considerations suggest that further research on herbaceous species' adaptive traits and life histories will significantly improve our successional modeling capabilities.

Implications for, and a Comment on, Park Management

Fire is undoubtedly the single most important determinant of the structure, species composition, and community characteristics of Glacier's forests. Natural fire periodicity may range from a few years to several hundred years, and fire intensity may range from a creeping ground fire with flame lengths of 2–3 cm to a crown fire with flames exceeding 100 m. The effects of fire on the forest communities therefore depend on the behavior of the fire, the communities' normal fire periodicity, and the location of these communities within the environmental hyperspace (see Figs. 10.5–10.8).

The low-elevation mesophytic forests are the least affected by fire, because of both the long fire periodicity and the high fuel moisture levels. Crown fires are rare; most of the infrequent burns destroy only the understory plants and some of the litter. The rare crown fire leads to seral communities dominated by *Picea* with less quantities of *Pseudotsuga* and *Larix occidentalis*. These seral species are soon replaced by those of the "climax" community.

More xerophytic low-elevation forests exhibit strong dominance by *Pinus contorta* during the early seral period, with replacement by *Pseudotsuga* and/or *Picea* (Camas and North Fork areas) or *Thuja* and *Tsuga* (McDonald drainage). *Larix* and *Pinus monticola* are also common. The most xerophytic low-elevation forests represent Glacier's "fire climax," a disclimax that may include *Pinus controta*, *P. ponderosa*, *Larix*, *Pseudotsuga*, *Populus tremuloides*, and/or *Betula papyrifera*, depending on site conditions.

Mid- and high-elevation mesic forests also experience mild fires, and even when the canopy is destroyed, the successional communities are

Figure 10.5. A lightning-ignited fire on Huckleberry Mountain, near the park's western boundary, burned several thousand hectares during 1967. Fire control is virtually impossible when multiple ignitions occur during drought periods. (Photograph by Mel Ruder for *The Hungry Horse News*)

composed of "climax" species, primarily *Picea* and *Abies lasiocarpa*. Drier sites will exhibit seral communities that include *Larix, Pseudotsuga, Populus tremuloides,* and/or the *Pinus monticola* × *albicaulis* complex. Most high-elevation forests do not include seral species, but fire recovery usually exhibits an increase in the density of *Pinus albicaulis*.

In earlier chapters we have noted that fires greatly increase understory cover immediately after the burn (until canopy closure) and also significantly increase species diversity (also discussed in Chapter 12). In Chapter 9 we noted several common understory species that were rare in mature forests but common in the seral communities. In addition, many tree species and community types can be maintained only by a regime of periodic fires. Another important role of fire, the reduction of flammable fuels, also was discussed in Chapter 9.

All of the results reported in this book clearly show that fire is a natural element of Glacier's ecosystems and is required if the park's natural communities are to be maintained. Suppression of natural fire in the park

Figure 10.6. Another naturally ignited fire, which occurred on Going to the Sun Mountain during 1971. The fire burned out of control until it reached natural fuelbreaks (solid rock). (Photograph by Mel Ruder for *The Hungry Horse News*)

for several decades of park management is altering, perhaps irreversibly, the park's ecosystems and is in direct conflict with the U.S. Department of the Interior National Park Service's responsibility for maintaining natural ecosystems.

What follows must be considered an "editorial," albeit, in my opinion, one that is greatly needed. Development of the gradient model brought me into intimate professional contact with a variety of National Park Service managers, administrators, and bureaucrats, and, with a few notable exceptions, I was appalled by what I found. Glacier National Park, and probably other National Park Service areas, is simply not meeting its legal mandate and public trust to protect and manage the environment using sound scientific principles. Examples are numerous; perhaps the best is the park's fire policy.

Like many other land management agencies worlwide, the park currently (early 1978) has a fire policy of total suppression. Several scientists, including Habeck (1970a, b, and in public statements and newspaper editorials) and myself (Kessell 1976c), have criticized this policy in Glacier Park. Several studies have recently been conducted in Glacier National

Figure 10.7. Early seral vegetation following intense natural fires provides crucial elk winter range. (Photograph by Mel Ruder for *The Hungry Horse News*)

Park, with National Park Service funding and sponsorship, on fire ecology relationships (Habeck 1970a, b; Singer 1975a; and the work reported here). The results of these studies show without doubt that maintenance of Glacier's natural ecosystems requires natural fires to run their course without human intervention. Fire suppression activities since the park's establishment in 1910 have reduced community diversity, reduced species diversity, reduced forage production, created unnatural fuel buildups (0.68 tons/ha/year), and created unnatural fuel bridges in the *Pinus ponderosa* savanna. The park management talked about the need for natural fire management in its *Master Plan, Management Plan,*[1] and other public documents, but the total suppression policy continued. Glacier funded several studies that showed the need for restoring fire, but the total suppression policy continued. Glacier gained the most refined wilderness fire behavior simulation system in existence, but still the total suppression policy continued. In fact, even a 0.1-ha training fire planned during the summer of 1976 under wet fuel moisture conditions in the ponderosa pine savanna was canceled by park management at the last minute because of

[1]U.S. Dept. of the Interior National Park Service public documents available from: Superintendent, Glacier National Park, West Glacier, Montana 59936.

Figure 10.8. Elk feeding in a seral *Pinus contorta* community on the park's west slope. These communities are widespread on the drier slopes and are maintained by natural fires. (Photograph by Mel Ruder for *The Hungry Horse News*)

fear of possible "adverse publicity" and "fire escape." (Fire expansion behavior predicted, by both computer models and experienced fire management personnel, was only 10 cm/min = 6 m/hr).

To the park's credit, it prepared (in 1977) an *Environmental Assessment* for a limited natural fire policy, but again the plan seems to offer more lip service than natural fire management. The plan will ultimately lead (in 1979) to natural fire zones in a portion of the park's interior.[2] Yet these are precisely the areas that have the lowest natural fire periodicity and lowest fuel loads. It is the heavily forested ridges along the park's west side that exhibit the heaviest unnatural fuel buildups because of fire suppression; these same areas are the most degraded by fire suppression in terms of both lowered diversity and potential for future holocausts caused by the fuel buildups. Yet there are, to my knowledge, no plans ever to allow natural fires in these areas, because of possible fire escape or "adverse public-

---

[2]Up-to-date information on the park's fire management policy is available from the Superintendent at the above address.

ity''—park managers shudder at the suggestion of prescription burning in these areas.

In the park managers' defense, I realize that the restoration of natural fires to a large, remote wilderness is a very difficult job, and that some elements of the public will oppose such a policy. However, many other national parks and wilderness areas have accomplished this fire management transition and have incurred neither bloodshed nor holocausts. Given the extremely refined simulation capabilities available to Glacier National Park, I think that it is time that the park quit stalling and started meeting its legal mandate in an aggressive, professional manner.

11

Other Terrestrial Communities

As is true for the species distributions of herbs and shrubs under a canopy, only a few obvious differences among the three major drainage units used in this study were noted for nonforested communities; the drainage areas therefore are combined for most of the following discussions. These communities are viewed first in terms of the primary succession mosaic, then in terms of the major component species, and finally in relation to the major disturbances that affect them.

The Role of Primary Succession

Chapter 5 described in detail the abundance of communities that either cannot support a forest overstory or have the potential to support a canopy but do not because of various disturbances. The nonforested communities presented a major problem in the development of the gradient model until the primary succession gradient was used to order the formation types found in the park. It must be realized that the primary succession gradient is primarily a construct that simplifies understanding of the vegetation mosaic. Ideally, following glacial retreat, a community progresses along the gradient, exhibiting first rock cover, then broken rocks and talus, then invasion by meadow, and finally a stable "climax" of forest or shrub. The rate at which a community progresses along the gradient depends on many other factors, and it is not possible to directly relate the gradient to a fixed time scale. In addition, other effects (high wind, slides, or hydric disturbances) may halt a stand's development at any point on the gradient or even cause it to retrogress to earlier primary succession stages. Examples of the latter include floods changing a forest community to a meadow, shrub, and/or talus community, fires destroying forest or savanna and maintaining meadow cover; and slides destroying a shrubfield or forest and maintaining meadow and talus cover. Despite these problems, however, the primary succession gradient serves as a useful tool in describing and understanding Glacier's terrestrial communities.

Another consideration in using the primary succession gradient concerns the discrete and mosaic communities that form it. One can clearly recognize "forest," "meadow," or "talus" as community types, but the continuous nature of the gradient implies intergradation among these communities. This intergradation is very dependent upon the size scale. For example, in examining a square plot with sides of at least 5 m, one can easily find a mosaic of rock–talus, talus–meadow, meadow–shrub, shrub–

forest, etc. Yet on a very small size scale (the size of individual plants), any point of the landscape is either rock, talus, meadow, or forest, etc. For the purposes of the Glacier model, consideration was given to landscape units as discrete communities down to an area of 25 m²—formation discontinuities smaller than this area were treated as a mosaic or transition type along the primary succession gradient.

The following discussion both views the major communities along the primary succession gradient as discrete units and considers changes in structure, species distribution, and community characteristics along the continuous gradient. Examples of nonforested communities are given in Figs. 11.1–11.6.

Distributions of Important Herb and Shrub Species

Figures 88–112 in Appendix 2 show the nomograms of herb and shrub cover, herb and shrub diversity, and the importance values (absolute cover on the 0–6% scale) for the Gramineae family, 13 genera and nine individual species on the primary succession and topographic-moisture gradients for (up to) five elevation classes of 305 m (1000 ft). All stands are mature (have not burned in the last 30 years).

The total herb and shrub cover nomogram shows increasing cover corresponding to higher primary succession types, with a frequent drop at

Figure 11.1 Typical mosaic of trees, shrubs, forbs, and gravel found along the park's major streams and rivers. (Photograph by Mel Ruder for *The Hungry Horse News*)

Figure 11.2. Floodplain of the North Fork of the Flathead River shows the typical forest and meadow mosaic. Flooding cycles and local drainage patterns determine this precarious balance. The Continental Divide is in the background. (Photograph by Mel Ruder for *The Hungry Horse News*)

the typical forest extreme (light-blocking canopy). Increasing elevation tends to decrease cover for any given primary succession or topographic type up to about 2200 m; above this elevation cover again increases. The slopes generally have the lowest cover, and mesic sites exhibit the highest cover.

Diversity is shown by the average number of species in one random 5 × 5 m plot. At the lowest elevations, diversity increases steadily with primary succession development and reaches a maximum in the bottomland forests.

Figure 11.3. A lower elevation, subalpine mosaic of forests, shrubs, meadow, rock outcrops, and avalanche areas is shown behind Avalanche Lake. The lake's elevation is only 1200 m. (Photograph by Mel Ruder for *The Hungry Horse News*)

At slightly higher elevations of 1220–1525 m (4000–5000 ft), the same relationship holds but peak diversity is achieved in the forested draws. Between 1525 and 2135 m (5000–7000 ft), earlier primary succession types exhibit higher diversity, maximum diversity is reached in the meadows, and diversity drops in the Krummholz and typical forests. The highest elevations (over 2135 m) show declining diversity for all communities, maximum diversity for the meadow–Krummholz mosaic, and very low diversity on the forested south-aspect open slopes, ridges, and peaks.

As also described for herbs and shrubs under a canopy, the species nomograms show a spectrum of individualistic responses. The difficulty of

Figure 11.4. The spectacular subalpine and alpine mosaic of Glacier National Park is shown in this photograph of upper Logan Creek valley. Note the numerous avalanche areas. The road is Going to the Sun Highway. (Photograph by Mel Ruder for *The Hungry Horse News*)

sampling alpine communities and the impossibility of returning to these areas when flowers were available often made it impossible to key grasses and sedges to the species level. *Carex* was most abundant on the xerophytic talus and meadows at the higher elevations and was absent from the high-elevation Krummholz and from almost all (excepting those experiencing hydric disturbance) low-elevation primary succession types. Grasses dominate the low-elevation talus and meadows but rapidly shift to somewhat more advanced primary succession types (meadow and Krummholz) at the high elevations. *Poa* spp. (primarily *P. alpina, P. gracillima,* and *P.*

Figure 11.5. The alpine Krummholz *Abies lasiocarpa* and rock mosaic provides vital mountain goat habitat. (Photograph by Mel Ruder for *The Hungry Horse News*)

leptocoma) are most abundant on the ridges and peaks's meadow–Krummholz cover at the highest elevations but may be found in most meadow and Krummholz communities above the middle elevations. *Agropyron* spp. (primarily *A. canium*) is common on the higher elevation southwest open slopes and ridges with mosaic meadow—Krummholz cover and extends to primary succession types as early as talus–meadow.

At the lower elevations, *Xerophyllum tenax* is a forest species and extends on to the meadows in low abundance only. At the higher elevations, however, it is a dominant species on the talus–meadow mosaic and in both meadows and Krummholz. Average cover ranges up to 25%, but localized cover in small areas may reach over 95%.

Salix is another genus not easily identified to the species during summer sampling. Low-elevation willows (primarily *S. candida, S. drummondiana, S. glauca,* and *S. scouleriana*) are common on all mesic and submesic sites from talus–meadow through forest cover. At the middle elevations of 1525–1830 m (5000–6000 ft), *Salix* (primarily *S. glauca* and *S. scouleriana*) is common in the ravines, draws, and sheltered slopes with meadow or shrub cover. High-elevation willows (*S. arctica* and *S. nivalis*) are abundant on the drier sites with talus, meadow, shrub, or Krummholz cover.

Of all the shrubs in Glacier, *Alnus sinuata* achieves the highest densities.

Figure 11.6. The devastating results of a severe avalanche that occurred in 1954. (Photograph by Mel Ruder for *The Hungry Horse News*)

At the lower elevations, it is present in low quantities in submesic meadows, shrubfields, and forests. It is the dominant species in shrubfields above 1200 m and often reaches a cover value exceeding 50% on the open slopes with all aspects except southwest. Unlike *Salix, Alnus* is absent from the earlier primary succession types.

Middle elevation meadows and shrubfields are often dominated by *Veratrum viride,* which reaches a maximum cover of 50% on many slopes. Its cover drops at the higher elevations, and it is displaced toward more xeric sites by the dominant *Alnus* and *Salix.*

Two other common meadow and shrubfield species are *Vaccinium* and *Rubus parviflorus. Vaccinium* is common on the drier meadows, shrubfields, and forests at all elevations and reaches peak cover at the midelevation open slopes and ridges with shrub cover, where it often forms a solid mat. *Rubus parviflorus* is a more mesophytic species and achieves peak cover in the mesic low-elevation shrubfields and "tall meadows." All except the highest shrubfields will include either *Vaccinium* or *Rubus parviflorus*, the balance between the two species being determined by topography and moisture conditions.

Fragaria vesca and *F. virginiana* are present at most elevations on the more xeric sites, reach peak cover in the low-elevation meadows, but also extend into the talus communities at the higher elevations. *Epilobium angustifolium* is also present at most elevations but achieves peak cover in the high-elevation early and midprimary succession types on the most xerophytic sites; these latter sites also often include *Epilobium alpinum* and *E. latifolium*. *Heracleum lanatum* is also present at all but the highest elevations but is common only on the low-elevation mesic meadows and shrubfields, where it is usually associated with *Rubus parviflorus*.

Middle-elevation meadows and shrubfields usually include *Achillea millefolium* and *Thalictrum occidentale*. *Achillea* is present on all but the most mesophytic meadows and reaches peak cover on the most xerophytic high-elevation meadows. *Thalictrum* is more common at the middle elevations, achieves up to 50% cover in meadow–Krummholz and meadow–shrub slope communities, and at high elevations extends as far as the rock–talus primary succession types. Another component of the colonial primary succession types (rock–talus) at the high elevations is *Penstemon* (several species).

Spiraea (mostly *S. betulifolia* but some *S. densiflora*) is common in subxeric meadow–shrub and forest communities at the lower elevations, whereas *Urtica dioica* is common but seldom abundant on the low- and mid-elevation sheltered slopes with meadow to shrub cover.

Three genera of Compositae—*Arnica, Erigeron,* and *Senecio*—are common in Glacier, especially at the higher elevations. At the lower elevations, *Arnica* is common on forested slopes and does extend onto the meadows; it is primarily *A. parryi* with some *A. latifolia*. Higher elevations (1830–2135 m) show a bimodal distribution for talus and meadow cover on the topographic-moisture gradient. The mesic mode is primarily *A. longifolia*, whereas the drier mode represents both *A. mollis* and the *A. diversifolia* hybird complex (Kessell 1974). At the highest elevations, *Arnica* achieves dominance on the drier slopes with meadow cover (mostly *A. mollis*), whereas *A. alpina* extends into the early (rock–talus) primary succession types.

Erigeron is present in meadows and forests at all but the lowest elevations. At the middle elevations, peak cover is reached in the most mesophytic meadows (*E. peregrinus*), whereas *E. speciosus* is present in the midelevation forests. Above 1830 m (6000 ft), *Erigeron* is common on talus–meadow, meadow, and Krummholz cover under all moisture conditions. The species *E. peregrinus* is dominant but meadow communities also include *E. simplex,* and *E. acris* and *E. leiomerus* extend to the early primary succession types.

Senecio is present in most midelevation slopes with forest or meadow–

forest cover and is usually a mixture of *S. triangularis* and *S. foetidus*. At the higher elevations in the mesic and submesic meadows *S. triangularis* is the dominant member of the genus, whereas *S. megacephalus* and *S. residifolius* extend to the rock and talus primary succession types. At the highest elevations, the early primary succession communities include a mixture of *S. fremontii*, *S. megacephalus*, and *S. residifolius*.

As a community progresses from early to later primary succession types, therefore, one or more complete turnovers of the species composition may be observed. Talus and rock fields will be invaded by *Penstemon, Senecio, Arnica*, and/or several other species (including *Polystichum lonchitis, Juncus drummondii, Luzula piperi, Eleocharis* spp., *Agropyron canium, Carex spectabilis, Salix arctica, Sedum* spp. *Epilobium alpinum, Aretemisia michauxiana, Oxyria digyna*, and *Erigonum flavum*). Once cover reaches the 10%–20% level, primary succession development will depend mainly on the macro- and micromoisture conditions. (A wealth of information on species composition by species too rare to be included in the nomograms is included in the original field data.[1] "Stable" meadow conditions are often altered by hydric or slide disturbances. Hydric effects often increase the density of *Juncus* (especially *J. drummondii*), *Luzula* (*L. piperi, L. hitchcockii*, and *L. spicata*), *Carex*, and *Castilleja*. Fellfields invariably lead to the establishment of several of the following species: *Woodsia scopulina, Juncus* spp., *Carex* spp., *Spiraea densiflora, Hypericum formosum, Arnica longifolia, Castilleja* spp., *Cryptogramma crispa, Mimulus lewisii*, and *Mimulus tilingii*. Further development to shrubfields will include *Alnus, Salix, Vaccinium*, and/or *Rubus parviflorus*, depending on moisture conditions and disturbances. Stands that ultimately reach forest cover have been described in Chapter 9.

The Role of Fire

Fire probably plays an important role in maintaining low-elevation meadow and prairie communities. Old photographs show that major forest invasions of prairie and meadow communities near Moose Creek and south of Polebridge (North Fork drainage) have occurred. However, Singer (1975a) determined the ages of *Pinus contorta* on the edge of Big Prairie (north of Polebridge) and found invasion rates to be low, with many trees on the periphery over 100 years old. Koterba (1968) and Habeck (1970a, b) reported the destruction of trees up to 40-years old invading prairie margins during the drought of 1967, and Singer (1975a) made similar observations during the summer of 1973. Habeck (1970a, b) and Singer (1975a) both

---

[1] Archived in Glacier National Park and with the author at Gradient Modeling, Inc., Missoula, Montana.

noted the role of fire in maintaining the *Pinus contorta* savanna communities adjacent to these prairie communities, and I have noted that fire control has led to the development of "fuel bridges" in the park's *Pinus ponderosa* savanna.

Singer's (1975a) work demonstrated that fire stimulated the reproduction of aspen (*Populus tremuloides*) stands in the North Fork as older stands had greatly reduced coverage by sprouts; he noted that the only deteriorating aspen stands that were regenerating were a few stands located next to grasslands. He also observed that fire exclusion had increased the density of *Artemisia tridentata* and was a factor in converting open riparian *Populus trichocarpa* stands into bunchgrass flats.

Singer's (1975a) results suggest that although fire is undoubtedly important in maintaining certain savanna, prairie, and meadow communities in the North Fork, it may not have the dominant role attributed to it by Habeck (1970a, b).

Fire has a minimal influence on the mesic unforested communities, because of both long fire periodicity and low intensities (a result of high fuel moisture). Fire does affect mid- and high-elevation xeric and subxeric meadows, as these often "cure out" (dessicate) during the summer fire season; however, most of these communities return to a "climax" state within a few years after the burn. Rock and talus primary succession types rarely have sufficient fuel to carry a fire.

One unusual situation relating to fire in unforested communities deserves special note here. A major portion of the Belton Hills near West Glacier which burned in 1929 with high intensity is still primarily shrub and meadow cover, whereas adjacent portions of the burn have returned to forest. The area is composed of steep, south and southwest open slopes and is a heavily used ungulate winter range. As discussed in Chapter 8, preliminary study suggests that grazing and recycling of the nutrient-limited *Acer* and *Ceanothus* shrubs may be prolonging the shrub community while retarding the reestablishment of *Pinus contorta* and *Pseudotsuga menziesii*. Fire may also combine with other disturbances, as discussed in the next section. The interaction of various disturbances to increase environmental and temporal habitat heterogeniety is further discussed in Chapter 12.

Other Causes of Secondary Successions

Both hydric and slide disturbances are also important determinants of the structure and species composition of Glacier's unforested communities. Each may either change species composition (but not formation type) or change both species composition and formation type.

Hydric disturbances that affect the forest communities along rivers were discussed in Chapter 9. Effects may also be more subtle; even when major changes do not occur in the canopy, the understory may exhibit species shifts, including an increase of *Salix, Juncus,* and *Carex.* Along their waterside edges, lake and pond bottomlands often exhibit a small hydric zone characterized by *Lycopodium, Potamogeton, Juncus, Carex, Sparganium, Typha,* and/or *Lemna.* Glacier Park also has small bogs and marshes that exhibit a hydric environment throughout the growing season and mesic meadows that flood during the spring runoff. Both are described in detail by Habeck (1970a).

Slides have a profound affect on Glacier's mid- and high-elevation unforested communities. Severe and frequent slides in ravines create a boulder and talus community with low (but often diverse) plant cover. Less frequent or severe slides permit greater meadow development, with occasional *Salix, Alnus, Ribes, Rubus,* and *Pinus albicaulis.* Areas with still less frequent and intense slides exhibit a solid shrub cover, which is partially or totally destroyed by the periodic slides. More subtle effects include a shift in species composition. Slide areas are virtually always more diverse than adjacent communities because of the increased environmental heterogeniety. Sometimes this effect is combined with fire succession to further augment species diversity—the most diverse community yet found in Glacier is a fellfield near Haystack Creek which also burned in 1967.

Distribution of Ground Fuels

As noted in Chapter 9, the main determinant of the distribution of fuels in unforested communities is the primary succession gradient. As was true with forest fuels, further separation of ground fuels on the elevation and topographic-moisture gradients did not usually offer a significant improvement in fuel predictions.

As also was true for the forest fuels discussed in Chapter 9, ground fuel distributions were skewed to the right, with most median–mean ratios ranging from 0.4 to 0.8. As a result, median rather than mean loadings are a better indication of the populations' central tendencies. Table 11.1 shows the changes in median fuel loads along the primary succession gradient. As expected, litter and the dead and down branchwood show an increase with stand development from talus to forest. Litter increases from 0.2 ton/ha in meadows to 2.2 ton/ha in typical forest, an 11-fold increase. Litter loads in typical forest are 20% higher than in shrubfields and 75% higher than in Krummholz. Grass and forb biomass is highest in prairies at 1.4 ton/ha; as sites become capable of supporting a shrub canopy, grass and forb loads drop to 0.4 ton/ha and drop again in upright forest to 0.2 ton/ha. Fine (1-hr time-lag) branchwood exhibits a 60% increase from the shrubfield to forest

Table 11.1. Median Fuel Loadings (by Fuel Types and Size Classes) and Packing Ratios (by Strata) Along the Formation Type (Primary Succession Gradient) Stratification. (Kessell and others 1978)

	Primary succession (formation) type								
	Glacier	Rock	Talus	Talus–meadow	Meadow	Prairie	Shrub	Krummholz	Upright forest
Sample size (N)	4	4	22	5	86	13	22	49	218
Fuel loads (ton/ha) for:									
Litter	0.00	0.29	0.04	0.00	0.10	0.62	1.88	1.28	2.23
Grass and forbs	0.00	0.28	0.11	0.05	0.76	1.36	0.39	0.41	0.25
Branchwoods									
1 hr	0.00	0.00	0.00	0.05	0.01	0.07	0.50	0.48	0.80
10 hr	0.00	0.00	0.00	0.15	0.00	0.00	1.13	1.17	1.34
100 hr	0.00	0.00	0.00	0.00	0.00	0.00	1.69	1.71	1.85
1000 hr	0.00	0.00	0.00	0.00	0.00	0.00	8.08	16.14	19.96
Shrubs									
Foliage	0.00	0.00	0.00	0.00	0.00	0.00	0.45	1.71	0.42
Branchwood	0.00	0.00	0.00	0.00	0.00	0.00	1.62	8.54	1.43
Packing ratios (dimensionless) for:									
Litter stratum	a	0.0094	0.0090	0.0079	0.0078	0.0172	0.0270	0.0233	0.0278
Downed woody stratum	a	a	a	0.0007	0.0017	0.0017	0.0056	0.0082	0.0101
Shrub stratum	a	a	a	a	a	a	0.0018	0.0033	0.0010

aPacking ratios are undefined for strata with zero fuel loads.

stands, whereas the 10- and 100-hr time-lag branchwood does not vary appreciably within shrubfield, Krummholz, and typical forest communities. Thousand-hour time-lag branchwood shows a notable increase from shrub to Krummholz to forest. Since all trees less than 5.0 cm dbh are considered part of the shrub stratum, moreover, shrub branchwood and foliage are highest in the Krummholz at 8.5 and 1.7 ton/ha, respectively. Stratal packing ratios follow trends similar to their respective fuel loading components.

12

Diversity Relationships

The research described in the last chapters provides a wealth of data for interpreting the response of species and community diversity to environmental heterogeneity, and for testing and comparing the various theories on dominance–diversity relationships and the various diversity measures. The following discussion is divided into three sections: First is a discussion of the diversity mosaic and how it responds to spatial and temporal habitat heterogeneity; next is a discussion of the theory supporting various dominance–diversity relationships and measures of species diversity in relationship to the Glacier community data; the final section is a discussion of the evolution and maintainence of Glacier's terrestrial diversity.

The Diversity Mosaic

In a discussion of how species populations divide community resources among themselves, Whittaker (1975b) notes that:

> The means by which a species population is controlled is a most important aspect of its niche. Control mechanisms and some aspects of niche are not really resources. Let us assume, however, that there is some correspondence among three things: the fraction of the niche hyperspace of the community that a species occupies, the fraction of the community's resources . . . that the species uses, and the fraction of the community's productivity that the species realizes. It may be clearer for our present purposes if we set aside niche characteristics that are not resources, and thus simplify the n-dimensional niche space to an m-dimensional resource space. We can then ask how this resource space is divided up among species, and what kinds of relative importances of species result.

By applying this approach, we may view the n-dimensional niche space of a species population in terms of the m-dimensional resource space within a community, and then we may view that m-dimensional intracommunity resource space in terms of a k-dimensional intercommunity resource space that is approximated by the gradient hyperspace model.

Let me first define the components of diversity that I wish to relate to the distribution and importance of species populations. *Alpha diversity* describes species diversity within a single homogeneous community. *Beta diversity* describes the turnover of species populations or their importance in response to environmental variation (such as is reflected by macro- or microhabitat gradients). *Gamma diversity* refers to the combined species diversity in a landscape or ecosystem (Whittaker 1975b).

Difficulties arise when these concepts are applied to natural communities. In the McDonald drainage area, for example, let us consider an apparently uniform 1-ha mature forest community on a southeast open slope. The overstory is primarily *Pseudotsuga menziesii* with an (apparently random) scattering of *Picea engelmannii* and *Abies lasiocarpa*. Now let us compare this community to a lower ravine community, which is composed of *Tsuga heterophylla* and *Thuja plicata*. From our diversity definitions, either stand should serve as an example of alpha diversity, whereas a comparison between the two communities should reflect beta diversity. The total diversities of all such stands in the McDonald drainage should reflect the drainage's gamma diversity.

If we view the forest canopy alone, our example may well reflect the diversity definitions. We can measure (by some appropriate method discussed in the next section) the canopy alpha diversity of the *Pseudotsuga* slope community and, if desired, compare it to the diversity of the *Tsuga–Thuja* ravine overstory. We can also compute some measure of environmental separation (perhaps in terms of gradient indices) and quantify the canopy species turnover (beta diversity) between the two communities. Finally, we can (theoretically at least) determine the canopy diversity of all such stands in the drainage and arrive at gamma diversity.

Our decision that the *Pseudotsuga* slope community is a homogeneous stand works well until we look under the canopy. Then we find a scattering of *Amelanchier;* if we divide the hectare into 20 × 20 m plots, we find some with almost no *Amelanchier* and a few with 50% coverage by *Amelanchier*. Clearly our homogeneous hectare community is no longer homogeneous. We may choose a 20 × 20 m plot (with uniform shrub cover) as our new homogeneous community, but if we sample and ordinate a series of 2 × 2 m plots within it, we find an obvious micromoisture gradient across the 20 × 20 m community; our 20 × 20 m community is therefore not homogeneous after all. Even the 2 × 2 m plot has exposed mineral soil at one corner, a clump of *Xerophyllum* covering half of it, and a single *Salix,* so we cannot even assume it to be homogeneous.

The point of this rather lengthy example is that although it may be useful to speak of within- and between-habitat species diversity, one must be extremely careful in defining the scale that is used to designate habitat homogeneity. A land management planner may consider several square kilometers to be homogeneous habitat, whereas a mycologist may describe habitat heterogeneity across a 2 m² plot.

Given these problems, we may still make useful application of the alpha, beta, and gamma diversity concepts. For the purposes of the following discussion, "within-habitat diversity," and the assumption of within-habitat homogeneity, refers to habitat differences that cannot be resolved by the gradient model; as in the distinction of uniform versus mosaic or intergrading primary succession types, this gives us a resolution down to about 25 m², even though some communities that are much larger reflect no variation on the macrogradients. I shall refer to this resolution as "macrohabitat heterogeneity," and the following discussion shows that it provides adequate resolution for most communities' overstories. It must be realized, however, that such an assumption of homogeneity does not hold for many of the species populations under the canopy. In the same way, I shall use the term "beta diversity" to apply to comparisons of communities with different locations on the macrogradients using this same scale of resolution, but it must be realized that the computed diversity measures for a single stand reflect primarily the overstory's alpha diversity but both the alpha diversity (in response to the macrogradients) and the beta diversity of the understory (in response to usually unmeasured microgradients).

Dominance–Diversity Relationships

An intelligent discussion of diversity requires an understanding of how the resource space of a community is divided among the component species. A useful approach is the construction of dominance–diversity curves, which relate species importance values to the species sequence (Whittaker 1965).

A number of theories have been proposed on the shape of these curves. Three major types of dominance–diversity curves have been recognized in natural communities: the geometric series (niche preemption), the MacArthur series (random niche boundary) and the log-normal distribution of Preston.

The geometric series assumes that each species occupies a fraction (k) of the available resources. Therefore, the first (most important) species receives k fraction of the resources, the second species receives $k(1 - k)$ fraction of the resources, and the ith species receives $k(1 - k)^{i-1}$ fraction (Motomura 1932, Whittaker 1969, 1972). Therefore:

$$n_i = Nk(1 - k)^{i-1} = n_1 c^{i-1} \qquad (12.1)$$

where N is the total of importance values for all species in the sample, n_i is the importance value of the ith species (in the sequence from most to least important), n_1 is the importance value of the most important species, and c is the ratio of the importance value of a species to that of its predecessor in the sequence (Whittaker 1975b). This general pattern has been observed for terrestrial plant communities with low diversities and for taxocenes or strata of communities with higher diversities (Whittaker 1965, 1969, 1972, Whittaker and Woodwell 1969, Reiners and others 1970).

The MacArthur series assumes that the boundaries between niche hypervolumes are set at random (MacArthur 1957, 1960, Vandermeer and MacArthur 1966); species are suggested to exhibit alternative modes of resource utilization. This yields the equation:

$$n_r = \frac{N}{S} \sum_{i=1}^{r} \frac{1}{S - i + 1} \tag{12.2}$$

where S is the number of species in the sample, N is the total importance values for all species, and n_r is the importance value of species r in the sequence of species from least to most important. This form has been approached by various animal taxocenes (MacArthur 1960, Hairston 1964, King 1964, Deevey 1969, Goulden 1969).

The log-normal distribution is assumed to apply to a large number of species not closely related in resource use (Whittaker and Woodwell 1969, Whittaker 1972, 1975b). As the number of factors controlling species distribution is expected to increase as the number of species increases, the resulting distribution is a Gaussian frequency distribution on a log scale (Preston 1948, 1962, Whittaker 1975b). The equation is:

$$s_r = s_0 e^{-(aR)^2} \tag{12.3}$$

where s_r is the number of species in an octave R octaves distant from the modal octave, which contains s_0 species, and a is a constant often equal to about 0.2 (Whittaker 1975b). Log-normal distributions have been observed for many different kinds of plant and animal communities (Preston 1948, 1962, Patrick and others 1954, C. B. Williams 1953, 1964, Whittaker 1965, 1969, Whittaker and Woodwell 1969, Batzli 1969).

Whittaker (1970, 1975b) has suggested that the geometric and log-normal distributions may be limiting cases for terrestrial plant communities, with most natural communities falling somewhere between these two extremes.

The use of dominance–diversity curves is important not only for observing how species divide the habitat resources, but also for the choice of appropriate diversity measures, as the effectiveness of the various diversity measures often depends on the dominance–diversity relationships.

Whittaker (1975b) groups diversity measures into three general classes: richness measures (such as total number of species), dominance measures (such as the Simpson index), and equitability measures (such as the Shannon–Wiener information index). [These indices are defined in Eqs. (12.4) and (12.5).] Alternatively, Peet (1974) considers both the Simpson and Shannon–Wiener indices to be heterogeneity measures; he distinguishes between type I heterogeneity measures (such as Shannon–Wiener), which place greater emphasis on species with low importance values, and type II heterogeneity measures (such as the Simpson index), which place greater emphasis on the dominant species. Peet (1974) suggests that such measures as Pielou's (1966) J, Lloyd's and Ghelardi's (1964) ϵ, or Hill's (1973b) ratios are the appropriate equitability measures.

The Glacier study used six diversity measures. The first, total number of species in plots of defined size, was used only for understory species. The next was the Shannon–Wiener H':

$$H' = -\Sigma p_i \log_e p_i \tag{12.4}$$

where p_i is the relative importance of the ith species. The next measure was the Simpson Index, C:

$$C = \Sigma p_i^2 \tag{12.5}$$

Whittaker's E_c was also used. It may be interpreted as the number of species per log cycle of the dominance–diversity curve:

$$E_c = s/(\log_{10} p_1 - \log_{10} p_n) \tag{12.6}$$

where s is the total number of species, p_1 is the relative importance of the most important species, and p_n is the relative importance of the least important species. The Simpson and Shannon–Wiener measures were also expressed as $1/C$ and $e^{H'}$, which indicate the number of equally important species required to produce the same level of heterogeneity.

An attempt was made to interpret the dominance–diversity curves for Glacier's overstory in terms of both dominance–diversity theory and the effect of macroenvironmental gradients on resource division. Curves were constructed for the overstory's density by combining from two to six field samples that had very similar gradient indices. Figure 12.1 shows a family of dominance–diversity curves for the canopy of mature forests in the McDonald drainage for varying elevations and topographies. The general trend is for the low mesophytic forests to show log-normal or nearly log-

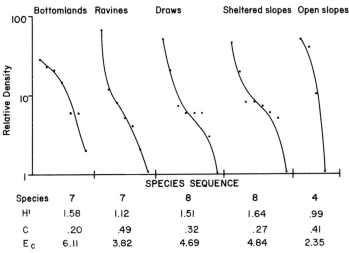

Figure 12.1 a–e. Overstory dominance–diversity curves and diversity measures for various topographies and elevations of mature forests in the McDonald drainage. **a** 1000 m. **b** 1300 m. **c** 1600 m. **d** 1900 m. **e** 2200 m.

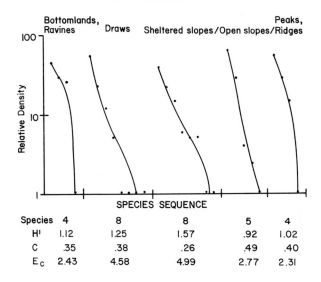

Overstory   Mature Forested Stands
McDonald drainage   1300 m

Species	4	8	8	5	4
H'	1.12	1.25	1.57	.92	1.02
C	.35	.38	.26	.49	.40
E_C	2.43	4.58	4.99	2.77	2.31

b

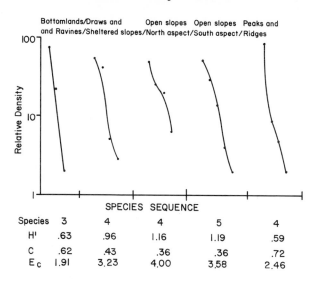

Overstory   Mature Forested Stands
McDonald drainage   1600 m

Species	3	4	4	5	4
H'	.63	.96	1.16	1.19	.59
C	.62	.43	.36	.36	.72
E_C	1.91	3.23	4.00	3.58	2.46

c

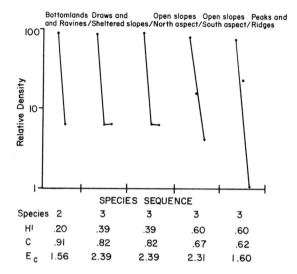

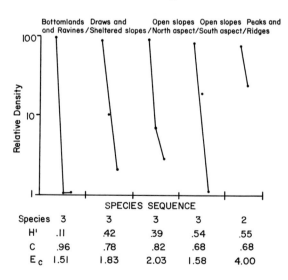

Figure 12.1 d and e. (For legend, see p. 152.)

normal distributions, whereas sites with more severe environmental conditions (xeric and/or high elevations) are nearly geometric in shape.

For 1000-m forests (Fig. 12.1a), all topographies except the open slopes exhibit a log-normal type of curve. At 1300 m (Fig. 12.1b), the mesic and submesic sites are nearly log normal, whereas the drier open slopes and ridges approach the geometric curve. At 1600 m (Fig. 12.1c), the draws, sheltered slopes, and north-aspect open slopes are the most log normal— really transitional between log normal and geometric—whereas the bottomlands, south-aspect open slopes, peaks, and ridges are nearly geometric. At 1900 m and 2200 m (Fig. 12.1d, e), the curves are all nearly geometric because of the small number of overstory species; note, however, that several communities with three overstory species show one species dominant and the other two with similar importance values. Also note that communities with E_c greater than 3.5 have generally log-normal species distributions, whereas those with E_c less than 2.5 have geometric distributions.

Dominance–diversity curves were also constructed for the overstories of successional forest communities (Fig. 12.2) and show that early and midseral communities are much more log normal (and more diverse) than are mature stands.

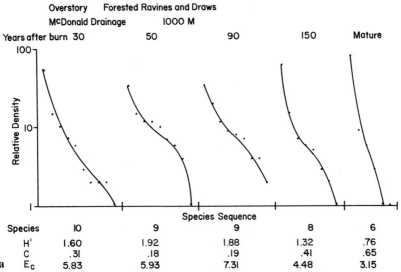

Figure 12.2 a–f. Overstory dominance–diversity curves and diversity measures for various stand ages and elevations of forested areas in the McDonald drainage. **a** 1000 m, ravines and draws. **b** 1000 m, sheltered slopes. **c** 1000 m, south-aspect open slopes. **d** 1400 m, sheltered slopes. **e** 1400 m, south-aspect open slopes. **f** 2000 m, south-aspect open slopes.

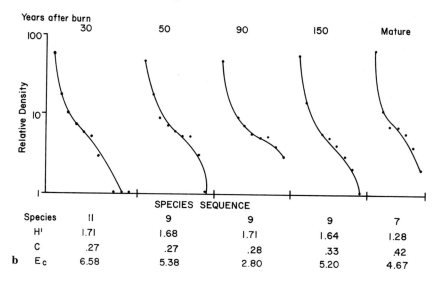

Overstory Mature Forested Sheltered Slopes
McDonald drainage 1000 m

Species	11	9	9	9	7
H'	1.71	1.68	1.71	1.64	1.28
C	.27	.27	.28	.33	.42
b E_c	6.58	5.38	2.80	5.20	4.67

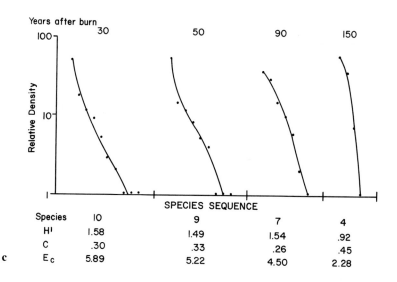

Overstory Forested Open Slopes, South aspect
McDonald drainage 1000 m

Species	10	9	7	4
H'	1.58	1.49	1.54	.92
C	.30	.33	.26	.45
c E_c	5.89	5.22	4.50	2.28

Figure 12.2 b and c. (For legend, see p. 155.)

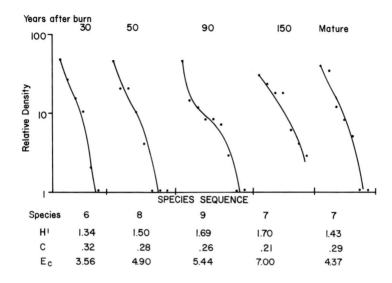

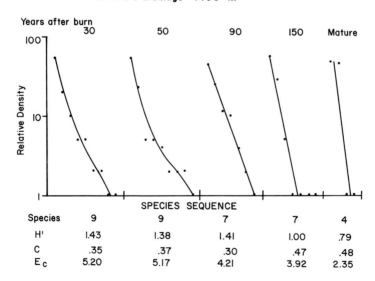

Figure 12.2 d and e. (For legend, see p. 155.)

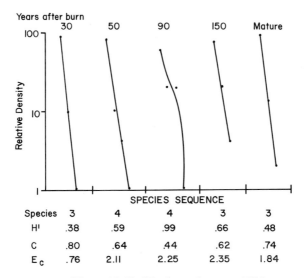

Figure 12.2f. (For legend, see p. 155.)

Figure 12.2a is the dominance–diversity curves for forested ravines and draws in the McDonald drainage; the 30-, 50-, and 90-year-old communities are clearly log normal, whereas the mature stands are nearly geometric and the 150-year-old stands are intermediate in form. The 1000-m sheltered slope communities (Fig. 12.2b), among the most diverse forests in the park, are log normal for all stand ages, but the most equitable and most diverse communities are the 30- and 50-year-old stands. The 1000-m south-aspect open slopes (Fig. 12.2c) are much more geometric than the more mesic sites, but again the most equitable communities are the 30- and 50-year-old stands.

At 1400 m, the midseral sheltered slopes (Fig. 12.2d) are the most diverse and most log normal, whereas the young and old stands are the least diverse and most geometric. The 1400-m south-aspect open slopes (Fig. 12.2e) show nearly geometric distributions for the older stands and intermediate dominance–diversity curves for the younger forests. At 2000 m (Fig. 12.2f), low overstory diversity causes communities at all ages to be nearly geometric. These results show that Glacier's overstory exhibits a range of dominance–diversity relationships, approaching the log-normal hypothesis at one extreme (favorable environments) and the geometric hypothesis at the other extreme (severe environments).

The results also explain why it was not possible to plot good nomograms of species diversity measures. Because of its insensitivity to rare species, the Simpson index is inappropriate when applied to equitable communities with a log-normal species curve, and attempts to show changes in the Simpson index across the gradients therefore reflect sample noise and variance more than species dominance. The Shannon–Wiener measure was better but its use was still hindered by other problems; because it was more sensitive to rare species, it incurred considerable noise from otherwise similar samples that differed on how well they represented rare species (relative density below 1%). E_c is also very sensitive to rare species, as it measures the linear slope of the dominance–diversity curve rather than the curve's shape, and may vary widely depending on the number of rare species included in a sample. The net result, then, was an inability to construct gradient nomograms of overstory species diversity; instead, it has been more instructive to interpret resource allocation in terms of the dominance–diversity relationships.

Similar attempts were also made to construct understory dominance–diversity curves and diversity measure nomograms. Here, beta diversity (in response to microhabitat variation) and sampling procedures caused even greater problems. Curves constructed from a single sample plot, even if the plot is representative of the community, turn out to be step diagrams because of the use of the 0–6% cover importance scale, and interpretation is therefore very difficult. The alternative is to combine several such samples or plots and thus approach a continuous importance value scale. However, this method combines samples from different microhabitats and so creates difficulty in separating the alpha diversity from the beta diversity. Various attempts with different combinations of techniques suggested that the understories of all but the most severe environments were generally log normal in shape. Future attempts to construct understory dominance–diversity curves or diversity nomograms should refine the field techniques used in this study, but are still likely to be plagued by the problems caused by microhabitat variation within a community.

Because of these problems, a special effort was made along the Boulder Pass alpine wind–snow gradient to reduce the effects of microhabitat variation. The overstory dominance–diversity curves (Fig. 12.3a) were constructed from two 0.05-ha samples for each point on the gradient; the pair of samples was selected to keep habitat differences at a minimum. The understory dominance–diversity curves (Fig. 12.3b) were constructed from four 2 × 2 m plots at each point on the gradient. The plots were all sampled during a 4-hr period, and a soil moisture meter was used to select four plots with nearly identical soil moisture conditions. The results suggest that the overstory becomes more geometric as environmental severity increases, but that the understory retains a nearly log-normal distribution even under the most severe conditions.

The results reported here from Glacier definitely support Whittaker's (1972, 1975b) suggestion that the geometric and log-normal distributions are the limiting cases for terrestrial plant communities, with most natural communities falling somewhere between the two extremes. Communities

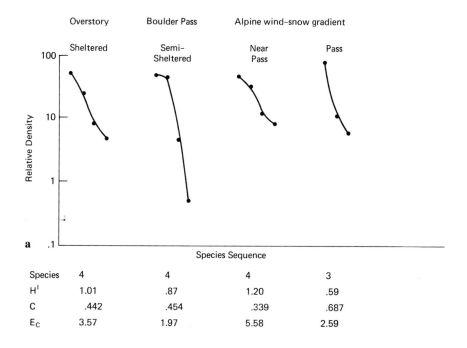

Species	4	4	4	3
H^1	1.01	.87	1.20	.59
C	.442	.454	.339	.687
E_C	3.57	1.97	5.58	2.59

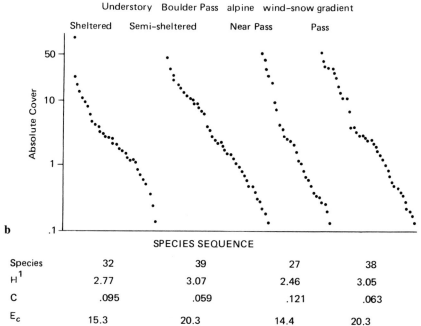

Species	32	39	27	38
H^1	2.77	3.07	2.46	3.05
C	.095	.059	.121	.063
E_c	15.3	20.3	14.4	20.3

Figure 12.3 a and b. Dominance–diversity curves and diversity measures for various positions on the alpine wind–snow gradient at Boulder Pass. **a** Overstory. **b** Understory.

that exhibit high diversity and an equitable distribution of species undoubtedly do so by complex division of the habitat space by the component species populations; most likely, different species are limited by different components of the habitat. In fact, one can argue that a geometric distribution is impossible with high species diversity. For example, with $k = 0.5$, the first species gets 50% of the resources, the second species receives 25%, the third species receives 12.5%, . . . , the 10th species receives 0.1%, . . . , the 15th species receives only 0.003%, etc. There is undoubtedly a minimum quantity of resources (call it k') that is required by a species population to allow it to successfully reproduce, compete, and maintain its population. If this minimum value of k' is 0.001, then a community exhibiting a geometric distribution (with $k = 0.5$) can support not over 10 species (different values of k exhibit little effect). The results from Glacier show that log-normal distributions are approached when the number of species exceeds seven or eight, and as k' is probably in the range of 0.005 (0.5%) for overstory species. k' probably decreases with smaller plant size (shrubs and herbs) and definitely decreases with increasing environmental heterogeneity (more microhabitats).

The correlation between environmental severity and the form of the dominance–diversity curves observed in Glacier suggest that k' must be higher under more extreme conditions; when resources are meager, a species needs a larger portion of those that are available. More moderate conditions permit niche and therefore habitat differentiation by more factors, so that species of equal abundance are not in direct competition for an individual limiting resource. In addition, a more diverse set of limiting factors undoubtedly operates in a heterogeneous or mosaic habitat and permits a greater diversity of specialized species to coexist. The trend toward more log-normal distributions and greater diversity in early and midseral communities supports this position.

With this discussion of alpha diversity and dominance–diversity relations in mind, let us now look at the mosaic diversity exhibited by Glacier's communities, caused by spatial and temporal habitat heterogeneity, and examine how habitats and species populations have evolved and maintain such diversity.

Evolution of Glacier's Terrestrial Plant Community Diversity

The above discussion has considered primarily alpha diversity, the product of division of resources by species populations within a "homogeneous" community. Yet we also have noted that, at least for the understory, microhabitats also greatly influence the division of a resource space and therefore the species richness of a community. Bratton (1976) stated that microtopography is a major influence on understory species diversity in the cove forests of the Great Smoky Mountains: "Microtopography and seasonal changes are responsible for much of the niche differentiation between the herbs and thus largely account for the species diversity of this rich herb stratum." In Glacier, microhabitats undoubtedly increase species diversity both through habitat (and thus niche) differentiation and by providing small "islands" of suitable habitat in an otherwise unsuitable environment. Examples of this latter effect include suitably mesic depressions in an otherwise xeric community, suitably sheltered spots on an otherwise exposed slope, and adequate soil development on an otherwise barren talus slide. These effects are especially evident in the micromosaic of the early primary succession types, where a 5×5 m plot may include bare rock, talus, meadow, shrub, and Krummholz communities, and is largely responsible for the high diversity of the alpine communities.

Aside from the microhabitat effects, the macrohabitat gradients described in earlier chapters are responsible for considerable (beta) species diversity. Species populations respond to each of these gradients, and so species importance, alpha and beta diversity, and other community characteristics all change along the gradients. In response to the elevation and topographic-moisture gradients, overstory beta diversity drops as the more severe environments (xeric or high elevation) are encountered. Overstory beta diversity is highest along the stand age gradient for midseral communities, lowest (nearly zero) for mature stands, and intermediate for the early seral forest, which extends through canopy closure.

The understory follows similar trends, but the situation is complicated by two factors: (1) Many rare understory species are detected in only some of the field samples and thus artificially increase the apparent beta diversity; and (2) suitable microhabitats in otherwise unsuitable stands can greatly extend a species' distribution range along the macrogradient and so decrease the beta diversity turnover rate. The latter effect is noted along all the macrogradients but is especially obvious along the primary succession, alpine wind–snow, and stand age gradients, again because of the incredible mosaic of microhabitats available to species populations.

However, the major diversity-producing agents in Glacier are the natural disturbances. Not only do natural fires create a mosaic of stand ages in otherwise homogeneous habitat, but variable fire intensity also greatly augments the diversity of habitat conditions. For example, a 4-km^2 area on Howe Ridge shows seven major burns with different stand ages and fire intensities in each burn, ranging from very mild ground fires to areas where the entire canopy, litter, and duff were destroyed. Natural slides cause retrogression of stands along the primary succession and stand age gradients, often rearrange or remove a portion of the substrate, differentially affect (in both space and intensity) the existing species populations, and create numerous microtopographies, microsubstrates, and micromoisture habitats. Hydric disturbances create similar habitat diversification, although the effect may not be so pronounced. Even the slow process of primary succession in the alpine communities produces numerous opportunities for invasion and colonization of pioneer species.

The net effect of these disturbances is to significantly augment beta and gamma diversity by creating a variety of available habitats, each of which exhibits a range of microenvironmental conditions. A large percentage of Glacier's landscape supports communities undergoing dynamic changes in response to these natural disturbances, and the landscape, community, and species population diversity is thus continually maintained by these natural processes.

The species populations that evolve and maintain themselves in this dynamic environmental mosaic reflect this variability. Species differentiated at the subspecies and variety level are very common in Glacier's flora (Kessell 1974). Work by Standley (1921), Harvey (1954a), and Habeck (1970a) and the results presented here suggest widespread ecotypic differentiation in many groups. Natural hybridization is obvious and widespread in many genera, including *Pinus, Picea, Populus, Salix, Castilleja,* and *Arnica.* Hybridization has also been documented for *Equisetum, Elymus, Sitanion, Dryas, Phyllodoce, Artemisia,* and *Antennaria* and is suspected but not documented for many other groups (Kessell 1974). This species variability is further augmented by the elastic nature of many non-hybridizing species.

A good summary statement applicable to Glacier's diversity comes from Whittaker (1975b): "Species diversity is a self-augmenting evolutionary phenomenon; evolution of diversity makes possible further evolution of diversity." Glacier's extraordinary landscape and community mosaic is the result of macroenvironmental, microhabitat, and species populations genetic variability, augmented and maintained by dynamic natural processes through evolutionary time.

SECTION III

IMPLEMENTATION OF THE GLACIER NATIONAL PARK GRADIENT MODELING INFORMATION SYSTEM

13

Gradient Models as the Base for a Dynamic Information System

As stated in Chapter 3, the gradient models of the vegetation and fuel were developed to provide: a description of the distributions of species populations and community characteristics; an understanding of the dynamic relationships among these species population, community characteristics, and the environment; and the basis for a terrestrial resource information system. The results of efforts to fulfill the first two purposes were described in Section II; this section describes the development of the resource and fire model that was built upon this gradient framework.

In Chapter 2 I described in some detail the needs for a detailed, quantitative resource inventory, modeling, and information system for sound management decisions in a natural or wilderness area. In that chapter, and in Chapter 3, I demonstrated that the traditional classification approaches to the landscape, such as "habitat types," offer a fairly good low-resolution (4-ha) system but simply cannot offer the high resolution (well under 1 ha) required for many management decisions. Because of the frequent lack of high resolution and other problems inherent in the classification approach, a new inventory and modeling system based on the gradient model approach was developed in Glacier National Park.

The Glacier National Park resource and fire management model, described in the following chapters, was developed to provide such an integration of the gradient models, inventory system, fire behavior models, and other necessary components. It was implemented in Glacier National Park in June of 1975 for an 8000-ha test area and became fully operational for a 40 000-ha area and operational (without precoded site inventory) for a 260 000-ha area in May 1976.

A Review of the Data Base

The gradient models described in Section II provide a detailed, quantitative description of species composition, fuel loadings, and community characteristics for any stand in the study area in Glacier if the location of the stand on each environmental gradient can first be determined; in other words, to use the graident models to predict composition for an individual, unsampled stand, one must first determine its elevation, topography, aspect, age, primary succession, drainage system, other disturbances, etc. To provide both high resolution and broad coverage of a large area, efficient use of such a system requires that these gradient indices be determined without first visiting the stand from the ground; any inventory system that indexes each stand therefore must determine the indices from aerial photographs, maps, or other remote sensing methods. Linking the gradient models with such a remote inventory system provides, stand by stand and within the limits of the inventory's resolution, a description of the physical site (directly from the inventory), a quantitative description of the vegetation (by providing to the gradient models the correct gradient indices for that stand), a quantitative description of the fuel (again, by providing the gradient indices), and a deterministic prediction of community development and species succession following a disturbance, such as fire (by varying the "stand age" gradient index while holding the other indices constant—note that the multiple-pathway succession model described in Chapter 10 was not developed until well after the completion of the Glacier model).

Furthermore, by linking other components with the gradient models and inventory system (such as fuel bed characteristics, fuel packing ratios, information on the spatial distribution of fuels, fuel moisture parameters, and wind vectors), the system provides all of the necessary inputs to current fire behavior models (spread rate, intensity, and flame length). The gradient models and inventory system therefore may be used to provide the required data base for fire behavior simulations and predictions.

The Gradient Modeling Information System

Gradient modeling met the needs for a resource information system, terrestrial resource inventory, and resource management package by linking four major components, as shown in Fig. 13.1:

1) A terrestrial resource inventory system
2) Gradient models of the vegetation and fuel
3) Weather and micrometeorology models
4) Fire behavior models

Its unique linkage of these components provides modeling capabilities previously unavailable from any other large-scale resource modeling system.

The *terrestrial resource inventory* system provides an efficient method of locating each stand on each environmental gradient, without the need to field sample every stand. The inventory information is derived from vertical black and white photographs, oblique false-color infrared photographs, topographic maps, and fire history maps. It is recorded hectare by hectare, offers 10-m resolution within a hectare, and also includes information on cultural and hydrologic features within each stand. By itself, it provides an extremely detailed site description, hectare by hectare. When linked to the

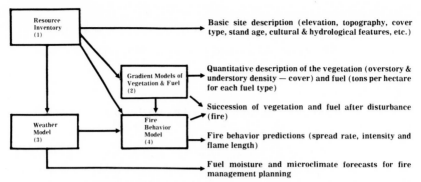

Figure 13.1. A simplified flowchart of the gradient resources and fire information system as developed in Glacier National Park. (Kessel 1976a)

gradient models, it provides the full quantitative description of the various site, vegetation, fuel, and community characteristics. The resource inventory system is described in detail in Chapter 14.

The *gradient models of the vegetation and fuel* were described in Section II. They predict vegetation, fuel, and community characteristics of the stand using the indices provided by the resource inventory.

The *weather model* provides a data base of fuel moisture, vegetation curing, wind vectors, and other meteorologic information and (on a preliminary test basis) uses the environmental gradients to extrapolate these weather parameters from base stations to remote sites using the gradient indices provided by the terrestrial resource inventory. The weather model is described in detail in Chapter 14.

The *fire behavior models* provide the appropriate algorithms to calculate fire behavior characteristics (spread rate, intensity, and flame length) from the site, fuel, and meteorologic parameters provided by the terrestrial resource inventory, gradient models, and weather models. The current deterministic ground fire behavior model (Rothermel 1972) is described in Chapter 14.

14

Structure and Components of the Resource and Fire Model

Terrestrial Resource Inventory (Remote Site Inventory)

The terrestrial resource inventory allows determination of each stand's gradient indices from maps, aerial photographs, and fire history information without field sampling or visiting each stand. These inventory data either may be recorded in advance for specified areas and stored on a computer or may be coded on the spot whenever resource information retrieval for a given site is required.

This resource inventory coding system offers the best approach we have found to efficiently match the resolution of the gradient models. Normally, if one wants to use the model immediately by entering only the site UTM coordinates, one codes the inventory in advance and stores it on either disk or tape. However, if one wishes to use the model for an area when the inventory has not yet been coded, one may locate the correct maps and photographs, code only those hectares for which information is desired, and immediately enter it to the system before the model is executed.

Description of the Inventory

The actual coding requires a set of preprinted forms, vertical black and white photographs of the area, oblique infrared photos (35 mm), a $7\frac{1}{2}$-minute topographic map, and fire history information. Ideally, the fire history information is plotted on the $7\frac{1}{2}$-minute map; it is often first necessary to trace fire perimeters on the aerial photographs, transfer them to the map, and determine ages from fire records, ground samples, etc.

Because fully corrected photographs usually are not available, it is also necessary to carefully transfer the hectare grid system to the vertical photographs. The major distortions in steep terrain are scale changes as a function of elevation, and these must be carefully determined. Once a square kilometer is "gridded," the information is coded using the map, fire history records, and photographs. The ideal photographs for this purpose would be vertical infrared pairs, but these frequently are not available. In Glacier, we obtained our own 35-mm false-color infrared obliques to supplement the vertical black and white photographs. One needs vertical photographs for accurate plotting and pinpointing of locations, but the infrared photographs offer much better vegetation cover interpretations, especially in areas of alpine mosaic where a small-scale mix of shrubs, meadow, rock, and talus is encountered.

After fire histories have been plotted and the hectares gridded, about 2–3 hr are required to completely code 1 km² (depending on the vegetation complexity). With a resolution of 10 m or better, coding the resource inventory represents nearly half the cost of developing the entire model; therefore considerable savings can be obtained by coding and storing on-line inventory records for areas of high management interest but not coding areas of lower interest unless and until such information retrieval is needed.

The system uses a set of four coding records, stores data in 1-km² (100-ha) units, and indexes stands hectare by hectare (but allows for and records discontinuities within a hectare). The type and number of records used depends on the complexity of the hectare. The coding and record formats are shown in Table 14.1.

Examples of Coding

Form H-101 records the basic site description and indicates whether other forms follow with either localized cover or special features data. If a hectare is entirely uniform in terms of site characteristics, age, and vegetative cover and includes no special (cultural or hydrologic) features, a one-line entry on the H-101 form completely describes the hectare to the model.

The UTM (Universal Transverse Mercator) coordinates of each hectare (in 100s of meters) are recorded in columns 6–14. Next, elevation (in feet, as all U.S. Geological Survey maps indicate contour lines in feet), topographic-moisture, and aspect are recorded in columns 16–22. If the stand has burned in the last 200 years, an entry is made in column 23, and the estimated age of the stand and intensity of the last fire are recorded in columns 24–27; if the stand has not burned in the last 200 years, columns 23–27 are left blank. Other disturbances are indicated in column 28; if no other disturbances are noted it is left blank. Vegetative formation (cover) type and primary succession are indicated in columns 29–31. If the stand appears to be affected by the alpine wind–snow gradient, an entry is made in column 32; otherwise it is left blank. A two-digit drainage area code is recorded in columns 33 and 34.

If the stand has totally uniform site conditions, columns 36–45 are left blank, a 1 is entered in both columns 1 and 2 (number of records for this stand and sequence number of this record), columns 3 and 4 are left blank (no localization or special records follow), and the inventory coding is completed for this hectare.

Examples of uniform stands that would be coded in this fashion are shown in Fig. 14.1, a and b. Stand a is a uniform forest and has cover type 7 (column 29) and primary succession code 99 (column 30–31). The other gradient parameters (elevation, aspect, topography, etc.) are recorded from the aerial photographs and 7½-minute topographic maps. The stand in Fig. 14.1b is a uniform shrub field, with cover type 9 and primary succession code 80; other parameters are recorded as for the forest.

Table 14.1.   Record Format Specifications for the Terrestrial Site Inventory.

Column	Appreviation	Information
FORM H-101: STANDARD HECTARE INVENTORY RECORD		
1	#C	No. of records to describe this hectare
2	S#	sequence No. of this record (within hectare)
3	L	1 if localization (H-111) record follows; otherwise 0
4	S	1 if special features (H-121) record follows; otherwise 0
	UTM	Universal Transverse Mercator coordinates
6–10	N	UTM North in 100s meters (no decimal point)
11–14	E	UTM East in 100s meters (no decimal point)
	SITE ACT	Actual site descriptors
16–20	EL	Elevation in feet MSL
21	T	Topographic-moisture, where:
		1 = Bottomland
		2 = Ravine
		3 = Draw
		4 = Sheltered slope
		5 = Open slope
		6 = Peak/ridge
		7 = Xeric flat
		8 = Slope (unspecified)
22	X	Aspect, where:
		1 = N
		2 = NE
		3 = E
		4 = SE
		5 = S
		6 = SW
		7 = W
		8 = NW
23	B	1 if burned, 0 if mature
24–26	YSB	Approx. years since last burn (omit if col. 23 = 0)
27	I	Intensity of last burn, where:
		1 = Full canopy (over 90%)
		2 = Mosaic canopy (over 50%)
		3 = Partial canopy (under 50%)
		4 = Understory only
		5 = Available fuel (areas without canopy
		(omit if col. 23 = 0)
28	OD	Other disturbances, where:
		2 = Fellfield/slide
		3 = Hydric
		4 = Animal grazing

Table 14.1. Record Format Specifications for the Terrestrial Site Inventory (*Cont.*)

Column	Appreviation	Information
FORM H-101: STANDARD HECTARE INVENTORY RECORD		
29	C	Vegetative formation (cover) type, where:
		0 = Glacier or permanent snow
		1 = Rock
		2 = Talus
		3 = Talus and meadow
		4 = Meadow
		5 = Mesic marsh/bog
		6 = Xeric or subxeric prairie
		7 = Typical forest
		8 = Krummholz forest
		9 = Shrub
30 and 31	PR	Primary succession, where:
		1 = Glacier or permanent snow
		10 = Solid rock
		20 = Broken rock
		30 = Boulders and talus
		40 = Talus
		50 = Talus and meadow
		60 = Meadow
		70 = meadow and Krummholz forest (or shrub)
		80 = Krummholz forest (or shrub)
		90 = Krummholz forest (or shrub) to typical forest
		99 = Typical forest
32	ALP	Alpine wind–snow gradient category, where:
		1 = Pass
		2 = Near pass
		3 = Other high wind area
		4 = Permanent snowfield
33 and 34	DR	2-digit drainage area code
	SPLIT	Location and coverage of this unit in hectare (omit if col. 1 = 1)
36 and 37	%	Percentage of hectare covered by this record
38–45	LOC	Grid location of this subunit on scale:
		7 8 9
		4 5 6 (block = 1 hectare)
		1 2 3 (north is up)
47–65	SITE	Readjustment of gradient site descriptors; used only
	FIT	when model predictions are inaccurate using true site data. Permits changing any site descriptor included in cols. 16–34 such that new descriptors give acceptable predictions. Format per cols. 16–34.

Table 14.1. Record Formal Specifications for the Terrestrial Site Inventory (*Cont.*)

Column	Appreviation	Information
FORM H-101: STANDARD HECTARE INVENTORY RECORD		
66 and 67	SPLR	Code No. of special branch record if used (otherwise 0)
69–77	LOC ID	UTM coordinates of an identical stand/unit. If used, cols. 16–67 are ignored. Used only when several hectares are absolutely identical, as in a lake, etc.
FORM H-111: STANDARD HECTARE INVENTORY LOCALIZATION RECORD		
1–9	UTM	Same code as cols. 6–14 of H-101
	FIRST	First localization type
11	C	Formation (cover) type (same code as col. 29 of H-101)
12 and 13	PR	Primary succession (same code as cols. 30 and 31 of H-101)
14 and 15	%	Percentage of area covered by this local type
16–19	CTG	Variance/mean contagion ratio on 20 × 20 m grid (default = 1.0)
	SECOND	Second localization type
21–29	As above	Same format as cols. 11–19 above
	THIRD	Third localization type
31–39	As above	Same format as cols. 11–19 above
	FOURTH	Fourth localization type
41–49	As above	Same format as cols. 11–19 above
FORM H-121: STANDARD HECTARE INVENTORY SPECIAL FEATURES RECORD		
1–9	UTM	Same code as cols. 6–14 of H-101
	FIRST	First special feature
11	T	Type of special feature, where: 1 = Lake or permanent pond 2 = Temporary pond 3 = Stream/river 4 = Trail 5 = Road 6 = Improvement/bldg(s). 7 = Campground 8 = Other
12–15	#	Four-digit identification number
16	DR	If col. 11 = 3, 4, or 5, direction of "flow" (same code as col. 22 of H-101)
17	DG	If col. 11 = 3, degree of stream (0 = temporary)
	SECOND	Second special feature
19–25	As above	Same format as cols. 11–17 above
	THIRD	Third special feature
27–33	As above	Same format as cols. 11–17 above
	FOURTH	Fourth special feature
35–41	As above	Same format as cols. 11–17 above

Table 14.1 Record Format Specifications for the Terrestrial Site Inventory (*Cont.*)

Line	Column	Information
FORM H-131: STANDARD MAPS, PHOTOGRAPHS, AND ACCESS RECORD		
1	10–80	ID codes of black and white aerial photographs
2	10–80	ID codes of infrared aerial photographs
3	10–80	ID code and name of 7.5-minute topographic map(s)
4	10–80	ID code and name of 7.5-minute vegetation map(s)
5–8	10–80	Verbal description of ground access information, including distance from nearest maintained trail and distance from nearest trailhead

Figure 14.1c shows a more complicated situation; here there exists a distinct discontinuity in the vegetative cover. Most of the hectare is forest, but a strip along the eastern border is uniform shrub cover. In this case, two H-101 records are recorded—one for the forested portion of the stand and one for the shrubfield. The first record has cover 7 and primary succession 99, whereas the second record has cover 9 and primary succession 80. In this case, we must also record the percentage of the hectare occupied by each cover type (columns 36 and 37) and the general location of each cover "block" on the nine-point system shown in the table. The forest cover occupies 75% of the hectare (columns 36 and 37) and positions 124578 (columns 38–45), whereas the shrub occupies 25% and positions 369. Finally, the correct total number of records and sequence numbers are entered in columns 1 and 2.

Note that the hectare was declared "blocked," and separate entries were made on the H-101 records, because of the discontinuity in cover. It is possible, but not necessary, for the two units to have different aspects, elevations, topographies, stand ages, other disturbances, etc. If this were true, the correct data would be entered for each cover "block." If the only differences were in vegetative cover, these other parameters would be identical for each line.

Figure 14.1d shows another blocked hectare; in this case, a ravine with shrub cover runs down the middle of a sheltered slope with forest cover. If we were to separate the forest block from the shrub block, therefore, we would also assign different topographies to each block. Because of the ravine, however, we have three different aspects for the hectare. The ravine has a south aspect, the strip of forest on the western edge has a southeast aspect, and the strip of forest on the eastern edge has a southwest aspect. Therefore, we must make three entries for the hectare and give the percent cover and relative location of each in columns 36–45.

Figure 14.1e shows another stand with continuous forest cover but including an old fire perimeter. The eastern portion of the stand burned 40 years ago, whereas the western portion burned 150 years ago. Two sepa-

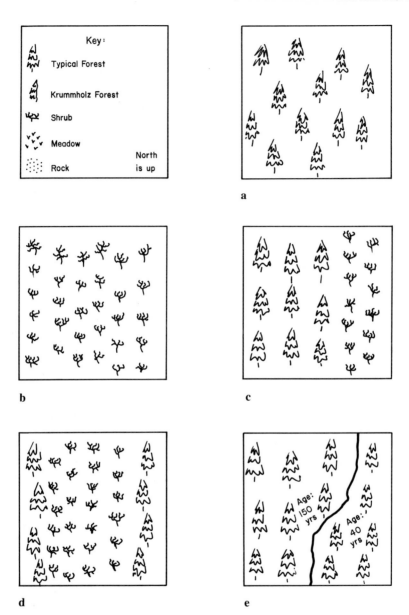

Figure 14.1a–k. Various hectare inventory examples. **a** Continuous forest cover. **b** Continuous shrub cover. **c** Blocked forest and shrub cover. **d** Blocked forest and shrub cover, and slope and ravine topography. **e** Blocked stand ages. **f** Blocked shrub, meadow, and forest cover. **g** Localized meadow on forest cover. **h** Localized

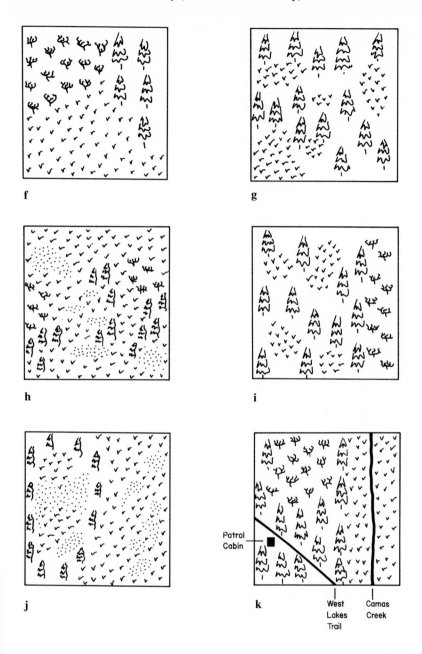

rock, shrub, and Krummholz on meadow cover. **i** Blocked and localized forest, shrub, and meadow cover. **j** Blocked and localized Krummholz, meadow, and rock cover. **k** Blocked forest, meadow, and shrub cover with both cultural and hydrologic features. See the text for an explanation of inventory coding procedures.

rate H-101 form entries are made, therefore, even though the only difference between the two blocks is stand age.

Figure 14.1f shows another stand that includes three distinct cover
types—forest, shrub, and meadow; Three H-101 records are recorded. Any
differences among the blocks in other gradient indices (elevation, aspect,
stand age, etc.) are of course also recorded.

The blocking method offers a convenient system for recording with
precision within-hectare discontinuities of cover or other stand characteristics. However, many cover discontinuities are on a size scale too small to
"block" (perhaps 10 m or less). This situation is very common in the alpine
areas, where a mosaic of Krummholz, shrub, meadow, talus, and rock
cover is observed.

In these cases, the stand is declared "localized" and the H-111 record is
used. The primary cover type (most common cover) and primary succession codes are recorded on the H-101 form. Then, for each localized cover
type (up to a maximum of four), the cover type code, primary succession
code, percent cover, and variance/mean contagion ratio on a 20×20 m grid
is recorded on record H-111. For example, Figure 14.1g shows a forested
stand with patches of meadow (too scattered for effective use of the
blocking system). On the H-101 form, therefore, cover type and primary
succession are recorded as 7-99. An entry of 1 is made in column 3 (to
indicate that a localization record follows). On Form H-111, the meadow's
cover type and primary succession codes are entered (4-60), along with the
meadow's percent cover and contagion value.

Figure 14.1h shows a more complicated situation, with several localized
cover types. Here, the most common cover is meadow; therefore, 4-60 is
recorded in columns 29–31 of the H-101 form. On the H-111 form, a
separate entry, which includes cover type, primary succession code, percent cover, and contagion value, is made for each of the localized covers
(rock, shrub, and Krummholz).

When a hectare is blocked into two or more major units, the localization
is a function of the block and not of the entire hectare. For example, in
Figure 14.1i, the hectare is blocked into forest cover and shrub cover. The
meadow is localized only within the forest. If the first H-101 record
describes the forest, therefore, it is followed by a localization (H-111)
record describing the localized meadow cover. However, the H-101 record
that describes the shrub cover is not followed by a localization record.

Figure 14.1j shows a hectare with two major cover types—Krummholz
and meadow. Each block has localized cover but different kinds, amounts,
and clumping of the localized cover. The first H-101 record therefore would
describe the Krummholz cover and would be followed by a localization
record describing the meadow and rock localized on the Krummholz. The
second H-101 record describes the meadow and is followed by a localization record that describes the localized rock.

In the simplest case, a hectare has uniform site conditions and is described by one H-101 record. At the other extreme, the system permits the hectare to be blocked into (up to) three major units, with up to four different cover types localized on each unit. By using such a variable resolution system, it is possible to rapidly code areas of uniform site characteristics but still preserve at least 10-m resolution for areas where a small-scale mosaic is found.

The system also permits recording of special (cultural or hydrologic) features within each stand or block by using the H-121 record. Whenever a block contains such a feature, a 1 is entered in column 4 of the H-101 record and the H-121 record is then completed. Special features are classified by type (column 11) and a unique four-digit name code (Columns 12–15). If the special feature is a stream, river, road, or trail, its direction of "flow" (or travel) is given in column 16. If the special feature is a stream or river, its order is also recorded (column 17). Up to four special features for each block may be recorded.

Figure 14.1k shows a hectare blocked into three units (forest, shrub, and meadow). The shrub portion has no special features. The meadow portion has a special feature (Camas Creek) of type 3 (column 11), name code 0012 (columns 12–15; determined from a master list), direction of flow 4 (south; column 16), and order 1 (column 17). The forested portion has two special features. The first is the West Lakes Trail (type 4; name code 0017; direction 4 or 8). The second is a patrol cabin (type 6; name code 0029).

Once all 100 hectares are coded for a square kilometer, form H-131 is completed; it records the aerial photograph ID codes for the square kilometer, the code number and names for the topographic and vegetation maps which include the square kilometer, and the ground access information to the square kilometer, including its distance from the nearest maintained trail and from the nearest trailhead.

Weather Models

Accurate knowledge of fire weather, specifically wind and the factors that determine fuel moisture, is crucial in describing or predicting fire behavior. Geiger (1965) noted that "since the influence of relief is so strong, the high mountain microclimate is a mosaic of vastly different conditions in the smallest of areas."

How does one determine "fire climate" boundaries within such a mosaic? Fosberg and Furman (1973) proposed a technique for defining large scale (300–500 km^2) zones of equal fire potential called "fire-climatic zones"; they are based on the mean equilibrium moisture content for fine fuels during the high fire danger season.

Another approach to the spatial and temporal prediction of fire weather, fire microclimatic zones, and real-time fire weather modeling is proposed here and in Mason and Kessell (1975). It determines the real-time fuel moisture for five fuel strata categories on environmental gradients as used by the vegetation and fuel models. The preliminary Glacier National Park weather study attempted to determine important factors of fire microclimate in the park and proposed a model for further development and refinement (Mason and Kessell 1975).

Required Wind Parameters

Rothermel's (1972) fire behavior model applies wind speed at midflame height (ϕ_w) as a multiplying factor in determining rate of spread (described in the "Fire Behavior Models" section). According to this model, wind speed significantly affects spread rate by bending the flame closer to the fuel and by altering the airflow characteristics through the fuel bed (and thus affecting the magnitude of the optimum packing ratio for a given fuel array).

The modified deterministic ground fire model emphasizes that each of the vertical vegetative fuel strata corresponds to a discreet bulk density and packing ratio. Because the particular optimum stratum through which the fire will spread is primarily a function of wind speed (Kessell and others 1978), wind velocity in each stratum is of extreme importance to behavior modeling. These strata include the compacted duff and litter, the loose array of dead and down woody fuels, and above the litter, grasses and forbs, the standing shrubs, the ladder fuels, and the crown fuel array.

Required Fuel Moisture Parameters

Rothermel (1972) also observes that the heat of ignition is a function of moisture content of the fuel particle (described in "Fire Behavior Models" section). He weights fuel by surface-area-to-volume ratios (σ) for determining spread rate, because the reaction velocity is a function of the availability of air and therefore of bed porosity. Therefore, the smaller σ fuels (larger diameter fuels—100- and 1000-hr time-lag fuels) are not important for predicting fire spread rate.

Dead and live fuels are not differentiated except with respect to moisture content in each stratum. Grasses, forbs, and shrub fuels are divided into their live and dead components. Dead fuels with a moisture content higher than their moisture of extinction are treated just as the live fuels, the moisture of which may not be a component for predicting reaction rate or intensity calculations but is in any case a heat sink.

The contagion of curing fuels should also be considered in predicting fire behavior. If cured plants are clumped and separate from live plants, fire spread may be fast and uninhibited through the cured portion; this heat sink is therefore unimportant if short pathways of dry fuels are available. For litter and duff moisture, depth to the moisture of extinction determines the fuel load available in this category for fire spread. Furthermore, in the prediction of fire ignitions, natural fire starts by lightning are a function of both topography and elevation (in determining lightning strike probability) and litter and duff moisture (which determine ignition probability) (Donald Latham, personal communication).

Management Requirements

In addition to the applied mathematical models described above, the manager needs readily accessible current and historical fire weather information from his base stations. These stations must provide the necessary information to predict fire weather in all of his "fire climate" zones within the management area.

Application: Analysis of 1975 Microclimate and Fuel Moisture

Routine fire weather observations were made during July and August of 1975 (Chapter 6) in the McDonald drainage of Glacier National Park's west slope at the stations shown in Table 14.2. On some rainy days, no data were collected at some stations; therefore the data presented here are biased toward nonrain days in which the 1400 hours MDT[1] 10-hr fuel moisture sticks[2] showed less than 20% moisture content.

Relative Humidity and Temperature at 1400 Hours MDT

The primary succession gradient described in previous chapters is useful to show relative humidity and temperature trends. Data from stations 1, 2, and 11 (Table 14.3) show that the low-elevation forest is cooler and wetter than the shrub or meadow cover. Geiger (1965) and others have documented similar relative humidity and temperature variations as a function of vegetation cover.

Table 14.3 also shows that day-to-day 1400 hours MDT variations in relative humidity are more extreme in the heavily forested cover (station 1) than in the meadow (station 11) in the low-elevation areas. This appears logical, because on calm, cloudy days, forest fuels exhibit minimum moisture loss after high moisture recovery from the previous night. On clear, windy days, however, the forest fuels will show considerable afternoon moisture loss. Meadows, conversely, will show a significant afternoon moisture loss even on calm or cloudy days by whatever wind, evaporation, or insolation is available. Meadows' 1400 hours MDT range and variance of relative humidities are therefore usually less than those observed under the

---

[1]Mountain Daylight Time

[2]As described in Chapter 6, section on "Fuel Moisture and Microclimate Study," fuel moisture sticks are physical analogs used to determine dead fuel moisture level

Table 14.2. Description of the Remote Weather Stations Established in the McDonald Drainage.[a]

Instruments[b]	Station no.	Elevation (m MSL)	Aspect	Topographic-moisture	Primary succession	Wind gradient	Slope (°)
A*	11	960	SW	Open slope	Meadow	None	1
A	1	1039	SW	Sheltered slope	Forest	None	2
C	2	1039	SW	Bottomland	Shrub	None	1
B	3	1097	W	Slope	Forest	None	1
B	4	1366	S	Slope	Open forest	None	50
C	5	1366	S	Open slope	Meadow, talus	None	4
C	6	1670	W	Ravine	Meadow, talus	None	50
B	7	1981	N	Open slope	Krummholz	None	2
A	9	2030	SE	Ridge	Rock and Krummholz	Near pass	Variable
						Pass	
C	10	2030	SE	Ridge	Krummholz and shrub	Pass	1
C	15	2569	SW	Peak	Rock	High wind	5

[a] Note that all stations were checked daily at about 1400 hours MDT with a belt weather kit containing a sling psychrometer, a hand wind meter, and a compass, except station 11; at staion 11, standard AFFIRMS observations were recorded.

[b] Key to Instruments: A, Standard U.S. Weather Bureau shelter, 1.4 m above ground, with recording hygrothermograph, commercial *P. ponderosa* 10-hr sticks, and Gradient Modeling, Inc. 1-hr sticks, set 25 cm above ground. A*, Same as A except no 1-hr sticks, but includes standard AFFIRMS anemometer and rain gauge. B, Recording hygrograph on white wooden platform with roof, about 20 cm above ground, and 1- and 10-hr sticks 25 cm above ground. C, 1- and 10-hr sticks only, 25 cm above ground.

Table 14.3. July 1975 Temperature and Relative Humidity Means and Standard Deviations for McDonald Drainage Stations.[a]

Station	Relative humidity (%)		Temperature (°C)		
	X̄	s	X̄	s	N
11	37.5	12.6	25.8	4.6	30
1	53.1	18.2	21.8	5.3	30
2	42.0	9.3	27.4	4.1	18
3	52.4	16.3	22.5	4.7	30
4	54.3	16.3	22.0	5.0	30
5	49.1	17.5	23.0	5.5	30
6	53.7	14.9	19.3	5.2	30
7	54.6	14.5	18.1	5.4	30
9	55.3	16.5	18.0	5.4	30
15	63.0	17.2	11.5	5.2	20

[a]All readings taken at approximately 1400 hours MDT.

canopy, although the diurnal variance is greater in the meadows. In other words, forests may or may not show significant moisture loss from night to the following afternoon, whereas meadows almost always show this loss; meadows are therefore much more sensitive than is forest cover to insolation and wind in modifying the cool, moist nighttime conditions. This may appear contrary to, but is actually quite unrelated to, the commonly stated axiom in microclimatology that forests are a moderating, more stable environment than open cover (Geiger 1965). Forest cover and meadows must therefore be separately monitored to observe this unequal day-to-day variance in relative humidity.

Temperature means also differed by over 5°C between stations 1 and 11. Relative humidity and temperature must therefore be taken carefully, close to the fuel, to be of use on the primary succession gradient representation. Relative humidity taken above the shrubs with a sling psychrometer was probably not indicative of fuel-level conditions.

The elevation gradient is also useful in describing fire microclimates within the park. Figure 14.2 shows the stations analyzed along this gradient at 1400 hours MDT. A line drawn through these stations yields a 0.7°C/100 m (3.5°F/1000 ft) lapse rate, equal to the often quoted normal lapse rate if stations 11 and 15 are neglected.

Relative humidity is inversely correlated with temperature on the elevation gradient, as normally expected (Fig. 14.2). Higher stations have consistently higher relative humidity, regardless of cover, then low-elevation stations do because of the inverse correlation between temperature and elevation. Because of differing day-to-day variations with cover (primary succession), simple linear regression of relative humidity among stations

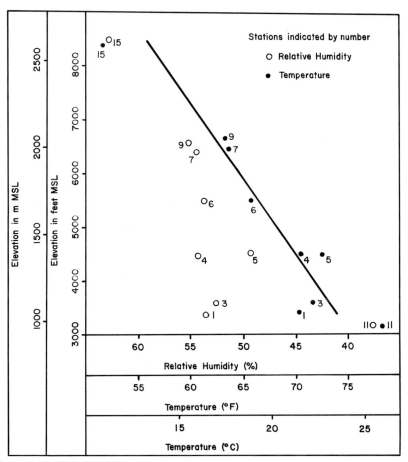

Figure 14.2. Relationship among temperature, relative humidity, and elevation for the McDonald drainage weather stations. Station numbers correspond to those in Table 14.2.

was better when cover types were similar (Douglas Mason, personal communication). Elevation therefore modifies and interacts with the effects of primary succession on temperature and relative humidity.

The topographic-moisture gradient implies greater moisture at the bottomland–ravines end of the gradient and lower moisture at the open slope– peak–ridge extreme, with northeast aspects the wettest and southwest aspects the driest. Although a range of topographic-moisture gradient positions is occupied by the stations (Table 14.2), these stations also vary on the elevation and primary succession gradients. One is therefore faced with the problem of too many independent variables and too few stations; more data are needed before any real conclusions can be drawn. Aspects among the stations were between 180° and 270° (true) except for station 7 (north aspect).

Figure 14.3 shows mean relative humidity at low elevations (July 1975) using primarily the data from Table 14.3. The contours are constructed with considerable extrapolation from the four appropriate stations and merely suggest tentative relationships among relative humidity, primary succession, and topographic-moisture. The collection of additional data should allow the determination of functional relationships among these variables and permit the replacement of absolute relative humidity values with the constants from functional, predictive equations.

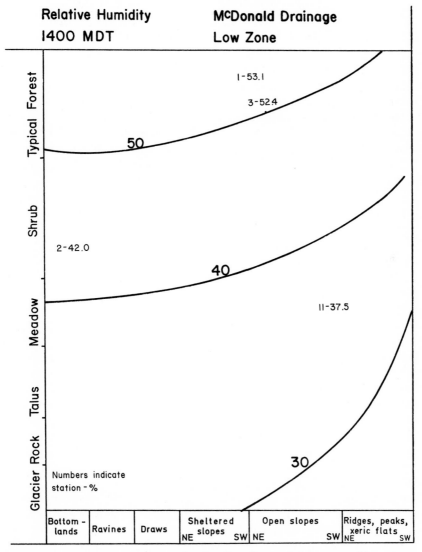

Figure 14.3. Tentative nomogram showing the effects of the topographic-moisture and primary succession gradients on mean relative humidity (July 1975).

Relative Humidity and Temperature Diurnal Variations

The elevation gradient plays a much more distinct role in fire weather at times other than 1400 hours MDT. Hayes (1941), Geiger (1965), Schroeder and Buck (1970), and many others describe nightly air inversions occurring at middle elevations in mountain valleys; this is a prominent feature in Glacier National Park (Mason and Kessell 1975). Recording hygrometers set at several stations showed that by 2200 hours MDT, all stations had relative humidities higher than at 1400 hours MDT. Humidity increases to a maximum just before sunrise.

Figure 14.4 shows a low zone extending up to about 1100–1300 m elevation; it contains cool, humid air at 2200 hours MDT and still cooler and more moist air by 0800 hours MDT. At the top of this zone, however, a sudden discontinuity of humidity is noted beginning at about 2200 hours MDT. Warm, drier (usually 55%–65% relative humidity) air extends toward the ridge tops while a thermal zone or inversion layer forms as colder air flows to the valley bottoms. Only in the high zone, near the high peaks, is colder air and high humidity again found. By 0800 hours MDT, all stations decrease to a minimum temperature and maximum humidity. Station 4, near the bottom of the thermal belt, shows the greatest change in humidity from 2200 to 0800 hours MDT, which suggests that the thermal zone may become unstable at this elevation as the night cooling progresses. The highest elevation station (station 15) varies very little in relative humidity from 2200 to 0800 hours MDT.

Day-to-day 2200 hours MDT relative humidity standard deviations are slightly (but not significantly) greater than the 0800 hours MDT standard deviations at all but station 15. Variations from day to day appear to increase with elevation, especially evident in the 0800 hours MDT data (Fig. 14.4).

Insufficient diurnal measurements recorded along the primary succession gradient prevent useful discussion. Diurnal data on the topographic-moisture gradient are too limited to show obvious conclusions. The principle of cooler, moist air flowing down the slopes and settling into depressions at night appears to be significant on this gradient (Geiger 1965, Schroeder and Buck 1970). High-elevation depressions (swamps, bottomlands, ravines, and draws) probably experience "miniinversions" at night (Douglas Mason, personal communication).

Ten-Hour Time-Lag Dead and Down Fuels at 1400 Hours MDT

Means for July and August in the low zone were closely related to vegetative cover or the primary succession gradient as shown in Fig. 14.5. Standard deviations decreased with the amount of cover, from 2.9% at station 1 to 1.3% at station 11. Greater day-to-day variation in the forest than in meadow cover supports the relative humidity conclusions discussed

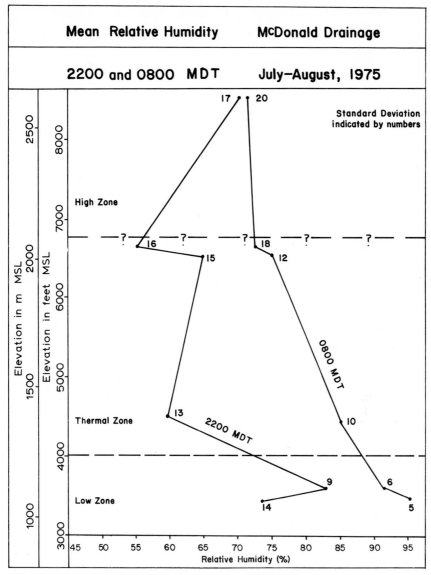

Figure 14.4. Relationship among elevation, 0800 hours MDT relative humidity, and 2200 hours MDT relative humidity suggest three fire weather zones in the McDonald drainage.

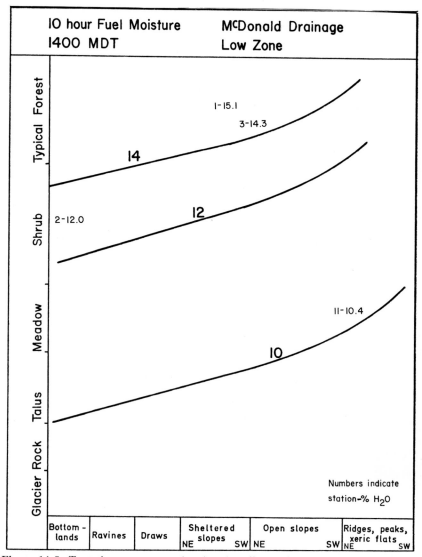

Figure 14.5. Tentative nomogram showing the effects of the topographic-moisture and primary succession gradients on 10-hr time-lag dead and down branchwood fuel moisture.

earlier. (Note that Fig. 14.5 is derived from minimal data and is, like Fig. 14.3, a tentative expression of relationships.) The greater variation of moisture at 1400 hours MDT in the forest shows that a forest cover base station is necessary since a wide range of moistures is possible here. Unfortunately, the park's AFFIRMS stations are all located in meadows, even those using AFFIRMS' "G" fuel model (coniferous forest model).

On the elevation gradient, 10-hr fuel moisture falls from a seasonal mean of 14% at 2570 m, to 12.4% at 2030 m, to 11.5% at 1670 m, to the 9.0%–10.3% range (depending on cover) at 1370 m. Fuel moisture rises again at lower elevations, with means of 11%–14% at 1000 m (depending on cover). This suggests a correlation with the night relative humidity–elevation profile (Fig. 14.4).

Because of the 10-hr time-lag characteristics of this fuel, the distinct night inversion on the elevation gradient probably plays a significant role in determining the 1400 hours MDT moisture. Therefore, correlations among stations within the same elevation zone are meaningful but correlations between zones are not.

Data on the topographic-moisture gradient were very limited; however, Fig. 14.5 does show a tentative relationship among topographic-moisture and 10-hr dead and down fuels.

Ten-Hour Time-Lag Dead and Down Fuel Moisture Diurnal Variations

On August 10 and 11, 1975, relative humidity, temperature, wind, 10-hr, and 1-hr time-lag fuel moistures were recorded at 3-hr intervals at several stations. These data indicate that the night 10-hr fuel moisture closely follows the night relative humidity curve along the elevation gradient (Fig. 14.4; and Douglas Mason, personal communication). Fuel moisture reached a minimum between 1800 and 2200 hours MDT and reached a maximum at about 0800 hours MDT.

At 2200 hours MDT, relative fuel moisture levels or differences between stations similar except for cover or primary succession (station 1 versus station 11; station 4 versus station 5) are similar to 1400 hours MDT moisture differences. At 0800 hours MDT, after several hours of little insolation, fuel moistures did not significantly differ along this primary succession gradient (Douglas Mason, personal communication). Geiger (1965) observes that dew-covered tree crowns may indicate drier fuels beneath; maximum dewfall in open areas is 1–2 m above the ground. Therefore, the primary succession gradient may be useful for fuel moisture prediction at night in relation to moisture recovery from dew.

The diurnal data are too limited for analysis along the topographic-moisture gradient.

One-Hour Time-Lag Dead and Down Fuel Moisture at 1400 Hours MDT

One-hour fuel moisture day-to-day means at 1400 hours MDT vary from station to station in a fashion similar to that of the 10-hr fuels. The two same environmental gradients, elevation and primary succession, are presumed to be the most influential. The short time lag of 1-hr fuels prevents the night elevation (moisture) zones from being an important influence on the 1400 hours MDT moisture. Therefore, correlations and regressions among any stations of similar cover are generally significant (Douglas Mason, personal communication). These fine fuels are probably more sensitive than larger fuels to insolation and therefore to the aspect component of the topographic-moisture gradient.

One-Hour Time-Lag Dead and Down Fuel Moisture Diurnal Variations

Based on the 10 and 11 August 1975 diurnal observations, 1-hr fuels appear to behave similarly to 10-hr fuels but show a quicker and a more extreme response to relative humidity changes. The discussion of 10-hr fuel moisture also applies to 1-hr fuels. Station 5 varied little from 6% moisture between 2200 and 0800 hours MDT, suggesting a stable thermal zone at this elevation. Maximum moisture at this station was reached earlier in the evening (2030 hours MDT), when cooling began but before downslope winds started to develop in the thermal belt.

Litter Moisture

Although the 1975 data are very limited, mean litter moisture appears to follow the same trend as 1- and 10-hr dead and down fuel moisture. Day-to-day variances are also greater in the low-zone forests than in meadows (as shown for relative humidity and 10-hr fuels).

Table 14.4 shows the wide variation in moisture content of litter on the primary succession gradient at 1525 hours MDT, 20 September 1975, in the area of station 11 (open slope, low-elevation zone, McDonald drainage).

Wind Speed at 1400 Hours MDT

The primary succession gradient is very useful in describing wind speeds, as both higher speeds and greater day-to-day variance appear to be associated with lower vegetative cover. As expected, for a given cover type, more open topographies corresponded to higher wind speeds.

Table 14.4. Litter Moisture Variation Along Primary Succession Gradient, Open Slope, Low Elevation Zone, McDonald Drainage.[a]

		Moisture (%)	
Primary Succession	Litter composition	Top 0.5 cm	Top 1.5 cm
Measow–forest transition, sparse grass	Sparse grass, pine needles, bark	11	—[b]
Meadow, thick grass	Sparse grass, mulch, moss	—	21
Dense *P. contorta* forest	*P. contorta* needles	79	108

[a]Readings at 1525 hours MDT on September 20, 1975.
[b]Mostly mineral soil below 0.5 cm.

Wind Speed Diurnal Variations

Daily wind patterns are dominated by down-slope, down-valley winds at night and by up-slope, up-valley winds during the day. Hygrothermograph traces indicate that unstable conditions occur from about 1000 to 2000 hours MDT. At high elevations (stations 4, 7, and 9), traces fluctuate throughout the night, as well as during the day, whereas the lower stations have only occasional nights exhibiting much wind variance. The daily late afternoon maximum is associated with the strongest convection occurring as the land surface warms to its maximum temperature.

Grasses, Forbs, and Shrub Moisture

The summer of 1975 was exceptionally wet, with July precipitation 1.42 cm greater than a 30-year July mean; August precipitation was 6.53 cm greater than the 30-year mean. Much of the lowland plants and some of the alpine grasses, forbs, and shrubs did not cure until September or later. The fire danger was generally low and the wet conditions precluded collecting much data on the curing process.

Proposed Future Refinement

As described above, preliminary work during the summer of 1975 indicated several important and unique approaches to fire weather modeling that are feasible as a result of the gradient model data base.

1) The first and most important object of proposed future work is to intensively *develop functional fuel moisture relationships along environmental gradients*. Elevation, aspect–topographic-moisture, and primary succession gradients are shown by the 1975 study as useful for determining fuel moisture gradients. Because these gradients have been intensively studied in application to vegetation and fuel distribution, much of the work of defining important characteristics of each gradient has already been completed. Field work should include:

a) establishing *recording hygrothermographs* at key elevations to determine the location and nature of the thermal belt
b) establishing *fire weather stations* at three or four key locations on two major gradients, primary succession and aspect–topographic-moisture (while holding constant the stands' location on other gradients)

These stations would measure relative humidity, temperature, insolation, precipitation amount and duration, wind, and moisture content of prepared 1-hr (fine fuel) and 10-hr time-lag fuel sticks. Measurements would be taken at approximately 1400 hours MDT (high wind and fire risk time), at 0700–0900 hours MDT (maximum fuel moisture), and at 2000–2200 hours MDT (minimum fuel moisture) each day. Results should then functionally relate the different positions on each gradient to one another so that only one base station would be required on each gradient to permit extrapolation and prediction of actual fuel moistures, temperatures, and humidities at all other positions at any time t. This would provide a diurnal, spatial remote fire weather analysis and model for the area and provide a methodology for developing similar systems elsewhere. This project would also provide an excellent opportunity to intensively test new fuel moisture analogs or other fuel moisture models.

2) An intensive *study of litter moisture along both the gradients described above* (item 1,b) *and in vertical profile* would be of considerable value of the fire behavior models described above, and for fire ignition predictions [see item 5 below]. Because of the large litter data resources available from the gradient model (depths, loading, packing ratios, and bulk densities have been sampled from 800 two-station stands and integrated into the model), functional relationships of litter moisture with any of these parameters may be realistically sought.

Field work would consist of determining litter moisture at 1-cm depth intervals at the stations described above using moisture meters such as the one McLeod (1974) has developed at Montana State University, at Bozeman. Results may also be used to test Fosberg's (1975) theoretical litter response calculations.

3) An intensive study has been proposed to *analyze the curing-out process of grasses, forbs, and shrubs in response to environmental gradients,* combining both seasonal trends and short-term (2- or 3-day) weather trends. Moisture levels of dead and live parts and percentage green would be measured for selected species at 2- or 3-day intervals at the stations described above. This would provide an excellent opportunity to test the "curing model" (Deeming and others 1977) of the National Fire Danger Rating System.

4) *Wind vertical profiles that included frequency distributions of velocity for different cover types* would be taken to provide needed information for the fire behavior model. Wind velocities recorded at stations described above may result in a preliminary analysis of wind velocity functional relationships in response to such environmental gradients as elevation, aspect–topographic-moisture, and primary succession and may provide data for a spatial pattern analysis.

5) A unique and useful proposed project is to *combine results of litter moisture studies with the park's base weather stations' historical fire and lightning strike occurrence records and those of fire seasons now being studied.* (The gradient model would allow extrapolation to remote sites of litter moisture at time of past strikes.) A natural fire ignition probability model could well result from this work.

Resulting Refinements for Managers

By understanding the spatial and diurnal characteristics of fire weather, it is hoped that fire managers can come nearer to making not only good spatial but also good temporal weather predictions. The quality of these predictions depends on understanding the dynamics of the interactive environmental gradients and spatial and temporal microclimates of the area.

Already operational in the *BURN* model is a fire weather information retrieval system; it provides a description of the new remote weather stations, the AFFIRMS stations, weather observations from these stations, plus the four manned lookouts, for the past 2 days and the 24-hr forecast for selected stations. Data for this system are updated nightly throughout the fire season.

Other drainages of regions with different climates that are of concern to fire managers should be studied by comparison with the McDonald drainage model. On the west side of the park, the same gradients should be applicable. As noted earlier, areas north and south of the McDonald drainage are drier; therefore, a base station in each elevation zone (low, thermal, and high) in each major drainage of concern may be necessary. In addition, hygrothermographs should be used to determine the elevation of these three zones in each drainage system because zone elevations are determined by local topographic features (Hayes 1941).

Conditional Forecasting

About a year after the completion of the Glacier model, another method of improving meteorologic inputs to fire models was tested; this evaluation of conditional forecasting was conducted by Larry Bradshaw and reported by Kessell and others (1977c).

Paegle (1974; also see Angulis 1969, 1970, Lund 1963) derived a synoptic climatology of eight unique 500-mb flow patterns over the western United States. Relative frequencies of precipitation, stratified by flow type, were computed; the resulting probabilities of precipitation, stratified by map type, offered significant improvement over nonstratified climatology. Paegle and Wright (1975) went on to investigate pattern recognition algorithms stratified by map type; this method resulted in further improvement of probability of precipitation forecasting.

Bradshaw (in Kessell and others 1977c) applied the techniques of Rasch and MacDonald (1975) to a select group of fire-weather stations in the northwestern United States. The resulting map-type frequencies of precipitation for each station were then tested against nonstratified climatology for forecasting skill. Persistence and transition values were also computed as a Markovian process.

Results showed that map-type stratification to predict probabilities of precipitation resulted in a 6.8% improvement over unstratified climatology for the entire summer season and a 19.7% improvement for the critical fire month of August. The study also produced matrices of persistence and transition probabilities for each map type.

The results of this brief evaluation of conditional forecasting suggest that it is an excellent method which can significantly improve summer fire weather forecasts. Instead of eliminating the need for microclimatic extrapolations of weather parameters along gradients, it complements that effort by improving the base-station forecasts. The approach is simple, easy to understand, and completely compatable with current U.S. National Weather Service (NOAA) products.

Fire Behavior Models

The Rothermel (1972) fire behavior model enables us to predict the spread rate and intensity of a fire if we know certain properties of the fuel matrix and the conditions under which it will burn. Once fire intensity is calculated, we may then derive other useful fire parameters, such as flame length, scorch height (the height above the ground where living foliage will be killed), and ignition height (the height above the ground where living foliage will be ignited). Three basic kinds of input parameters are required by the Rothermel model. First are parameters describing the fuel array, such as fuel loads and packing ratios; these are quantified from field samples, as described in Chap. 6 (Section on "Fuel Sampling") or by inference from sampled to unsampled areas. Next are time- and space-dependent parameters that describe the conditions under which the fire will burn. These include fuel moistures for each fuel size class; the proportion of live and dead fuels in the grass, forb, and shrub strata; wind velocity; and the site's slope steepness. The third set of inputs includes fuel properties, such as the mineral content, silica-free mineral content, low heat of combustion, and average surface area to volume ratios, for each fuel category and size class. These parameters vary little from one fuel array to another, and usually standard constants (e.g., from Rothermel 1972) are assumed. A simple derivation of the current Rothermel (1972) fire behavior model follows, based on Kessell (1977, Kessell and others 1978).

The Rothermel Model

How does a fire spread from one burning volume to another that is similar in composition but not yet burning? Consider two unit volumes of fuel adjacent to each other within a larger fuel array. The two volumes are as small as possible, with the condition that they exhibit the same bulk density, packing ratio, and fuel particle array as the larger fuel matrix within which they are located. That is, they have the smallest cubic volume possible that still maintains the bed's physical characteristics. How does a propagating fire spread from one of these unit volumes to the next, adjacent volume?

Frandsen (1971) derived a formula for this fire spread between volumes, based on a consideration of the law of conservation of energy. The burning volume will produce a certain reaction intensity, I_R. Some of this total heat flux will become available to pass into and raise the temperature of the fuel in the second unit volume. This available heat is termed the propagating intensity, I_p. The fire will spread into the second volume if a critical amount of the fuel, the effective bulk density (ρ_{be}), is raised to the level of heat required for ignition, Q_{ig}. The effective bulk density is the density of the fuel array heated to ignition (g/cm³), and Q_{ig} is the unit heat requirement for ignition (kcal/g). Together, they represent the volume heat requirements (kcal/cm³) to raise the second unit volume to ignition.

From this treatment, fire spread rate R can be viewed as the ratio of the propagating intensity from the burning volume to the volume heat requirement posed by the second volume:

$$R = \frac{I_p}{\rho_{be}Q_{ig}} = \frac{\text{Heat flux (kcal/m}^2\text{/min)}}{\text{Heat sink (g/cm}^3\text{)(kcal/g)}} \tag{14.1}$$

This relationship alone will not allow for a prediction of fire spread rate unless the propagating intensity can be determined prior to ignition.

The reaction intensity can be expressed as the product of the load loss rate in a burning fuel array times the heat content per unit mass:

$$I_R = h\left(\frac{-d\omega}{dt}\right) \tag{14.2}$$

where h is measured in kcal/g and $-d\omega/dt$, in g/m²/min. This is approximated by:

$$I_R = \frac{h[-(\omega_0 - \omega_f)]}{\tau_R} = h\omega_0\frac{[(\omega_f - \omega_0)/\omega_0]}{\tau_R} \tag{14.3}$$

where ω_0 and ω_f are the initial and final fuel loadings (g/m²) and τ_R is the reaction time (minutes). Note that the expression $[(\omega_f - \omega_0)/\omega_0]/\tau_R$ is the proportion of mass consumed per unit time, or the reaction velocity, γ.

What controls the reaction velocity? For dry wood with no mineral content, it is solely a function of the availability of air. This, in turn, depends on the fuel bed porosity and the exposed surface area of the fuels through which pyrolysis occurs. For a perfect fuel particle, therefore, the maximum reaction velocity γ' is:

$$\gamma' = f(\beta, \sigma) \tag{14.4}$$

where β is the packing ratio (the proportion of the fuel array that is actually occupied by fuel) and σ is the fuel particle surface area to volume ratio. Both may be determined empirically.

However, we generally do not have a perfect fuel, so γ' must be decreased according to coefficients for the fuel's moisture content (η_M) and mineral-dampening (η_s) coefficient:

$$\gamma = \gamma' \eta_M \eta_s \qquad (14.5)$$

The moisture-dampening coefficient is a function of how close the moisture content is to the moisture content of extinction M_x (the moisture content at which the fuel will not burn).

We have arrived at Eq. (14.3) for the reaction intensity I_R based only on the characteristics of the static fuel bed. Because these parameters can be measured before a fire occurs, predictions of fire behavior are possible. The fire intensity formula is now:

$$I_R = h \omega_i \gamma' \eta_M \eta_s \qquad (14.6)$$

Rothermel (1972) actually constructed a large number of fuel beds of different, but known, ω_i, h, β, and σ. He ignited the beds, used a weighing system to determine reaction intensity through mass loss, and recorded the rate of spread. From these empirical studies the heat flux could be calculated from Eq. (14.1) if R, ρ_{be}, and Q_{ig} were known for a specific fuel bed. The heat flux was then related to reaction intensity (I_R) by the formula:

$$I_p = \xi I_R \qquad \text{or} \qquad \xi = \frac{I_p}{I_R} \qquad (14.7)$$

where ξ is an empirically derived constant. This constant ξ changes for each fuel bed but it has been measured for many types and mixes of fuels. These studies found that:

$$\xi = f(\beta, \sigma) \qquad (14.8)$$

Now, to go back to our original question, we can calculate the heat flux and therefore the rate of spread from parameters that can be measured in a forest before a fire, as:

$$R = \frac{I_p}{\rho_{be} Q_{ig}} = \frac{\xi I_R}{\rho_{be} Q_{ig}} = \frac{\xi h \omega_i \gamma' \eta_M \eta_s}{\rho_{be} Q_{ig}} \qquad (14.9)$$

This approaches the final Rothermel equation. However, we need to make a few final adjustments. We know that the heat to ignition Q_{ig} is virtually the same for all fuels except for the moisture content M_f that must be vaporized to raise a fuel particle to ignition temperature. Therefore:

$$Q_{ig} = f(M_f) \qquad (14.10)$$

Now, what about ρ_{be}, the amount of fuel that must be supplied with Q_{ig} units of heat per unit mass? This was taken into account by expressing it as a ratio, ϵ, where:

$$\epsilon = \frac{\rho_{be}}{\rho_b} = f(\sigma) \tag{14.11}$$

where ρ_b is the bulk density of the bed, defined as the initial wood mass divided by the vertical depth of that wood mass.

Slope and wind are introduced as multiplying factors ($\phi_w + \phi_s$) on the rate of spread. If the flame is brought closer to the fuel (as by a higher wind), the rate of spread increases. Because these two factors also alter the airflow characteristics through a fuel bed, they also must depend on the fuel's σ and β, as well as on the magnitude of the wind and the slope.

Thus we finally arrive at the Rothermel mathematical model of fire spread, but we have done so using parameters that can be measured in the field:

$$R = \frac{\xi h \omega_i \gamma' \eta_M \eta_s}{\epsilon \rho_b Q_{ig}} (1 + \phi_w + \phi_s) \tag{14.12}$$

The equations in the fire spread model were empirically evaluated using fuel beds of uniformly sized and uniformly spaced fuel particles. To account for mixtures of fuels of various size classes, the complete model provides a method of weighting the fuel parameters by surface area. Fine fuels therefore dominate the prediction of fire spread rate because of their high surface area to volume ratios, thereby satisfying the axiom that "fine fuels carry the fire." For example, if a bed is 75% 1-hr fuels (by surface area) and 25% 10-hr fuels (by surface area), the weighted $\bar{\sigma}$ is:

$$\bar{\sigma} = 0.75\sigma \ (1 \ \text{hr}) + 0.25\sigma \ (10 \ \text{hr}) \tag{14.13}$$

We may also calculate flame length (FL) by:

$$FL = \alpha I_\beta^b \tag{14.14}$$

where I_β is the intensity per unit length (not area) of fireline per unit time and α and b are constants. From here we may calculate crown scorch height, crown ignition height, and (if we know the forest crown dimensions) the proportion of the canopy that will be killed by the fire.

A limitation of the Rothermel model described above is that it assumes a steady-state rate of spread and assumes that fuels are uniform and continuous in both the horizontal and the vertical dimension. These assumptions of uniformity are often violated by natural fuel arrays. For example, many fuel types seem randomly distributed, whereas large branchwoods often exhibit clumped distributions because of fallen snags, etc. Furthermore, most forest ground fuels include the three distinct vertical strata (litter, branchwood + grass and forbs, and shrub layer) noted in the analysis of Glacier's fuels; each stratum exhibits a characteristic fuel load, packing ratio, and moisture content often very different from the characteristics of the other strata. For example, the litter may be sufficiently dry to propagate a fire although the grass, forbs, and shrubs are too green to burn. Alternatively, dessicated shrubs occurring on a steep slope may spread a fire much faster than either of the other two strata. However, recent research has revealed ways to overcome these uniformity assumptions of the model, as discussed next.

Vertical and Horizontal Discontinuity of Fuels

Although the Rothermel model assumes a horizontal and vertical uniformity of fuels, no scale is specified; the user may choose whether the horizontal uniformity assumption will apply to 1 ha or to 1 m², and whether the vertical uniformity assumption will apply to the entire fuel array or to each discreet stratum. The user therefore may refine the model by limiting the scale of the uniformity assumptions.

For example, Frandsen (1974) has developed an elaboration of the Rothermel model that superimposes a hexagonal grid upon the forest floor (hexagons are the maximal polygons that form an interlocking network). Typical hexagonal diameters are 1–4 m. Frandesen's model assumes uniform fuel loads and depths within a hexagon but allows fuel loads and depths to vary from one hexagon to another. Basic fire behavior computations, except for the rate of spread, use the Rothermel model. The Frandsen model therefore permits a fire to burn irregularly through such an array, accelerating through hexagons with heavy fuel loads and pausing where it encounters areas with little or no fuel. The method of assigning fuels to each hexagon is left to the potential user. The assignment can be random, based on field-measured contagion (clumpedness) values (as suggested in Chap. 6), or based on actual stand by stand field measurements (as performed for slash fuels by Frandsen). If fuels are uniformly distributed, the results of the hexagon model converge with those of the basic Rothermel fire behavior model.

Another approach to nonuniform horizontal fuels is to apply the basic Rothermel model to the entire distribution of fuel parameters instead of to mean or median fuel loads. With this approach, the probability distribution of natural fuels (which is a function of natural variations) produces a similar probability distribution of fire behavior patterns (rather than mean, median, or "typical" fire behavior). We hope to compare the results of this method to those produced by Frandsen's hexagonal model.

The problem of the vertical uniformity assumption has been alleviated by the development of a multi-strata elaboration of the Rothermel fire model (Bevins 1976, Kessell and others 1978). This model distinguishes three distinct vertical strata of ground fuels; each stratum has a distinct load, packing ratio, surface area to volume ratio, and moisture content. The fire spreads through the stratum that permits maximum spread rate. Note that the fastest stratum is not fixed but varies with changes in wind speed and slope steepness, because the optimum packing ratio (that which produces the maximum reaction velocity) is a function of wind speed, as shown in Fig. 14.6. The final spread rate is calculated by weighting for discontinuity in the strata. If the fastest spread rate is through a discontinuous shrub stratum, for example, the fire spreads through the next fastest stratum when shrubs are absent, etc.

Integration of the Model's Components

The Glacier National Park Basic Resource and Fire Ecology Systems Model required an information retrieval and simulation model that could project site characteristics, cultural and hydrologic features, full quantitative species composition, successional development, community characteristics, and quantitative fuel loadings by a user's simply entering a stand's geographic coordinates; furthermore, it was desirable to calculate fire behavior parameters from this information by simply entering either the current weather or a weather forecast. Such a system requires a very high level of integration of the components described earlier in this chapter.

The key to this integration was the conversion of the environmental gradient hyperspace to a gradient matrix hyperspace. The study initially considered storing species and fuels distributions as continuous functions of the gradient hyperspace (as shown in the species nomograms in Appendix 2), but this approach was rejected because of excessive costs. Such a procedure would require a single but expensive construction of the equations (high-degree polynomials) that describe these relationships. How-

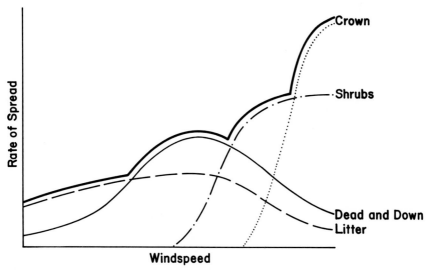

Figure 14.6. Idealized relationship between rate of fire spread and wind speed for various fuel strata.

ever, it would also be necessary to solve these equations each time the model was executed for every species and fuel property, repeated for every hectare selected for analysis. Such a process was estimated to increase operating costs by an order of magnitude. Furthermore, as discussed in Chapter 16, the storage of species and fuels in external files permits subsequent revision of this data base without making changes in the source program, job control procedures, or user procedures. Therefore the matrix method was selected.

Each complex gradient was treated as a vector composed of several individual elements. The elevation gradient was converted to a vector of 18 elements, each with a range of 101.6 m (333.33 ft). The topographic-moisture gradient was converted to another vector of 18 elements, each including two or three aspects within a single topographic type. The stand age gradient was converted to a vector of 21 elements, each with a span of 10 years. The primary succession gradient was converted to a vector of 20 elements, each including five points on the 0–99 primary succession scale. The drainage gradient was treated as a three-element vector corresponding to the three major units described in Chapter 8. The alpine wind–snow gradients, as well as the disturbance types, were treated not as vectors but as special community types that could trigger various user messages (the messages were also dependent on the location of a stand on the other gradients). Thus the five gradients:

Elevation × Topographic-moisture × Stand age
$$\times \text{ Primary succession} \times \text{Drainage}$$

may be represented as a matrix:

$$18 \times 18 \times 21 \times 20 \times 3$$

of 408 240 elements. This represents the maximum gradient resolution possible with the system—the equivalent of 408 240 community types. For practical purposes, however, the size of the matrix was significantly reduced, as shown below.

As described in Chapter 8, five stand age classes are adequate for describing forest communities and these communities, of course, occupy only one extreme of the primary succession gradient. For species composition and community characteristics in forest stands, therefore, the gradient matrix may be reduced to:

Elevation × Topographic-moisture × Stand age × Drainage

and represented in a matrix:

$$18 \times 18 \times 5 \times 3$$

of 4860 elements.

For the characteristics and species composition of nonforested communities the model requires the primary succession gradient but may drop the stand age and drainage gradients (Chaps. 8–11) and so use a matrix:

Elevation × Topographic-moisture × primary succession

which has vectors of:

18 × 18 × 20

for a total of 6480 elements.

Forest fuels, as described in Chapter 9, may be represented in a matrix of:

Stand age × Fire intensity

of size:

21 × 2

whereas nonforested fuels (Chap. 11) require only the first 18 elements of the primary succession gradient.

Once this gradient matrix space is created, it is necessary to construct the individual gradient models for each species, community characteristic, and fuel type. For all plant species (including trees), importance values stored in the matrix were on the 0–6% scale described in Chapter 6. Total herb and shrub cover was also on this scale. Species diversity measures used integers to represent the number of species expected in sample plots of specified size. Fuel loadings remained in metric tons per hectare.

Once such models are stored on a digital computer, appropriate software is required to decode the site inventory records and assign matrix indices to each block of a site inventory record. Such an assignment locates a unique element of the gradient space that is associated with that site inventory block. Displaying the contents of that element therefore associates an importance value (such as 2.1 ton/ha or 5%–25% relative density) with the appropriate community attribute (such as litter loading or *Picea* density). Additional software provides a method of varying the stand age indices and so allows display of deterministic successions for all forested stands. Further software and dictionaries display an English-language version of the coded inventory records and allow reportlike formats for all model outputs (see Chap. 16). Finally, additional software permits entry of weather parameters and computation and display of fire behavior predictions, using the ecologic attributes (such as vegetative cover, fuels, and slope steepness) retrieved above.

A more complete description of the source program and data files used to accomplish these results is given in Chapter 16 and in Kessell (1976d); a description of use and operation of the model by personnel not familiar with ADP methods is also given in Chapter 16 and in Kessell (1975b).

15

Testing the Model

An important part of the process of judging the utility and effectiveness of the gradient modeling package is quantitative testing of its various components. Because the model is a system in which inventory errors could cause errors in vegetation and fuel prediction, which in turn would affect fire behavior predictions, its four major components have been tested individually; these include site characteristics, vegetation, fuel, and fire behavior.

Site Information

The resource site inventory codes site features; the location of the stand on each gradient; cultural features; hydrologic features; maps, photographs and access data; and disturbance history from aerial photographs, topographic maps, and field information. The great bulk of the coding is done by remote methods without visiting the stand. We decided to judge site information resolution by two methods: (1) how well the system provided a physical stand description as required by managers, and (2) how well it recorded the gradient indices that are used by the model to compute vegetation and fuel, and thus fire behavior. However, Glacier's managers stated, as soon as the prototype model was developed, that the model's resolution greatly exceeded their requirements; therefore testing was conducted to see how well the site inventory records gradient indices.

Such parameters as elevation; aspect; gross landforms; vegetative cover; cultural features; hydrologic features; maps, photographs, and access information; and "other disturbances" are recorded with virtually 100% accuracy. Topographic types are more than adequate for most management purposes (bottomlands, ravines–draws, slopes, ridges, peaks); in terms of gradient parameters, some stands show topography that ranges from ravines to draws, or from sheltered slopes to open slopes, and the index parameter therefore may be in eror by one-half index point (on a six-point scale) for perhaps 10%–15% of the stands and may be in error by one index point for about 2%–3% of the stands. Stand ages are determined by first locating burn perimeters on the aerial photos and then dating the burn by: (1) fire records if available (usually available for large fires since 1910); (2) stand ages recorded in any field samples taken within the burn perimeter if method 1 is not available (ages determined in the field from increment cores); and (3) estimates from interpretation of infrared aerial photographs if methods 1 and 2 are not available. Method 1 dates the stand age to the nearest year, and method 2 gives stand age errors of 5%–10% (trivial for the purposes of the model). Method is used only when the other methods cannot be applied (perhaps 15% of the 40 000-ha site inventory used the third method).

The blocking and localization of cover and other site characteristics has been checked both in the field and from low-altitude flights in light aircraft. The method gives virtually 100% resolution on a 10-m scale and, for open alpine environments (localized Krummholz, shrub, meadow, talus, and rock cover), it can often resolve down to 2–4 m.

Vegetation

The model predicts the quantitative composition of the overstory (all species) and common understory species (about 45 species total) using the 0–6% importance value scale described in Chapter 6. The original gradient models of the overstory were derived before the 1974 field season from about 400 field samples (plus the published data of other researchers); these preliminary models were then used to predict the species composition (overstory) of 312 stands sampled during the 1974 field season. For each stand, the actual overstory species composition was compared to the predicted composition. For 284 of these stands (91%), the percent similarity between the actual and predicted values was at least 70%. The value of 70%–80% is the measured internal association value for Glacier's forest overstories. It is the percent similarity between replicate samples taken in the same stand and therefore represents the best predictions possible for small areas (that is, predict the composition of half the stand by sampling the other half). After these tests were completed, the gradient models were revised to include the results of these 312 new samples.

Similar quantitative testing of herb and shrub species composition predictions has not been completed, because the work which produced the herb and shrub gradient models was not completed until the fall of 1975 and field work on this project has not been possible since June 1976. At any time in the future new samples may be taken to compare actual and predicted species composition. Preliminary work suggests that results will be comparable to the accuracy of the overstory predictions.

Fuel Loadings

As with the testing of herb and shrub species composition, it has not yet been possible to use the gradient fuel models to test predict new stands prior to sampling. However, the standard errors for fuel loadings ranged from 20% to 30%, and the standard deviations (from natural variation in fuel distributions) were usually about equal to the mean fuel loadings

(Chaps. 9 and 11; Kessell and others 1978); the prediction errors should average only a fraction of a standard deviation (perhaps one-third). Most fuel modelers hope to achieve results where the prediction is within one standard deviation of the mean (Richard Rothermel, personal communication). At any rate, the fuel resolution of the Glacier model is such that weather inputs, rather than fuel inputs, are the limiting factors affecting fire behavior predictions (Chap. 14; Kessell and others 1978).

Fire Behavior

Despite unusually wet weather and low fire incidence during the summers of 1975 and 1976 (very frustrating when one is trying to test a fire model), the full model package using both the single-stratum and multistrata fire models was tested on three fires—the 1976 Larson fire, the 1975 Redhorn fire, and the 1974 Curley fire.

The Larson fire occured in mid-July 1976 on a south-aspect forested slope on Apgar Mountain which had also burned in 1929. The weather was showery, just before and after the fire was reported, with site relative humidities ranging from 55% to 99%; prediction therefore required the use of a range of fine fuel moistures. The fire behavior prediction method was to code the site inventory and let the model calculate vegetation, fuel loadings, and fire behavior. Results appear in Table 15.1 and show that actual behavior was roughly at the midpoint of predicted behavior ranges.

The Redhorn fire occured in late July 1975 in the upper Belly River drainage on an east-aspect mature slope. Vegetation cover was shrub, with localized forest *(Picea)* covering about 10% of the stand. Wind ranged from 0 to 5 mph (measured), fuel moisture was predicted by extrapolation using the weather model described in Chapter 14, and the gradient model calcu-

Table 15.1. Comparison of Predicted and Actual Fire Behavior for the Larson Fire, 1976.

Actual spread rate	9 cm/min
Predicted spread rate (gradient model and modified single-stratum Rothermel fire behavior model)	2–16 cm/min[a]
Predicted spread rate (gradient model and three-strata fire behavior model)[b]	6–19 cm/min[a]

[a]Range caused by uncertainties in fuel moisture.
[b]Described in Chapter 14.

Table 15.2. Comparison of Predicted and Actual Fire Behavior for the Redhorn Fire, 1975.

Actual spread rate	0–2.5 cm/min[a]
Predicted spread rate (gradient model and modified single-stratum Rothermel fire behavior model)	0–4.0 cm/min[a]
Predicted spread rate (gradient model and three-strata fire behavior model)[b]	1.0–6.0 cm/min[a]

[a]Range caused by natural variation in wind and fuel moisture.
[b]Described in Chapter 14.

lated vegetation, fuel, and fire behavior from the site inventory. Results appear in Table 15.2 and again show very good predictions.

Whereas both the Larson and Redhorn fires were small (less than 1.0 ha), the Curley fire burned 43 ha in the Divide Creek drainage from 29 July to 31 July 1974. The fire was on a steep (25°–45°) mature forested slope with east aspect (1830-m elevation), which gave way to rock, talus, and localized trees at about 2150-m elevation. Overstory was mostly *Abies lasiocarpa* and *Pinus albicaulis*. Wind was from the east at 10–20 mph, and fuels were rather dry (10-hr dead and down fuel moisture was 4%–6%). Although the model was not completed until several months after the fire, it was used as described above (site inventory was input and the model calculated vegetation, fuel, and fire behavior) using the St. Mary weather (7 km to the north at 1370-m elevation). The results appear in Table 15.3, and Fig. 15.1 compares the actual and predicted perimeters on the two runs of the fire.

Table 15.3. Comparison of Predicted and Actual Fire Behavior for the Curley Fire, 1974.[a]

	First run of fire	Second run of fire
Actual spread rate	18.4 m/min	60.9 m/min
Predicted spread rate (gradient model and modified single-stratum Rothermel fire behavior model)	16.8 m/min	38.0 m/min
Predicted spread rate (gradient model and three-strata fire behavior model)[b]	24.0 m/min	61.5 m/min

[a]See also Fig. 15.1.
[b]Described in Chapter 14.

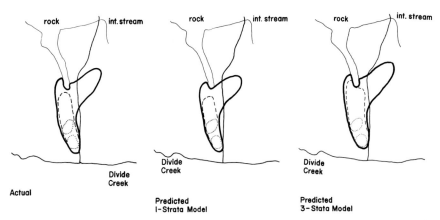

Figure 15.1 a–c. a Actual perimeters of the 1974 Curley fire in Glacier National Park. **b** Perimeter predicted by the one-stratum fire model. **c** Perimeter predicted by the three-strata fire model. Dotted lines show perimeters when the fire was first spotted, dashed lines are perimeters after the first run, and solid lines show perimeters after the second run. See Table 15.3 for quantitative data on the fire.

The results show that the single-stratum model was adequate for the first run; in fact, it gave somewhat better predictions than did the three-strata model. However, the single-stratum model was inadequate for predicting the second run (38.0 m/min predicted, compared to 60.9 m/min observed). The reason for this is that the fire, on its second run, spread through the partially cured shrub stratum (because of the increased slope and wind); the single-stratum model cannot handle this kind of behavior, but the three-strata model was designed for just this kind of circumstance (see Chap. 14).

In addition, the model was also run on 12 smoke reports called in to headquarters by fire lookouts. All of the smokes were extinguished naturally before fire crews reached them. For 11 of the smokes, the model predicted zero rate of spread. For one of them it predicted a spread rate of 0.5–3.0 cm/min (apparently because of an extrapolation error in fuel moistures).

This somewhat limited testing of the fire behavior component of the model indicates that it performs much better than anticipated. In addition, the results suggest that under low to moderate fire conditions the single-stratum model is adequate, but that the three-strata model is highly desirable for circumstances where a fire spreads through the upper strata.

16

Model Computer Program, Operation, and Utility

Source Program and File Structure

The entire gradient modeling fire and resource management package is executed by a single FORTRAN IV G1 mainline program and its 24 FORTRAN IV G1 subroutines. The source program is part of an integrated package that includes data files, Job Control Language, and Command Lists and it is operational on the Boeing Computer Services IBM 370/168 VS digital computer system located in McLean, Virginia. The package is designed for interactive operation from a remote low-speed terminal located at Glacier National Park headquarters, but it may be run in a batch environment if desired. The following discussion is abridged from the model's *Systems Manual* (Kessell 1976d).

I wrote source program during the period January 1975 to June 1976, except for subroutines FIRMOD and FOURW. FIRMOD was adapted by Jane Kapler, Glacier National Park, from the FIREMOD package developed at the Northern Forest Fire Laboratory, Missoula, Montana, by Frank Albini. Ms. Kapler also wrote the FOURW subroutine. Program logic is as follows:

MAINLINE PROGRAM

Fifteen options or specification parameters are read from the user specification input file.

Subroutine RDINVD is called to read dictionaries.

Subroutine RDINV is called to read a square kilometer site inventory file.

Subroutine RTRGRA is called to read the tree gradient matrices file if either overstory gradient analysis or overstory succession gradient analysis is requested.

Subroutine HERBR is called to read the understory cover and diversity gradient matrices file if either understory cover and diversity or understory succession cover and diversity is requested.

Subroutine RDFUEL is called to read the fuel gradient file if either fuel gradient analysis or fire behavior is requested.

Subroutine WEAIN is called to read the weather input file if fire behavior is requested.

Subroutine MPA is called to print maps, photographs, and access records for the square kilometer if requested.

The mainline program then reads the UTM coordinates for the first hectare requested for modeling. The slope override value is also read at this time if it is used. If UTM North equals 0, the program branches to stop. The mainline will branch back to this point to read new UTM coordinates for the next requested hectare after all requested options for the first hectare are completed.

Subroutine DPINV is called to locate and decode the correct inventory records for this hectare. This subroutine will call subroutines DP101, DP111, and DP121 if required. If automatic slope steepness is requested, it also calls subroutine SLOPE1.

Subroutine CAGRIN is called to compute gradient matrix indices if any gradient analyses are requested.

Subroutine TRGRA is called to compute and print the overstory gradient analysis if requested.

Subroutine REGROW is called to compute and print the overstory succession gradient analysis if requested.

Subroutine HERBC is called to compute and print the understory cover and diversity if requested.

Subroutine HERBS is called to compute and print the understory cover and diversity succession if requested.

Subroutine HERSP is called to compute and print the understory species gradient analysis if requested.

Subroutine FUGRA is called to compute the fuel gradient analysis if either fuel analysis or fire behavior is requested. It also prints the fuel gradient analysis if requested.

Subroutine FIREB is called to compute and print the fire behavior analysis if requested. It calls subroutine FIRMOD, which calls subroutine FOURW.

The mainline then branches back to read the new hectare UTM coordinates for next hectare.

A very simplified flowchart of the program is shown in Figure 16.1.

SUBROUTINES

Subroutine RDINV reads an entire square kilometer inventory partition from a permanent file. Four record formats are used, corresponding to the H-101, H-111, H-121, and H-131 inventory records (Table 14.1, Chap. 14). All variables are returned to the mainline. A welcome message is printed.

Subroutine RDINVD reads two dictionaries—the basic model dictionary and the special features name–ID code dictionary. All variables are returned to the mainline.

Subroutine MPA prints the maps, photographs, and access for the square kilometer.

Subroutine DPINV searches the square kilometer inventory matrices to find the records for the requested hectare. If a print of inventory

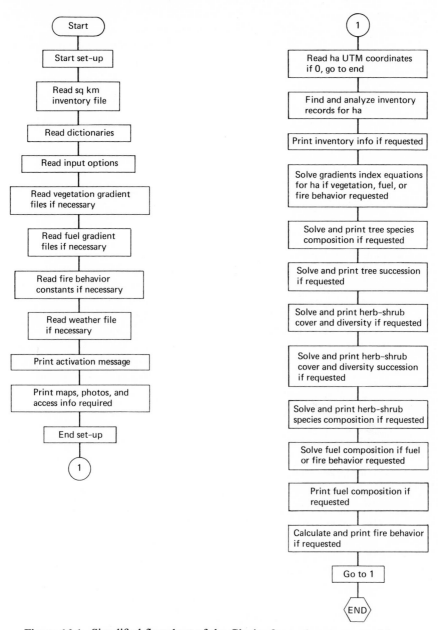

Figure 16.1. Simplified flowchart of the Glacier fire and resource model.

analysis is requested, it calls subroutine DP101, and it calls subroutines DP111 and DP121 if H-111 and/or H-121 records, respectively, are recorded for this hectare. It also calls subroutine SLOPE1 if an automatic slope steepness calculation is requested. All variables are returned to the mainline.

Subroutine DP101 matches the hectare site codes with the appropriate dictionary codes to print a text of the basic inventory analysis.

Subroutine DP111 matches the hectare site localization codes (if present) with the appropriate dictionary codes to print a text of localized cover.

Subroutine DP121 matches the hectare site special features codes (if present) with the appropriate dictionary codes to print a text of special features.

Subroutine RTRGRA reads all overstory gradient matrices from the permanent input file. All variables are returned to the mainline.

Subroutine CAGRIN utilizes the gradient site parameters for the specified hectare and converts them to the matrix indices used by all gradient analyses. Any specified site-fit indices replace the actual indices. All matrix indices are returned to the mainline.

Subroutine TRGRA matches the gradient indices with the overstory gradient matrices to compute and print the overstory gradient analysis.

Subroutine REGROW matches the gradient indices with the overstory gradient matrices and, by varying the age indices, computes and prints the overstory succession gradient analysis.

Subroutine RDFUEL reads all fuel gradient matrices from the permanent input file. All variables are returned to the mainline.

Subroutine FUGRA matches the gradient indices with the fuel gradient matrices to compute and print the fuel gradient analysis. The fuel of any localized cover is averaged into the block fuel loadings.

Subroutine WEAIN reads all user-specified weather input parameters to be used in fire behavior calculations from the user input file. All variables are returned to the mainline.

Subroutine FDICTO reads the fire behavior model constants from the permanent input file. All variables are returned to the mainline.

Subroutine FIREB sets up data arrays and serves to switch and loop for the fire behavior computation subroutine FIRMOD. If the user specifies a real fire run, the output is also written to a permanent file. This subroutine calls subroutine FIRMOD.

Subroutine FIRMOD computes fire behavior using the single-stratum Rothermel model from the fuel parameters and weather parameters in subroutines FUGRA, WEAIN, and FIREB. It calls subroutine FOURW to calculate the wind vectors.

Subroutine FOURW computes the wind vectors for four-directional fire spread from the wind and aspect parameters determined in subroutines DP101 and WEAIN. Vectors are returned to subroutine FIRMOD.

Subroutine HERBR reads all understory cover and diversity gradient matrices from the permanent input file. All variables are returned to the mainline.

Subroutine HERBC matches the gradient indices with the understory cover and diversity matrices to compute and print the gradient analysis of understory cover and diversity.

Subroutine HERBS matches the gradient indices with the understory cover and diversity matrices and, by varying the age indices, computes and prints the understory cover and diversity succession gradient analysis.

Subroutine HERSP matches the gradient indices with the understory species matrices read by this subroutine from the permanent input file to compute and print a gradient analysis of understory species composition.

Subroutine SLOPE1 searches the permanent inventory matrices to calculate the matrices of elevation and aspect used for automatic slope steepness calculation. It then calls subroutine SLOPE2 to compute slope steepness for hectare blocks.

Subroutine SLOPE2 calculates slope by blocks from the data provided by subroutine SLOPE1. The calculated slopes are passed to the mainline through subroutines SLOPE1 and DPINV.

FILES

In addition to the source program, its object code, and Job Control Language, up to 14 additional files may be used by the model for execution, and many additional files are used by the Command List interactive instructions. Detailed documentation of the model files is included in Kessell (1976d).

Operation of the Model

The model was designed for use by personnel with absolutely no automatic data processing training. Every effort was made to make use of the model simple, straightforward, and virtually foolproof. This was accomplished by designing a system in which the user saw the model only through the various "canned" Command Lists (CLISTs). CLISTs permit the user to answer various questions, specify options, enter site coordinates, and so forth, using simple English-language commands; the user never need see the FORTRAN or Job Control Language (JCL) involved, as the latter is automatically generated and edited by the CLISTs.

The model uses the acronym "BURN," which is used for the following basic CLISTs:

HELPBURN	This command provides a brief introduction to procedures for running the model.
BURN	This command provides detailed directions for executing command RUNBURN.
BURNAIDS	This command briefly describes each model CLIST command available.
WEATHER	This command allows retrieval and printing of the last 1 or 2 days' AFFIRMS weather data and/or remote weather station data.
RUNBURN	This command executes the model. It provides selection of areas, options, and weather data; permits assignment of priority and run time; and automatically logs the run. (A simplified flowchart is shown in Fig. 16.2.)
GETBURN	This command retrieves and prints the results of the model run and provides an option to cancel the job.
SAVEBURN	This command copies the model output to a permanent file.

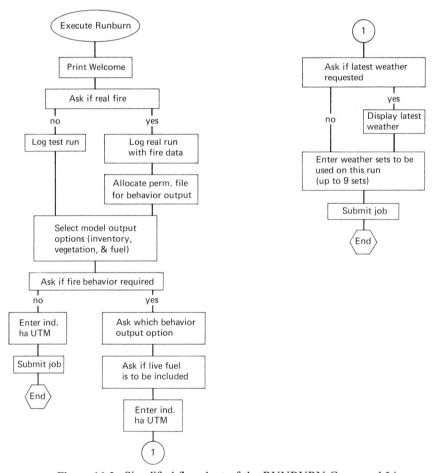

Figure 16.2. Simplified flowchart of the RUNBURN Command List.

Other CLISTs are described in the system's User's Manual (Kessell 1975b).

Output from the Model

As noted above, the model is designed to display the results of a run on the user's low-speed terminal. However, if the output is not needed immediately, it may be routed to a high-speed printer and mailed to the park by giving a single one-line command.

Regardless of where the output was to be displayed, it was deemed very important to present output that was easily interpreted by personnel not trained in computer use (the same persons who run the model). Therefore I made a major effort to use somewhat detailed and elaborate formatting so the results are either an English-language text or a simple table. Graphic display of fire behavior would be very desirable, but it was not available for this study. Sample output for a single unblocked hectare requesting all model options is shown in Fig. 16.3; a blocked hectare will repeat each analysis for each hectare block. Much more extensive sample output is available from the Superintendent, Glacier National Park; the National Park Service Division of Systems Design (Washington); and Gradient Modeling, Inc.

Welcome to the Glacier National Park Basic Resource and Fire Ecology Systems Model. The model has been activated for the square kilometer located at UTM 5394. North, 288. East.

Maps, photo and access information for kilometer at UTM 5394. North. 288. East —
KM is located on 7.5 Min topo map(s) 20 Camas Ridge East.
KM is located on GNP Vegetation Map(s) 20S Camas Ridge East (South).
KM is located on BLACK AND WHITE PHOTOS 18—A-2,3.
KM is located on infrared photo(s) 720605NW, 720606NW, 720701NW.
Ground access information to the kilometer is as follows — kilometer is 3.4KM directly north of the Lake McDonald Ranger Station. It is 2.4 KM from the McDonald Creek Trail.

Inventory analysis for Hectare 5394.0 North, 288.1 East —
This hectare has uniform site conditions, and is therefore not blocked into subunits.
This stand has an elevation of 1640. meters, which is equal to 5380. feet (MSL).
It is an open slope, with south aspect.
The slope steepness is 27 degrees.
Its vegetative cover type is typical (upright) forest, and its primary succession (soil development condition) is Code 90 transition from Krummholz (stunted) forest (or shrub) to typical forest.
The stand is mature.
No other distrubances are recorded.
The stand is not affected by the alpine wind-snow gradient.
No special branch record is recorded for this stand.

Figure 16.3. Sample model output for a single unblocked hectare. The run is for a forested stand (1 ha) located on Mount Stanton just north of Lake McDonald. Fire behavior predictions are based on typical early August weather; Weather Set 1 is the driest afternoon weather, and Weather Set 2 shows projected overnight weather (slightly wetter fuel because of some humidity recovery). The results were obtained within 5 min of the time the user requested the run. The cost of generating this sample run was less than $5; additional hectares would have cost about 80¢ each.

No gradient adjustment (site fit) parameters have been recorded for this stand.
Finally, the stand is located in the McDonald drainage.

This stand includes localized cover types. These include —
Meadow
It has a primary succession code of 70, which is mixed meadow and Krummholz (stunted) forest (or shrub).
Its contagion value (variance / mean cover ratio on a 10 x 10 meter grid) is 3.0.
This stand has no recorded special features.

Gradient analysis of tree species for Hectare 5394.0 North, 288.1 East —
(Species with a relative density of less than 1 percent are not printed).
Hectare is not blocked, and has a primary cover type of typical (upright) forest.
Tree species in this cover type include —

Abies Lasiocarpa (Subalpine Fir)	With a relative density of 25 to 50 percent
Picea Hyrbrid Complex (Spruce)	With a relative density of 5 to 25 percent
Pseudotsuga Menziesii (Douglas-Fir)	With a relative density of 1 to 5 percent
Pinus Albicaulis X Monticola Complex (White Pine)	With a relative density of 25 to 50 percent

Localized cover types include —
Localized meadow
Trees are not present in this localized cover type

Secondary (Fire) Succession Model of the canopy for Hectare 5394.0 North, 288.1 East —
This portion of the gradient model predicts the successional development of the stand if it is burned by a fire which destroys at least 50 percent of the crown (if a forest) or top stratum (if shrub or herb cover.)

These predictions are —
Tree composition 1 to 20 years after burn:
Cover will be primarily meadow, with trees invading from 1 to 8 years after the fire.

Tree composition 20 to 50 years after burn:

Abies Lasiocarpa (Subalpine Fir)	With a relative density of 50 to 75 percent
Picea Hybrid Complex (Spruce)	With a relative density of 5 to 25 percent
Pseudotsuga Menziesii (Douglas-Fir)	With a relative density of 5 to 25 percent
Pinus Albicaulis X Monticola Complex (White Pine)	With a relative density of 1 to 5 percent
Pinus Contorta (Lodgepole Pine)	With a relative density of 1 to 5 percent
Larix Occidentalis (Western Larch)	With a relative density of 1 to 5 percent
Populus Tremuloides (Quaking Aspen — Localized)	With a relative density of 5 to 25 percent

Tree composition 50 to 90 years after burn:

Abies Lasiocarpa (Subalpine Fir)	With a relative density of 25 to 50 percent
Picea Hybrid Complex (Spruce)	With a relative density of 1 to 5 percent
Pseudotsuga Menziesii (Douglas-Fir)	With a relative density of 25 to 50 percent
Pinus Albicaulis X Monticola Complex (White Pine)	With a relative density of 1 to 5 percent
Pinus Contorta (Lodgepole Pine)	With a relative density of 25 to 50 percent
Larix Occidentalis (Western Larch)	With a relative density of 1 to 5 percent

Tree composition 90 to 150 years after burn:

Abies Lasiocarpa (Subalpine Fir)	With a relative density of 25 to 50 percent
Picea Hybrid Complex (Spruce)	With a relative density of 5 to 25 percent
Pseudotsuga Menziesii (Douglas-Fir)	With a relative density of 5 to 25 percent
Pinus Albicaulis X Monticola Complex (White Pine)	With a relative density of 5 to 25 percent
Pinus Contorta (Lodgepole Pine)	With a relative density of 5 to 25 percent

Tree composition 150 to 300 years after burn:

Abies Lasiocarpa (Subalpine Fir)	With a relative density of 25 to 50 percent
Picea Hybrid Complex (Spruce)	With a relative density of 5 to 25 percent
Pseudotsuga Menziesii (Douglas-Fir)	With a relative density of 1 to 5 percent
Pinus Albicaulis X Monticola Complex (White Pine)	With a relative density of 25 to 50 percent

Gradient analysis of herb & shrub cover & diversity for Hectare 5394.0 North, 288.1 East —
Herb and Shrub cover — 75 to 95 percent
Average Vascular Plant Species diversity in a pair of random 2 x 2 meter quadrats — 8 species.

Secondary (Fire) Succession Model of the understory for Hectare 5394.0 North, 288.1 East —
This portion of the gradient model predicts the herb & shrub cover & diversity successional development of the stand after a ground fire.
These predictions are —

YEARS AFTER BURN	COVER (PERCENT)	DIVERSITY (#SPECIES IN 2 2X 2 METER QUADRATS)
1 — 20	50 — 75	8
20 — 50	75 — 95	8
50 — 90	50 — 75	7
90 — 150	75 — 95	9
150 — 300	75 — 95	8

Figure 16.3. *(Continued).*

Gradient analysis of common herb and shrub species for Hectare 5394.0 North. 288.1 East —

SPECIES	ABSOLUTE COVER (%)
Xerophyllum tenax	5 — 25
Thalictrum occidentale	1 — 5
Spirea betulifolia	1 — 5
Pachistima myrsinites	1 — 5
Menziesia ferruginea	5 — 25
Vaccinium sp.	1 — 5
Arnica sp.	5 — 25
Sorbus scopulina	5 — 25

Gradient analysis of fuel loadings for Hectare 5394.0 North. 288.1 East —
The calculated average loadings for each block include the effects of all localized cover types.

Litter	Average loading — 13.00 metric tons per hectare.
Grass and forbs	Average loading — 6.13 metric tons per hectare.
1 hour dead and down	Average loading — 0.61 metric tons per hectare.
10 hour dead and down	Average loading — 3.07 metric tons per hectare.
100 hour dead and down	Average loading — 3.10 metric tons per hectare.
Greater than 100 hour dead and down	Average loading — 30.75 metric tons per hectare.

Fire behavior predictions for Hectare 5394.0 North. 288.1 East —
The model uses the Rothermel Behavior Model, excludes live shrubs. and assumes uniform spatial distribution of fuels.
Site slope — 27. degrees, tangent — 0.51 thus percent slope — 51.

Behavior predictions using Weather Set 1

WIND PARAMETERS							FUEL MOISTURE PARAMETERS				
Direction	Minimum		Maximum		Increment		G F	Litter	1 Hour	10 Hour	100 Hour 100 Hour
	KPH	MPH	KPH	MPH	KPH	MPH	200.00	6.00	4.00	6.00	10.00 10.00
240.00	0.0	0.0	32.2	20.0	16.1	10.0					

Date: 751021, Time/Run: 2304

Fire Behavior Predictions. Block 1

Wind at Midflame Height (KPH — MPH — M/Min)			Direction of Spread		Rate of Spread (Meters/Min)	Flame Length (Meters)	Byrams Intensity (KGCAL/Min/Fireline M)
0.0	0.0	0.0	Upslope	(North)	2.24	1.36	7283.15
			Across	(East)	0.38	0.61	1249.78
			Downslope	(South)	0.0	0.0	0.0
			Across	(West)	0.38	0.61	1249.78
16.1	10.0	268.17	Upslope	(North)	6.82	2.27	22191.55
			Across	(East)	13.79	3.14	44855.40
			Downslope	(South)	0.0	0.0	0.0
			Across	(West)	0.0	0.0	0.0
32.2	20.0	536.34	Upslope	(North)	19.99	3.73	65045.34
			Across	(East)	52.31	5.80	170198.12
			Downslope	(South)	0.0	0.0	0.0
			Across	(West)	0.0	0.0	0.0

Output complete for Subroutine Firmod. 1-Blk HA

Behavior predictions using Weather Set 2

WIND PARAMETERS							FUEL MOISTURE PARAMETERS				
Directions	Minimum		Maximum		Increment		G F	Litter	1 Hour	10 Hour	100 Hour 100 Hour
	KPH — MPH		KPH — MPH		KPH — MPH		200.00	8.00	6.00	7.00	10.00 10.00
240.00	0.0	0.0	32.2	20.0	16.1	10.0					

Date: 751021, Time/Run: 2304

Fire Behavior Predictions. Block 1

Wind at Midflame Height (KPH — MPH — M/Min)			Direction Of Spread		Rate of Spreak (Meters/Min)	Flame Length (Meters)	Byrams Intensity (KGCAL/Min/Fireline M)
0.0	0.0	0.0	Upslope	(North)	2.13	1.31	6664.79
			Across	(East)	0.37	0.58	1143.67
			Downslope	(South)	0.0	0.0	0.0
			Across	(West)	0.37	0.58	1143.67
16.1	10.0	268.17	Upslope	(North)	6.50	2.18	20307.42
			Across	(East)	13.14	3.02	41047.02
			Downslope	(South)	0.0	0.0	0.0
			Across	(West)	0.0	0.0	0.0
32.2	20.0	536.34	Upslope	(North)	19.06	3.58	59522.78
			Across	(East)	49.88	5.57	155747.81
			Downslope	(South)	0.0	0.0	0.0
			Across	(West)	0.0	0.0	0.0

Output complete for Subroutine Firmod. 1-Blk Ha
End of Output

Figure 16.3. (*Continued*).

SECTION IV

SUBSEQUENT
APPLICATIONS
OF GRADIENT MODELING
SYSTEMS

17

Gradient Models and Multiple Resolution Levels

The completion and implementation of the gradient modeling system in Glacier National Park in 1976 demonstrated several things. First, a gradient analysis, as opposed to a classification, approach can produce a sound resource management information system. Gradient analysis had not previously been used for management purposes on this scale.

Second, the concept of gradient modeling, with its linkage of vegetation and fuels models with a remote site inventory and appropriate computer software, is a viable and useful method for constructing resource management information systems. The Glacier system has capabilities still not offered by any other resource information system.

Third, the results in Glacier Park suggested numerous other geographic areas and resource management problems where similar concepts might be employed.

However, several researchers and managers—even including some who understood gradient modeling—were critical of the Glacier Park model. Their criticisms usually centered on two points: the system's extremely high resolution ("unnecessary"), and the system's cost compared to available low-resolution but less effective methods ("too expensive"). These critics assumed that gradient modeling is a viable method only when a very accurate, 10-m resolution system is wanted. My response was that Glacier Park got a very accurate, 10-m resolution system solely because Glacier Park wanted a very accurate, 10-m resolution system; I believed that gradient modeling would still be a viable and cost-effective method at lower resolution levels. I soon had an opportunity to test that belief, as the U.S. Department of Agriculture Forest Service wanted both a general fire management information integration system for coniferous forest ecosystem, and a site-specific, variable resolution (100–1000 m) terrestrial information system for southern California chaparral wildlands. Before I descibe those projects, however, let us take another brief look at the accuracy and resolution requirements posed by different resource management applications and needs.

Management Information Systems

As noted in Chapter 2, different resource management problems and different land uses pose different accuracy and resolution requirements. Additional constraints are imposed by fiscal considerations; perhaps a manager would like 25-m resolution, but would 100-m resolution be good enough if that choice allowed modeling a much larger area? To answer this question the manager must know just what information and accuracy is lost if the resolution is dropped to 100 m (or 500 m, or 1000 m) and until recently researchers have been unable to answer that question. Fortunately, the completion of the two projects noted in the first section of this chapter brought us a step closer to the answer.

Table 17.1 highlights six operational models and systems that use gradient modeling techniques[1]; each was developed and implemented during the period of mid-1975 through early 1978 by Gradient Modeling, Inc., my nonprofit research foundation. As we look at these systems, note not only how each builds upon the capabilities of the preceeding models, but especially how we have attempted to generalize and apply our findings to broader problems and/or geographic areas.

Fire Behavior Information Integration System

Work on the *Fire Behavior Information Integration System (FBIIS)* for southern California chaparral was initiated in October 1976 and completed in March 1977. Its purpose was to provide a site-specific, general terrestrial (and especially fuels) inventory for a 140-km^2 prototype area located primarily in the Angeles National Forest. The area is predominantly a mosaic of chaparral, grass, and oak (*Quercus* spp.) enclaves, with forest intrusions at the higher elevations. We used the basic gradient modeling methodology developed in Glacier Park, but on a variable (rather than fixed) resolution level of 50–1000 m. Full details are provided in Kessell and others (1977a) and Kessell and Cattelino (1978).

Because a major purpose of the study was to compare the accuracy lost (and money saved) at lower resolution levels, we designed a gradient framework to describe communities and fuels over this entire resolution range. We chose a very simple two-gradient system: a vegetative cover gradient analogous to Glacier's primary succession gradient, which arranged major cover types in a nearly continuous sequence, and a stand age gradient. Existing vegetation and fuels data were then stratified on these two gradients, and an appropriate remote site inventory (coded from aerial photographs and fire history maps) was developed and coded for the 140-km^2 area.

The basic *FBIIS* used 1-ha area (100 m linear) resolution but recorded secondary cover types down to 0.25 ha (approximately 50 m). The computer software not only allows site by site data retrieval and inference, as does the Glacier Park model, but also provides statistical summary capabilities over large areas.

---

[1] Several other systems are under development in the United States, Canada, and Australia, but are not yet operationally available to managers.

Table 17.1. Operational Models and Systems that use Gradient Modeling Techniques.

Name of model	What it does for the resource manager
Glacier Park model *(BURN)*	Detailed terrestrial inventory; predicts fire behavior and fire effects on plant succession
Fire Behavior Information Integration System *(FBIIS)*	Terrestrial inventory and fuels data base for southern California
Fire Management Information Integration System *(FMIIS)*	Provides accuracy estimates of fire behavior models based on: (1) resolution, (2) inference method, (3) fire model inputs' accuracy
GANDALF	Fire management system designed to simulate scheduled activities and fire impacts on lodgepole pine communities through time
TAROT	Fire hazard evaluation system that helps managers evaluate potential effects of specific fuel complexes on resource management objectives; integrates the combined effects of fuel buildups and weather
*FOR*est *P*lanning *LAN*guage and Simulator *(FORPLAN)*	Incorporates all of the above, plus other components, into a unified system; simulates fire effects on plants, animals, and fuels; simulates fire behavior, danger, and hazard; simulates other management activities; is programmed by common English words and phrases

Once the 1-ha system was developed, we recoded portions of the area on new resolution levels of 4 ha (200 m), 11.1 ha (333 m), and 100 ha (1000 m). This allowed us to compare, for each resolution level, the accuracy of vegetation inference, fuels inference, and costs. The results (Kessell and Cattelino 1978) showed that, in this rather uniform vegetation type, the lower resolution levels were suited for many resource management purposes; it further determined development and implementation costs for each resolution level (Table 3.3, Chap. 3).

Fire Management Information Integration System

Near the completion of the southern California *FBIIS* in early 1977, we entered into a cooperative agreement with the Forest Service's Northern Forest Fire Laboratory, Missoula, Montana, to apply some of the techniques developed in Glacier Park and southern California to the evaluation of the *Fire Management Information Integration Systems (FMIIS)* for coniferous forest ecosystems. Our first problem was to define an *FMIIS;* I did so in the following way (Kessell and others 1977c):

> Fire occurrence, behavior, and effects, and thus fire's impact on resource outputs, are estimated, predicted, or simulated by combining and integrating various data bases, models, and other inference techniques. These data, models, and inference methods may be simple or sophisticated, crude or refined, computer-based or "seat of the pants," accurate or not, and appropriate or not. All of these systems which link data, models, and inference techniques for the purposes of simulating fire behavior and consequences are hereby grouped into the general class of *Fire Management Information Integration Systems (FMIIS).*

Next we reviewed and evaluated the state of the art for each major component of an *FMIIS,* including land inventories (terrain, vegetation, and fuels), fire occurrence information, fire weather information, fire effects knowledge, fire behavior models, fire effects models, and fire risk information. Our purpose was:

1) To determine what was available
2) To determine weaknesses, limiting factors, and missing links
3) For some components, to attempt to improve on current state of the art

Let me highlight some of these activities (based on Kessell and others 1977c). Review of existing land inventories showed the tremendous range available—from "seat of the pants" to 10-m resolution! What the review did not show was a determination of what resolution would be needed for different purposes or what would be lost at successively lower resolution levels. A major part of our work, therefore, was the development of land inventories (using gradient models and inference techniques) for successively lower resolution levels (10 m to 1000 m); we then used these inferences of fuels, at each resolution level, to predict fire behavior (using state of the art fire behavior models). Results (for primarily lodgepole pine communities in western Montana) showed that although *mean* fire behavior can be adequately predicted with 10- to 25-ha resolution levels, *extreme* (95th percentile high and low behavior) fire behavior prediction requires at least 4-ha resolution.

Our weather review led to the application of conditional forecasting as a fire management tool; that work, highlighted in Chapter 14, complements the microclimate extrapolation work we had completed in Glacier Park and is detailed in Kessell and others (1977c). The fire modeling work included the development of elaborations on the basic fire behavior model and the development of an extensive sensitivity analysis of the fire model; the results are reported in Kessell and others (1978). We found fire occurrence information and predictive models lacking; that has since become an area of active work at the Northern Forest Fire Laboratory.

The greatest need demonstrated by the evaluation of *Fire Management Information Integration Systems* is the ability to simulate fire effects on ecosystems in general and resource outputs in particular. We met this need by developing two specific simulators, *GANDALF* and *TAROT,* aimed at two particular problems—fire effects on stand (timber) yield, and the integration of fuels and weather information to predict fire hazard. These two systems are highlighted in the next section of this chapter. More importantly, we initiated the development of *FORPLAN,* the *FOR*est *P*lanning *LAN*guage and Simulator, as a new resource information system and disturbance simulator; *FORPLAN* has significantly advanced the state of the art and is described below, in Chapter 19, and by Potter and others (1979).

GANDALF AND TAROT

Every family, every corporation, has its loser, and I am about to describe ours—*GANDALF.* Perhaps that is too critical—*GANDALF* was completed on schedule, and on budget, and did exactly what it was intended to do; therefore it could be considered a success. The "loser" stigma is that in early 1977, we thought that *GANDALF* would be the pioneer of a new line of simulators; instead, cost, capabilities, and computer constraints made it a "dead end."

GANDALF was developed to build upon the excellent stand prognosis work of Stage (1973) for lodgepole pine communities by adding new management (especially fire) capabilities. It is a forest stand simulator that projects the effects of timber harvest activity and fire on the stand's stocking, mortality, natural fuels, activity fuels (fuels generated from harvest or thinning), understory vegetation, and fire hazard. *GANDALF* accomplishes this by linking:

1) A stand simulator [derived from Stage (1973)]
2) A stand cruising simulator
3) A harvest and thinning simulator
4) A fuels—fire behavior—fire impacts simulator

Two versions of *GANDALF* were developed. The first deals with the forest, on a tree by tree basis, and the forest floor, on a cell by cell basis; the second uses distributions of trees, fuels, etc. (Kessell and others 1977b). Although *GANDALF* meets its development requirements, it is a difficult and expensive type of system to develop, has very large computer requirements, and is limited to lodgepole pine communities; its completion showed that our original intent to build a separate simulator for each major community type was ill conceived. We therefore abandoned *GANDALF* as a valuable lesson in systems design and humility and looked toward *FORPLAN* as our general system. However, in the meantime, we had developed *TAROT,* a simple yet elegant fire hazard simulator.

The fire hazard evaluation system *TAROT* is based on the tenet that fire hazard should be interpreted not in terms of the fuel array characteristics per se, but in terms of potential fire behavior as presented by that specific fuel array under varying weather conditions (Kessell and others 1978). This type of approach answers questions about fire hazard that knowledge of fuel loads alone cannot address.

The fire hazard evaluation design consists of two parallel processes. The first process converts historical fire weather records into cumulative distributions of temperature, wind speed, and the four size classes of dead and down branchwood fuel moistures. The second process converts raw fuel data into fuel loads, percent vegetation cover, and packing ratios within randomly sampled fuel plots. These results are input to *TAROT*'s fire behavior model, which then computes fire behavior (rate of spread, intensity, flame length, scorch height, and ignition height) for each fuel sample at each successive fifth percentile fuel moisture level.

The results show cumulative percentiles of fire behavior for each moisture level; alternatively, cumulative percentiles of mean or extreme behavior may be observed across the weather spectrum. The user can determine, for example, that fire under 50th percentile weather will destroy only 10% of the forest (or alternatively, that any given stand has a probability of only 0.1 of being destroyed), but that under 95th percentile weather there is a 0.9 probability overstory trees will be destroyed. Another use is to predict the number of "risk days" (the days when a fire would destroy the overstory) and observe changes in this fire hazard or risk from fuel buildup. For example, a certain 100-year-old forest has only 6 risk days during the average fire season, but after an additional 50 years fuel buildup the number of risk days increases to 18. This gives the manager a concrete, easy to interpret measure of increased fire hazard from fuel buildup. More detailed information is provided in Kessell and others (1978).

Other Related Work

Concurrent with the development of the systems and simulators just described, we (Gradient Modeling, Inc., personnel) have been involved in several other related research projects. Some are outlined below to show the kinds of considerations necessary for the development of integrated resource management systems.

A major concern and interest continues to be the evaluation of community stratification techniques. Many of the advantages of gradient modeling methods have been described; however, for some applications, we have been content to use traditional classification methods. Our ongoing research on the appropriate uses of the various techniques centers on two areas. The first is the quantitative evaluation of ordination techniques using real field, rather than simulated, data from Glacier National Park, Mount

Rainier National Park (in cooperation with Dr. Jerry Franklin, Principal Plant Ecologist, U. S. Forest Service, Pacific Northwest Forest and Range Experiment Station), and Kosciusko National Park, NSW, Australia. The second is the side by side construction and comparison of both a habitat type classification and a gradient modeling system from the same vegetation data bases. These projects will be completed in 1980.

We are especially interested in the elaboration and evaluation of fire behavior models, as not only do they constitute the heart of many of our systems but they also simply have not received much field evaluation. In cooperation with the Canadian Forest Service, Northern Forest Research Centre, Edmonton, Alberta, we are testing some of the fire models on prescribed (research) fires and hope to expand the multistrata fire model to include forest crowns. Similar work comparing fire models in Australia is being conducted in cooperation with Dr. Malcolm Gill, Div. of Plant Industry, Commonwealth Scientific and Industrial Research Organization, Canberra, A.C.T., Australia.

Another interest is the improvement of computer software packages available to resource managers. Our current work includes the development of broad integrated systems, such as *FORPLAN;* the development of software for minicomputers; and a recently completed project that has developed fire behavior and fire danger modeling software for the Texas Instruments TI-59 programmable pocket calculator.[1]

Our main effort for the past 2 years has been the integration of inventory systems, fire behavior simulators, fire effects simulators, and other resource management simulators for land management planning. This led to the development of *FORPLAN. FORPLAN* (and the related system, *PREPLAN*) integrate many of the capabilities described in earlier chapters, plus several new components, into single, easy to use systems. Before I describe *FORPLAN,* however, let me briefly review some recent and exciting developments in simulating vertebrate habitat utilization and disturbance effects on vertebrates.

---

[1]Conducted under cooperative agreement with the U.S. Forest Service, Intermountain Forest and Range Experiment Station, Research Work Unit 2107.

18

Models of Vertebrate Habitat Utilization

As noted in Chapter 3, Singer (1975a, b) used a modification of our gradient scheme in Glacier National Park to model the habitat distributions of common ungulate species. During subsequent work, we attempted to improve upon this scheme by expressing both plant species' abundance and their relative palatability to different animals as gradient functions (Kessell and others 1977b). In the meantime, Thomas and others (1977) have completed an extensive study of vertebrate habitat utilization in the Blue Mountains of Oregon for all terrestrial vertebrate species; their methods have since been applied in Montana, California, and several other areas; the results of the Montana and Blue Mountains work have been incorporated into *FORPLAN*. This chapter reviews both Singer's work in Glacier National Park and the Thomas and others research in the Blue Mountains.

Gradient Models of Ungulates

During the period 1973–1975, Singer (1975a) conducted an extensive study on ungulate and fire relationships in the North Fork drainage of Glacier National Park. Although he related the distribution of species and their seasonal ranges to vegetation and habitat classification units rather than to gradients, it soon became apparent that: (1) Many of his classification units in fact represented a series approaching one or more environmental gradient; and (2) many of his classification units could be precisely defined within the environmental gradient hyperspace.

Singer then initiated a study of mountain goats *(Oreamnos americanus),* elk *(Cervus canadensis),* mule deer *(Odocoileus hemionus),* and moose *(Alces alces)* in relation to the proposed reconstruction of U.S. Highway 2 where it runs through Glacier Park in the Middle Fork drainage (Singer 1975b). In his analysis of winter ranges for these ungulates, he chose the gradient nomogram approach; he used three of the gradients defined by our vegetation model (elevation, topographic-moisture, and primary succession) and two new gradients (snow depth and slope steepness). Time since burn was not used because most of his study area was burned by a single large fire in 1910.

Figures 18.1–18.3 show gradient nomograms for mule deer, elk, moose, and mountain goat winter ranges on the elevation versus topographic-moisture and primary succession versus topographic-moisture gradients. The former include all primary succession types; the isodens connect areas of equal relative density, where relative density is defined as the percentage of all ungulate observations. The nomograms clearly differentiate the winter ranges of the four species.

Moose are restricted to the bottomland forests below 1600 m elevation. Mountain goats predominate in the very early (rocks, ledges, and permanent snowfield) primary successions on the 1250–1800 m slopes, with much lower densities in the middle primary succession (talus, meadow, and shrub) communities. Mule deer show a bimodal distribution, preferring the 1500–1800 m elevation draws with shrub, meadow, or forest cover and the 1750–2100 m elevation south-aspect open slopes and ridges with meadow, shrub, or forest cover. Elk also show a bimodal distribution and are predominant on the low-elevation forested bottomlands, draws, and sheltered slopes and on the midelevation, south-aspect open slopes and ridges with meadow, shrub, or forest cover.

Singer also found good habitat differentiation on the snow depth and slope steepness gradients, as shown in Figs. 18.4–18.6. On the unforested open slopes, mountain goats prefer the 40°–60° slopes with minimal snow depth. Mule deer show peak density at the middle snow depths (70–90 cm) on gentle slopes (25°) but also utilize the low-snow habitats on either side of the steepness range dominated by the mountain goats. Elk also show a bimodal distribution, preferring the 20°–30° slopes with 30–50 cm of snow and exhibiting another mode on the 40°–45° slopes with high snow depths (100–110 cm).

Somewhat similar distributions are observed for the unforested peaks and ridges. Mountain goats predominate the 40°–50° slopes with low snow depth, mule deer utilize slightly gentler slopes with fairly low snow depth, and elk use both the 40° slopes with high snow depths and the gentler slopes with low to moderate snow depth.

It would be very interesting to compare the distribution of winter ranges on these gradients to the distribution of browse species, but because Singer's Middle Fork study area was not part of the vegetation gradient analysis study area, this is not yet possible.

Vertebrate Habitat Utilization Model

Thomas and others (1977) recognized timber management as the dominant land management activity of the Blue Mountains Province of northeastern Oregon and southeastern Washington. They note that timber management dramatically affects wildlife habitat, affects large areas, and is well funded, whereas wildlife habitat management has little financing, affects small areas, and has little overall impact. They concluded that broad-scale wildlife goals must be accomplished through timber management goals and developed a system to determine the wildlife effects of timber and other management activities.

Thomas first grouped the 327 resident terrestrial vertebrate species of

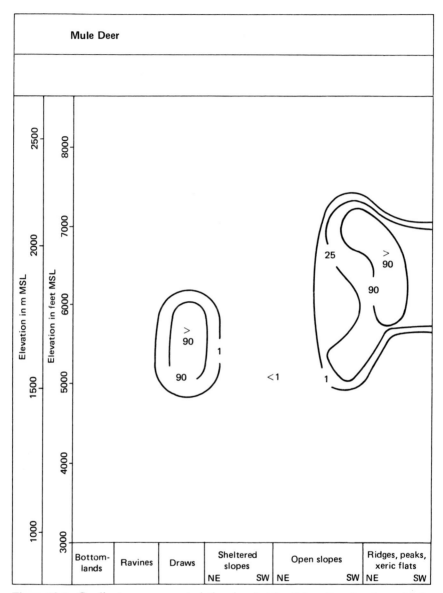

Figure 18.1. Gradient nomograms (relative density) for *Odocoileus hemionus* (mule deer) on the elevation, topographic-moisture, and primary succession gradients. (Singer 1975b).

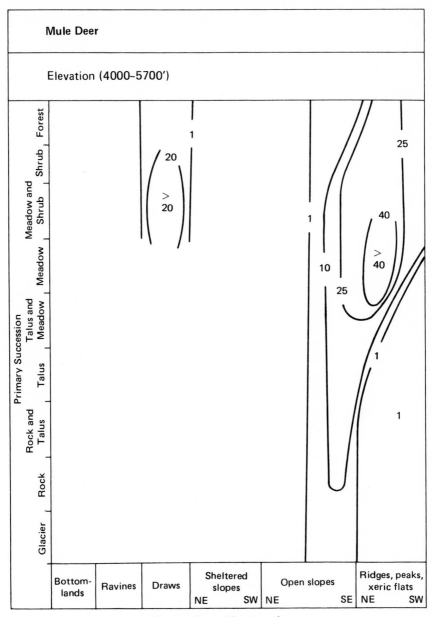

Figure 18.1. *(Continued).*

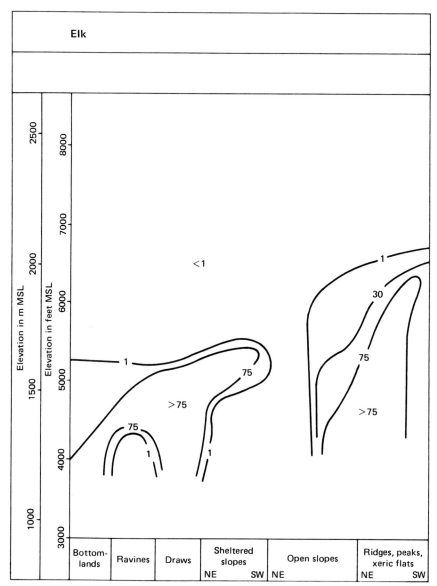

Figure 18.2. Gradient nomograms (relative density) for *Cervus canadensis* (elk) on the elevation, topographic-moisture, and primary succession gradients. (Singer 1975b).

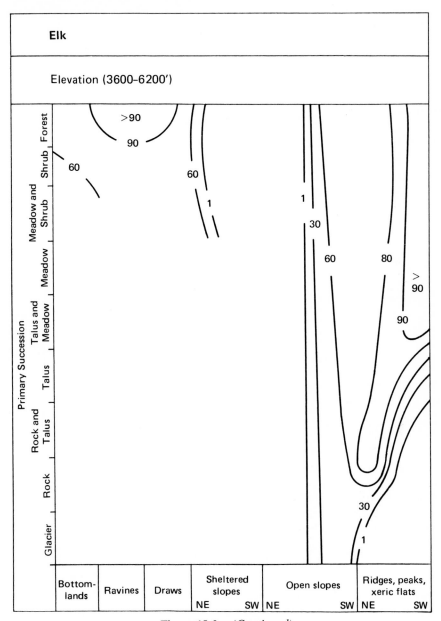

Figure 18.2. *(Continued).*

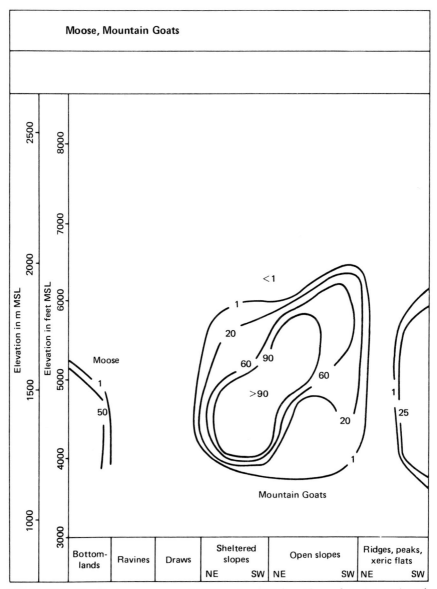

Figure 18.3. Gradient nomograms (relative density) for *Alces alces* (moose) and *Oreamnos americanus* (mountain goats) on the elevation, topographic-moisture, and primary succession gradients. (Singer 1975b)

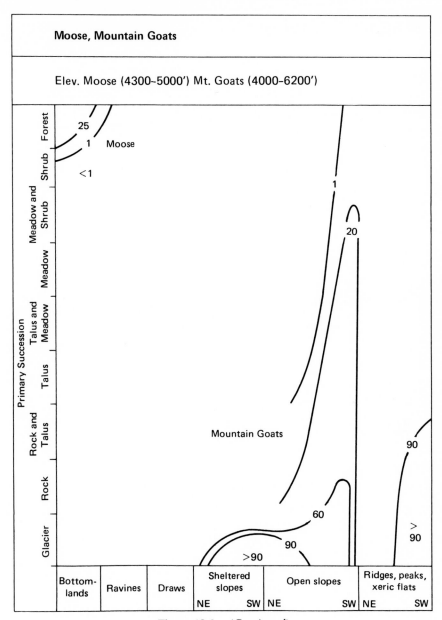

Figure 18.3. *(Continued).*

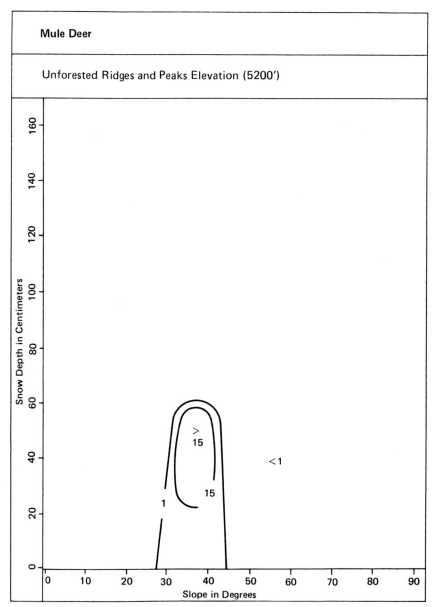

Figure 18.4. Gradient nomograms (relative density) for *Odocoileus hemionus* (mule deer) on the snow depth and slope steepness gradients. (Singer 1975b)

Mule Deer

Unforested Open Slopes Elevation (4000–5700′)

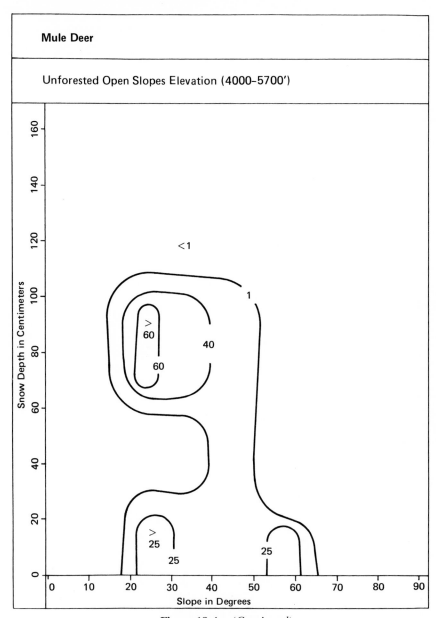

Figure 18.4. *(Continued).*

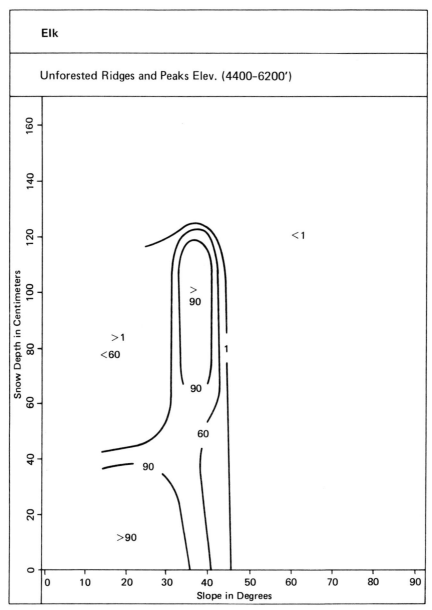

Figure 18.5. Gradient nomograms (relative density) for *Cervus canadensis* (elk) on the snow depth and slope steepness gradients. (Singer 1975b)

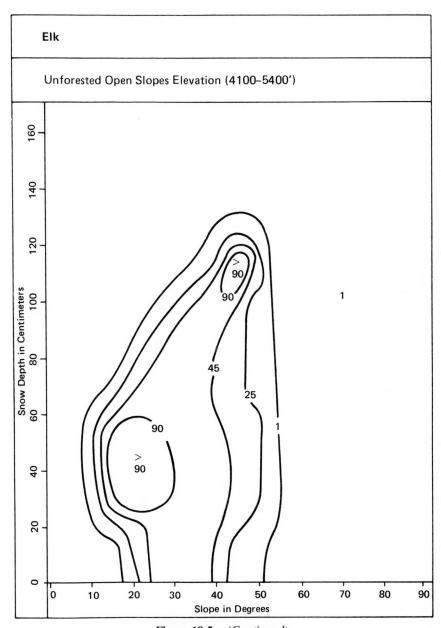

Figure 18.5.   *(Continued).*

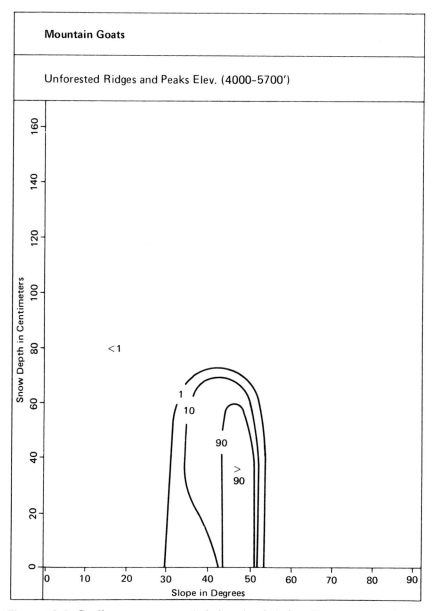

Figure 18.6. Gradient nomograms (relative density) for *Oreamnos americanus* (mountain goats) on the snow depth and slope steepness gradients. (Singer 1975b)

the Blue Mountains into 16 "life forms", which group species with similar habitat requirements for feeding and reproduction (Table 18.1). He next stratified the area into 15 major community types and recognized (up to) six seral stages for each community type.

Next, for each vertebrate species, Thomas and others determined which

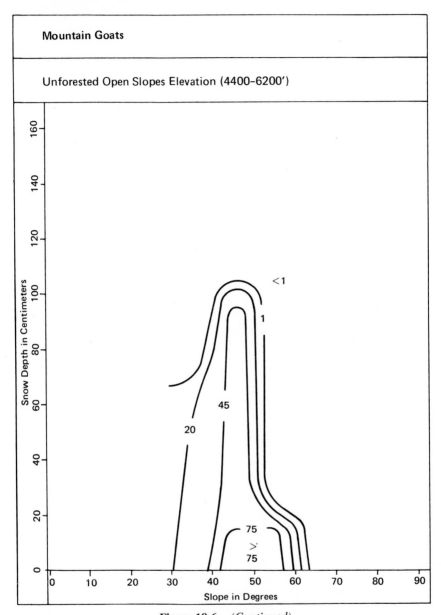

Figure 18.6. *(Continued).*

of the 15 community types could be utilized for feeding and which for reproduction. An example is shown in Table 18.2 for life form No. 13. The species in this life form can utilize seven of the 15 community types; differences among species are shown in the table. They next determined, for each species, which successional stages of the community could be utilized for feeding and reproduction. The example for life form No. 13 is

Table 18.1. Description of Vertebrate Life Forms Occurring in the Blue Mountains. (Thomas and others 1977)

Life form number	Reproduces	Feeds
1	In water	In water
2	In water	On ground, in bushes, and/or in trees
3	On ground above water	In water, on ground, in bushes and trees
4	In cliffs, caves, rims, and/ or talus	On ground or in air
5	On ground without specific water, cliff, rim, or talus association	On ground
6	On ground	In bushes, trees, or air
7	In bushes	On ground, water, or air
8	In bushes	In bushes, trees, or air
9	Deciduous trees	In bushes, trees, or air
10	Coniferous trees	In bushes, trees, or air
11	In trees	On ground, bushes, trees, or air
12	On very thick branches	On ground or in water
13	Excavates own hole in tree	On ground, bushes, trees, or air
14	In hole made by another species or natural hole	On ground, water, or air
15	Underground burrow	On or above ground
16	Underground burrow	In water or air

shown in Table 18.3. In this example, none of the species can utilize the early (fewer than 40 years old) seral stages for reproduction, and only two species can utilize them for feeding. Thomas and others (1977) provide full tables for all 16 life forms and 327 resident species.

Thomas used these results for several purposes. The first and obvious use was to provide a model of habitat utilization by vertebrates; they can be used to determine which species use each successional stage of each community type. This also allowed calculation of a "versatility index" in the following way. The habitat utilization tables provide a quantitative measure, for each species, of the number of community types and successional stages that they may use for feeding or reproduction. The sum of these numbers is the versatility index—the lower the index, the less adaptive and less versatile, and therefore the more vulnerable, is the species.

Thomas and his co-workers also determined how several different management actions affect successional changes, as shown in Table 18.4. Although many of these activities retard succession, some advance it; examples include cold burns of certain seral stages, cattle and sheep

Table 18.2. Species Orientation by Plant Communities for Life Form No. 13 in the Blue Mountains[a] (Thomas and others 1977)

Species	Juniper dominant	Aspen	Riparian/ deciduous	Ponderosa pine	Mixed conifer	Lodgepole pine	White fir	Subalpine fir
Common flicker	× O	× O	× O	× O	× O		× O	
Pileated woodpecker				× O	× O		× O	
Lewis' woodpecker			× O	× O	× O			
Yellow-bellied sapsucker		× O	× O	× O	× O			
Williamson's sapsucker				× O	× O		× O	
Hairy woodpecker		× O		× O	× O	× O	× O	× O
Downy woodpecker		× O	× O					
White-headed woodpecker				× O	× O			
Black-backed three-toed woodpecker					× O	× O	× O	O
Northern three-toed woodpecker						× O		× O
White-breasted nuthatch				× O	× O			
Red-breasted nuthatch					× O	× O	× O	
Pygmy nuthatch		O		× O	× O			

[a] ×, reproduction; O, feeding.

Table 18.3. Species Orientation by Successional Stages for Life Form No. 13 in the Blue Mountains.[a] (Thomas and others 1977)

Species	Grass–forb	Brush–seedling (0–10 yr.)	Pole–sapling (11–39 yrs.)	Young (40–79 yr.)	Mature (80–159 yr.)	Old (160 yr.)
Common flicker	O	O	O	×O	×O	×O
Pileated woodpecker				O	×O	×O
Lewis' woodpecker	O	O			×O	×O
Yellow-bellied sapsucker				×O	×O	×O
Williamson's sapsucker				×O	×O	×O
Hairy woodpecker				×O	×O	×O
Downy woodpecker				×O	×O	×O
White-headed woodpecker					×O	×O
Black-backed three-toed woodpecker				×O	×O	×O
Northern three-toed woodpecker				×O	×O	×O
White-breasted nuthatch					×O	×O
Red-breasted nuthatch				×O	×O	×O
Pygmy nuthatch					×O	×O

[a]×, reproduction; O, feeding.

Table 18.4. Anticipated Changes in Forest Community Structure from Management Action.[a] (Thomas and others 1977)

Management action	Successional stage of condition					
	Grass–forb	Shrub–seedling	Pole–sapling	Young	Mature	Old growth
Brush control						
Herbicides	→	→	→	←	←	←
Mechanical control	←	←	—	←	←	←
Controlled burn						
Cold burn	←	←	→	→	←	←
Hot burn	←	←	←	←	←	←
Fertilization	—	→	→	→	→	—
Grazing						
Cattle and sheep	→	→	—	—	—	—
Goats	←	←	—	←	←	←
Planting						
Trees	→	→	—	→	→	—
Shrubs	→	—				
Grasses–forbs	→	←	—	→	→	—
Regeneration cut						
Clear cut				←	←	←
Shelterwood				→	←	←
Seed tree				←	←	←
Salvage			←	←	←	←
Thinning (including single tree selection)		→	→	→	→	—

[a]Symbols are: → advances succession; ← retards succession; — no effect on succession.

grazing of pioneer communities, and thinning of young stands. Still other activities do not affect animal succession. The Thomas group also provide, for each species, information on the special and unique habitats that it utilizes (marshes, cliffs, caves, talus, snags, logs, etc.), information on its utilization of ecotones, and an extensive bibliography.

Their results provide an extremely valuable data base and simulator for determining the effects of management actions and natural succession on animal habitat utilization. Furthermore, although we have incorporated their results into *FORPLAN,* use of their model requires no computational facilities of any kind; all of the capabilities of their system can be realized from (admittedly tedious) manual use of their tables.

Similar vertebrate habitat utilization research has been conducted in California and Montana, and I anticipate the Thomas and others (1977) example to become the standard approach soon; it is undoubtedly the most significant vertebrate habitat utilization research that has been conducted in several decades.

19

The Next Generation of Gradient Modeling Systems: *FORPLAN* and *PREPLAN*

In designing the *FOR*est *P*lanning *LAN*guage and Simulator *(FORPLAN)*, we were acutely aware not only of the diverse information needs of land managment planning, but especially of the confusion and difficulty that have been created by the recent exponential proliferation of systems, simulators, and data bases. We wanted to include the unique capabilities of such systems as the Glacier Park model, *FBIIS, FMIIS, GANDALF, TAROT,* and the Thomas and others (1977) vertebrate habitat models; yet we were determined to build a system that would invoke automatically those subroutines appropriate to the scope and resolution of the user's input data. In this way, if the user could indicate what was available for data and what was wanted in the way of information, *FORPLAN* would provide the best possible level of simulation and hence the most reliable output. The user would not have to select from dozens of possible programs and simulators; the selection and linkage of modules appropriate to specific data and requests would be provided by *FORPLAN*.

  FORPLAN therefore can be viewed as: a concept in which several systems and several levels of resolution are handled simultaneously; a language that both employs and recognizes common English words; and a discrete-time simulator that models major forest disturbances. This chapter is devoted to describing the language, the data base structure, and the subroutine modules, and finally *FORPLAN*'s potential use as a simulator for land management planning; this material draws heavily on work reported by Potter and others (1979).

Structure of the Language

What makes *FORPLAN* so easy to use is that *FORPLAN* recognizes and interprets common English words. For instance, a manager in western Montana wanting to know the effect of a fire on plants in a 180-year-old Douglas-fir stand would type:
  AREA MONTANA
  AGE 180 YEARS
  VEG DOUGFIR
  PRINT PLANTS

    DISTURB FIRE
    STATISTICS
    AFFECT ON PLANTS
    END

The *FORPLAN* simulator scans the user input stream for recognized vocabulary. It then interprets each line of user input for use by the subroutines, which are designed to simulate the various activities associated with forest management.

There are three levels of vocabulary in *FORPLAN*: key words, attributes, and states. KEYWORDS are the primary words that signal branches to appropriate subroutines and submodules. KEYWORDs either stand alone or are followed by an ATTRIBUTE that makes the key word more specific; some attributes are further modified by STATES. *FORPLAN* recognizes the following key words: AREA (recognized synonym: GET), SPECIFY, DISTURB, AFFECT (recognized synonym: PRINT), STATISTICS, COST, and END. Each key word is described in detail below with its associated attributes and states. A diagram of the key word–attribute–state relationships is given in Fig. 19.1; a complete list of key words, attributes, and states (for the 1978 version of *FORPLAN*) appears in Table 19.1.

AREA is the word that signals the data base for which the simulation is to be invoked. It currently has three attributes associated with it: BLUE

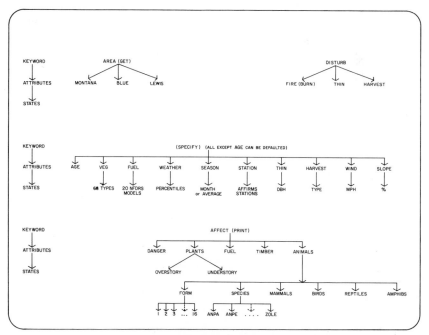

Figure 19.1. *FORPLAN* language structure.

Table 19.1. *FORPLAN* Keywords, Attributes, and States (1978 Version)

Keyword	Attributes	States
AREA (GET)	BLUE	
	MONTANA	
	LEWIS	
SPECIFY	AGE	Age in years
(may be omitted)		
	VEG	type optional; see Table 19.2 for list of sample recognized states
	FUEL	NFDRS fuel model optional
	THIN	DBH (diameter at breast height) (cm) followed by an integer
	HARVEST	
	WIND	Integer (mph); default 5 mph
	WEATHER	Integer percentile; default 90th
	STATION	AFFIRMS weather stations
	SEASON	Month of year may be specified
	SLOPE	Integer percent; default 30%
DISTURB	FIRE (BURN)	
	THIN	
	HARVEST	
AFFECT (PRINT)	DANGER	
	PLANTS	OVERSTORY or UNDERSTORY
	FUEL	
	TIMBER	
	ANIMALS	FORM
		FORM *n* (an integer between 1 and 16)
		SPECIES (prints all species)
		SPECIES *aaaa* (*aaaa* is species code)
		MAMMALS
		REPTILES
		BIRDS
		AMPHIBS
		MANAGED
STATISTICS		
COST	(Not implemented)	
END		

(Blue Mountains Provinces of northeastern Oregon), MONTANA (west of the Continental Divide), and LEWIS (Lewis and Clark National Forest located in Montana). Examples:

GET MONTANA

AREA IS THE *BLUE* MOUNTAINS

GET THE *LEWIS* AND CLARK DATA BASE

Note that *FORPLAN* scans each line for the key word and associated attribute ignoring any extra words that may appear in the line. Italicized words in the examples above are the words recognized by *FORPLAN*. So the user may add words to make the line more readable or meaningful without altering the *FORPLAN* processor interpretation. Additional attributes are being implemented as *FORPLAN* incorporates the data bases of more forests. Each attribute provides for the attachment of specific fuel and weather data bases. If no attribute is listed, the simulation defaults to a National Fire Danger Rating System (NFDRS) fuel model (Deeming and others 1977) and a default weather set.

SPECIFY is a key word that may be omitted; that is, if a user input line contains no key word, it is assumed to contain an attribute of the key word, SPECIFY. Hence SPECIFY AGE 10 can be abbreviated to AGE 10. The attributes associated with SPECIFY are: AGE, VEG, FUEL, STATION, WEATHER, SEASON, THIN, HARVEST, WIND, and SLOPE. Each of these attributes has associated states that set the appropriate pointers to fuel and data base arrays, parameters to the fire model, etc.

The states of AGE are integers from 1 to 999 that specify the stand age in years since previous major disturbance. The SPECIFY VEG and SPECIFY FUEL commands are used to select individual plant community types and the NFDRS fuel models, respectively. For each AREA, VEG recognizes up to 68 community types as states. States for the areas BLUE and MONTANA are listed in Table 19.2, and states for LEWIS have been derived for individual habitat types as developed by Pfister and others (1977).

The recognized states for the FUEL attribute are the 20 letters associated with the 1978 NFDRS fuel models. Note that this specification allows for user option in determining resolution level. The specification of an AREA, a stand age (SPECIFY AGE), and a plant community type (SPECIFY VEG) provides automatic linkage to an appropriate fuel data base. In the cases of MONTANA and LEWIS these data bases are of sufficiently high resolution to invoke a fire model that utilizes locally collected fuels information (Kessell and others 1978). In the case of AREA BLUE, stylized NFDRS models based on stand types but futher stratified by stand age are invoked. These models are of lower resolution and branch to a basic fire model (Rothermel 1972), but because of typing and age stratification they can provide linkage to appropriate plant and animal data bases. Alternatively, the SPECIFY FUEL command overrides the above linkage and provides the lowest resolution input to the basic fire model.

The STATION attribute allows the user to stipulate the AFFIRMS weather station to be used in computing fuel moistures. Default weather stations are provided for each area. Weather SEASONS are divided into individual months (MAY–OCTOBER inclusive) and a seasonal AVERAGE (default).

The WEATHER attribute recognizes an integer from 1 to 99 as the percentile weather desired. The 90th percentile weather indicates that 90%

Table 19.2. Recognized STATES of the *FORPLAN* command SPECIFY VEG for AREA MONTANA and AREA BLUE

Area Montana (Montana west of the Continental Divide)

State name	Community name
TALUS	Talus–Meadow
MEADOW	Meadow
PRAIRIE	Prairie
SHRUB	Shrub
KRUMMHOLZ	Krummholz
CEDAR	Cedar–Hemlock forest
SPRUCE	Spruce forest
LARCH	Larch–Douglas-fir forest
PONDEROSA	Ponderosa pine savanna–forest
LODGEPOLE	Lodgepole pine forest
DOUGFIR	Douglas-fir forest
WHITEBARK	Whitebark pine forest
SUBALPINE	Subalpine fir forest

Area Blue (Blue Mountains province of northeastern Oregon)

State name	Community name
DRY	Dry meadow
MOIST	Moist meadow
OTHER	Other grass dominant
SAGE	Sage–bitterbrush dominant
MAHOGANY	Mahogany dominant
JUNIPER	Juniper dominant
ASPEN	Aspen
RIPARIAN	Water–riparian–deciduous
BRUSH	Other brush and shrub dominant
PONDEROSA	Ponderosa pine savanna–forest
WHITE	White fir forest
MIXED	Mixed conifer forest
LODGEPOLE	Lodgepole pine forest
SUBALPINE	Subalpine fir forest
ALPINE	Alpine meadow

of all historical weather recorded at that AFFIRMS station for the specified season was wetter. The 1st percentile weather is therefore the wettest recorded and the 99th percentile weather is the driest. The default value for WEATHER is the 90th percentile.

THIN has a minumun dbh state associated with it. HARVEST signifies a clearcut. WIND recognizes an integer from 1 to 99 as a state (speed in mph); default value is 5 mph. SLOPE recognizes an integer percent state from 1 to 100 and its default value is 30%.

DISTURB is the key word that invokes a branch to subroutines to simulate various requested activities or disturbances. The following attributes are recognized: FIRE (recognized synonym: BURN), THIN, and HARVEST. States for these activities are set by SPECIFY commands, including the time at which the activity or disturbance is to occur (set by the AGE attribute). The STATISTICS key word initiates print routines associated with the above activities, and COST (when implemented) will produce cost–benefit analyses for the above simulations.

AFFECT and the recognized synonym PRINT have the following associated attributes: DANGER, PLANTS, FUEL, TIMBER, and ANIMALS. These activities are subroutines that incorporate their own print routines for tables and graphs as follows.

PRINT DANGER causes a National Fire Danger Rating simulation to occur and prints graphs of three familiar fire danger rating statistics (spread component, energy release component, and burning index) as functions of the moisture percentile. PRINT TIMBER is being implemented to incorporate a stand prognosis module developed for *GANDALF* (Chap. 17; Kessell and others 1977b) utilizing equations and techniques developed by Stage (1973). Output will consist of stand yield tables.

The attribute PLANTS has states OVERSTORY and UNDERSTORY associated with it. The output format for PRINT PLANTS differs by area as a result of diferences in available research data bases. By way of example, for AREA MONTANA, PRINT PLANTS OVERSTORY causes the printing of a list of overstory species composition by percent relative density of each species present to be printed. PRINT PLANTS UNDERSTORY for the same area provides percent total herb and shrub cover, herb and shrub diversity as measured in two 2×2 meter sample plots, as well as a species list by percent absolute cover. If neither OVERSTORY nor UNDERSTORY is specified, the user receives a printout of both.

States for the PRINT ANIMALS command allow the user some selectivity in quantity and type of information received. PRINT ANIMALS FORM produces a table of the number of species occurring (for both reproduction and feeding) grouped by the vertebrate life forms of Thomas and others (1977) (Chap. 18). A selection of PRINT ANIMALS FORM n (where n is an integer from 1 to 16) produces similar information for the specific life form n. PRINT ANIMALS SPECIES (or AFFECT ANIMALS SPECIES) produces a listing of all species utilizing (for reproduction or feeding) the stand resulting from the previously specified *FORPLAN* commands. For example, if the AFFECT ANIMALS SPECIES command were preceded by a DISTURB FIRE command, the printout would include the animals in the postfire community. The user can also restrict output to a particular taxonomic group by specifying one of the states: MAMMALS, REPTILES, BIRDS, or AMPHIBS. To restrict the output to an individual species, the user may request: AFFECT ANIMALS

SPECIES *aaaa,* where *aaaa* is a four letter abbreviation for the individual species. An additional option allows the user to specify the forest's managed animals. The user can then PRINT ANIMALS MANAGED to obtain information on these selected species throughout the simulation.

PRINT FUEL (or AFFECT FUEL) provides the user with current fuel loadings by fuel types, packing ratios, and percent cover by strata in the case of high-resolution data.

FORPLAN Modules

FORPLAN is modular in structure and consists of three categories of modules: data bases, disturbance and activity simulations, and print routines. The vocabulary that causes these modules to be invoked has been described above. Let us now examine the internal structure of the *FORPLAN* simulator.

For each AREA currently implemented, data bases have been developed for fuel, vegetation, weather, and animals. *FORPLAN* weather data are currently available for one to four representative weather station within each area. Moose Creek station represents western Montana, whereas Meacham ranger station represents the Blue Mountains. Four additional stations are available for the Lewis and Clark National Forest. Fuel moistures for each fuel size class and category, and for each cumulative percentile level, were obtained from the programs FIREDAT (Furman and Brink 1975) and *TAROT* (Kessell and others 1978). These parameters were augmented by percent cured data (for both woody shrubs and nonwoody understory vegetation) for each percentile level, cover type, and area based on algorithms of the NFDRS (Deeming and others 1977). These data are accessed by the *FORPLAN* user simply by requesting the desired station, percentile level, and season.

The Blue Mountains province is represented by 15 major plant community types. Each type is stratified (by Thomas and others 1977) into six stand age classes of 0, 1–10, 11–39, 40–79, 80–159, and over 159 years since the last major disturbance. Fuel models were derived from the basic 1978 NFDRS fuel models to correspond to each community type. By inference from fuel buildup curves for western Montana (Kessell and others 1978; none is available for the Blue Mountains province specifically), these fuel models were refined to include estimated effects of stand age. All Blue Mountains meadow communities use the NFDRS T model. Early (0–10 years) and midseral (11–79 years) coniferous forests use a modified H model; late seral (80–159 years) coniferous forests use the standard H model; mature (over 159 years) coniferous forests use the G model. Mature ponderosa pine is represented by the NFDRS C model; seral ponderosa pine communities use modifications of that same model. Mature mahogany and juniper (over 80 years) employ the NFDRS F model; younger stands use a modification of that model. Aspen and riparian–deciduous communities use the NFDRS R model for older stands (over 80 years) and modifications of that model for younger communities.

Montana west of the Continental Divide is represented by 13 major plant community types that are further stratified into five age classes of 0–20, 21–50, 51–90, 91–150, and over 150 years. The fuels data reported in Chapters 9 and 11 were included in *FORPLAN* without modification; an appropriate algorithm matches the community type and stand age with the correct fuel model. Furthermore, these fuel data are stratified by the three vertical strata (Chap. 14 and Kessell and others 1978) and hence represent the highest level of resolution.

Fuels data for the Lewis and Clark National Forest are stratified by habitat type and stand age and have been derived from over 400 field samples.

Montana vegetation data were derived from the Glacier National Park data base and stratified along spatial and temporal environmental gradients (Sections II and III). However, for the purposes of *FORPLAN,* the gradient analysis was greatly simplified by grouping the communities into 13 major community types, identical to the fuels stratification noted above. This classification produced eight major forest types: cedar–hemlock, spruce, Larch–Douglas-fir, ponderosa pine, lodgepole pine, Douglas-fir, whitebark pine, and subalpine fir. These forest types were further stratified by stand age into the five age classes described for the fuels data. Five major nonforested types were also derived: talus–meadow, meadow, prairie, shrub, and Krummholz. These communities were not stratified by stand age because of the minimal and/or short-term nature of fire effects on these communities in western Montana. For each community type and age class, quantitative species composition was determined (from Section II) and included in *FORPLAN*. These data included all overstory (tree) species quantified by relative density (of stems over 7.62-cm dbh), and 29 common understory shrubs and herbs quantified by total percent ground cover. Further, the age stratification of forests permits the deterministic prediction of postfire forest succession (trees, shrubs, and herbs) from this data base. The multiple-pathways succession model of Noble and Slatyer (1977, Cattelino and others 1979; Chaps. 3 and 10) is being implemented by major forest types into *FORPLAN*.

Blue Mountains Province vegetation data were derived from Thomas and others (1977), F. C. Hall (1973), and Franklin and Dyrness (1973) in the following fashion. Thomas and co-workers stratified the area's plant communities into 15 major types for the purpose of relating vertebrate species distributions to plant communities. These types as implemented into *FORPLAN* include: dry meadow, wet meadow, other grass dominant, sage–bitterbrush dominant, mahogany dominant, juniper dominant, aspen, water–riparian–deciduous, other brush and shrub dominant, ponderosa pine, mixed conifer, white fir, lodgepole pine, subalpine fir, and alpine meadow. This stratification is a simplification of 44 community types recognized by F. C. Hall (1973) and generally (but imperfectly) corresponds to the habitat types defined and/or reviewed by Franklin and Dyrness (1973). Unfortunately, none of the classification systems provides full quantitative data on overstory species or common herbs and shrubs; published quantitative data on seral stages of the communities are virtually nonexistent. The quality and quantity of the existing species composition data are also very variable, and for some communities *FORPLAN* is limited to providing the brief descriptions furnished by F. C. Hall (1973). At present, published quantitative succession data are also not available for any of the communities, and even solid qualitative predictions of fire succession are unavailable for some of them.

The difficulties in determining plant succession in eastern Oregon are more than compensated for by the very detailed and complete animal succession modeling capabilities derived from Thomas and others (1977) as described in the last chapter. In addition to providing habitat utilization information species by species, *FORPLAN* also allows access by taxonomic groups and life forms (groups with similar ecologic requirements for food and reproduction). The same species and groups are also recognized by any special or unique habitats they utilize (snags, logs, talus, caves, for example) and by major ecotones that they utilize. For any Blue Mountains community type and stand age, therefore, full occurrence information for all species, taxonomic groups, and/or life forms is available.

This application of the Thomas data also, of course, predicts the effects of fire on the vertebrates, as fire affects the community's stand age index. For example, following each *FORPLAN* DISTURB command, *FORPLAN* predicts the effects of that disturbance on the vertebrates. In addition, the Thomas data as adapted by *FORPLAN* also permit alternative succession pathways for different components of the community. Some disturbances, such as hot fires and clear cutting, force animal successions that follow the plant and fuel successions (that is, a return to age zero and a pioneer community, followed by seral stages leading to maturity). Yet other disturbances, such as a cool burn or thinning, actually "advance" the succession in terms of the animal communities but cause it to retrogress in terms of plants. These multiple pathways for the different components are included in *FORPLAN*.

Originally, work of Thomas and others (1977) was inferred for MONTANA and LEWIS by matching community types and known successional pathways. More recently, the Forest Service has developed a similar vertebrate data base for Montana; this has been incorporated in *FORPLAN* for AREA LEWIS.

Let us now examine the modules that serve as disturbance and activity simulations. A module to respond to the DISTURB FIRE command was originally adapted from the multistrata version (Bevins 1976, Kessell and others 1978) of the Rothermel (1972) fire behavior model. However, in order to obtain fire behavior statistics for the lower resolution NFDRS fuel models and their adaptations, the basic Rothermel model (one stratum) was also needed. Hence, the module invoked by the DISTURB FIRE command currently incorporates both capabilities. By means of a flag, it operates as either the Rothermel model or the multistrata version. In addition, new algorithms for weighted rate of spread have been incorporated. A recent modification to the weighting factors for fuel loadings and relevant equations provides NFDRS danger rating capabilities.

The important point to be noted about this fire disturbance module is that user input data are assessed automatically by *FORPLAN,* and the appropriate fire model is determined as a result of the resolution level of this input data. Not only does the user not have to select which fire model to use, he or she is automatically prevented from making an inappropriate selection. In addition, DISTURB FIRE and PRINT DANGER commands perform two very different tasks, although both invoke the same fire module. A DISTURB FIRE command simulates an actual disturbance to the stand by resetting the stand age, altering successional pathways, and making resultant changes in fuel, animal, and plant data bases. If followed by a STATISTICS command, the rate of spread, Byram's fireline intensity, flame length, and scorch height for the specified simulation are printed. In contrast, PRINT DANGER neither alters the simulated ecosystem or any of its components nor resets the stand age. It provides a current measure of "disaster potential" should such a disturbance occur as well as probable spread component, energy release component, and burning index. This allows the user to assess the potential for catastrophic disturbance without even simulating that disturbance. Again, the manager is prohibited from falling into the trap of selecting a fire danger model to predict fire behavior, because *FORPLAN* makes the distinction and the subsequent model selection.

The effects of DISTURB THIN and DISTURB HARVEST on animals have already been noted. In addition, they currently modify fuels and other parameters to the fire module. These disturbance activities are currently being augmented to incorporate some of the specific features of these activities that were developed by *GANDALF* specifically for lodgepole pine communities.

The final category of modules to be discussed consists of the print routines. These routines were developed to provide the user with specific stand information at any discrete point in time during the simulation. Most of the PRINT commands have been described above in other contexts, and examples of each routine appear in Table 19.3 (located at the end of the chapter). The PRINT TIMBER routine currently provides stand information to the user as specified: area, age, and community type. As more stand prognosis capabilities are implemented from *GANDALF* and other models, this event will be augmented to include a cruise table.

FORPLAN as a Managerial Tool

A manager is concerned about a 300-year-old ponderosa pine *(Pinus ponderosa)* stand. Because fire has been suppressed in this stand, there is considerable fuel buildup. Should lightning strike this stand in late August, does the manager have any hope of saving it? The manager types:

AREA MONTANA
AGE 300
VEG PONDEROSA
SEASON AUGUST
WEATHER 95 PERCENTILE
PRINT FUEL
DISTURB FIRE
STATISTICS
END

He frowns as he studies the computer output. The fire behavior statistics confirm his fears. If lightning should strike on a hot day in late summer, no amount of control would be likely to save the stand. Clearly, he must consider some fire management alternatives. Again he sits down at the computer and types:

AREA MONTANA
AGE 300
VEG PONDEROSA
PRINT FUEL
SEASON MAY
DISTURB FIRE
PRINT FUEL
AFFECT ON ANIMALS SPECIES CECA
END

This output tells him not only the fuel reduction to be expected from a prescribed burn, but also the effect on elk *(Cervus canadensis —* code name CECA) habitat, another of his management concerns.

Yet he also has another management alternative. He requests the following simulation:

AREA MONTANA
AGE 300
VEG PONDEROSA
SPECIFY THIN DBH 12 CM
DISTURB THIN
 SEASON OCTOBER
WEATHER 50 PERCENTILE
DISTURB FIRE
SEASON AUGUST

```
WEATHER 95 PERCENTILE
DISTURB FIRE
STATISTICS
AFFECT ON ANIMALS SPECIES CECA
END
```

This run tells him whether a precautionary thin, followed by an October prescription burning of the resulting slash, will reduce his fire hazard sufficiently without affecting his elk habitat.

Armed with probable outcomes of these management alternatives, the manager can now plan a strategy for the forest. The major advantages to the manager of using a system such as *FORPLAN* include:

1) The ability to *compare* management strategies and alternatives before making a final decision
2) The ability to *consider* a decision without either the cost or the consequences of implementing that decision
3) The ability to *predict* probable effects of a decision at future points in time

FORPLAN makes no recommendations to the manager. Instead, it is designed to display the probable effects on plants, animals, and fuels of various activities and disturbances. It aids the manager in making enlightened decisions, but the decisions are still up to the manager.

Continued Development and Availability of *FORPLAN*

At present, *FORPLAN* has been successfully applied to and evaluated on the Lewis and Clark National Forest by fire management personnel. Using fuel, plant, animal, and weather data bases specific to that area, simulation runs were evaluated to ensure that they were both scientifically credible and acceptable as a managerial tool. Based upon the results of this validation, a set of guidelines that instruct managers in preparing data to implement *FORPLAN* on other forest ecosystems was developed (Kessell and others 1979). These guidelines were evaluated, by both researchers and managers, as they were used to implement *FORPLAN* on the Helena National Forest (Montana). A *User's Guide* (Potter and others 1978) is available, and new user guides are being developed as *FORPLAN* expands to include new capabilities and is implemented in additional areas. A *Systems Manual* is being developed and will be available in September 1979. I expect the U.S. Forest Service will, in the near future, reach a decision on the national implementation of *FORPLAN* or a *FORPLAN*-like system.

PREPLAN. The Kosciusko National Park Model

As this book goes to press (February 1979) I am developing a prototype resource management system for the Australian National Parks and Wildlife Service and the National Parks and Wildlife Service of New South Wales. The system is called *PREPLAN* (*PR*istine *E*nvironment *P*lanning *LAN*guage and Simulator) and is being implemented on a test and demonstration basis in Kosciusko National Park, N.S.W.[1] Major goals of the modeling effort include the development of a grid-based resource inventory and fire behavior and effects modeling capabilities. *PREPLAN* is mentioned here because it provides a good example of several points emphasized in this book.

First, *PREPLAN* is designed to include the best features of both the Glacier National Park gradient modeling system and *FORPLAN*. Specifically, it includes a grid-based remote inventory linked to gradient models, yet includes the English language vocabulary and simplicity of use introduced by *FORPLAN*.

Second, *PREPLAN* allows greater flexibility than is available from any previous system. For example, the user may SPECIFY the community in three distinct fashions: 1) by specifying the geographical coordinates and automatically linking to a precoded remote inventory (as does the Glacier National Park model); 2) by specifying the community type and disturbance history (as does *FORPLAN*); or 3) by stating (in English words) the stand's location on major gradients at the time of model execution (an option previously available from no other system).

Third, *PREPLAN* utilizes the technology of previous models, such as Glacier's linkage of gradient models and the remote site inventory, and *FORPLAN*'s vocabulary and flexibility, while including locally available and acceptable components, such as the McArthur fire models (1977), Costin's (1954) extensive vegetation data base, and the park's own resource information (Roger Good, Graeme Warboys, and Mark Butz personal communication).

Fourth, and to my mind of greatest value, *PREPLAN* demonstrates the desire of land managers in Australia to utilize the best available technological tools to help them manage the national parks in a competent and professional manner.

As the *PREPLAN* system is being developed, other land management agencies in the United States, Canada, Australia, and South Africa are evaluating *FORPLAN*- and *PREPLAN*-like systems for application to diverse ecosystems and management problems.

---

[1]Kosciusko National Park includes 600 000 ha of the Snowy Mountains (Australian Alps), and contains montane through alpine environments.

Table 19.3. Interactive Operation and Sample Output from *FORPLAN* Demonstrate Some of the Simulator's Capabilities.

In the first example, manager Dennis is concerned about the effects of fire suppression on spruce communities in western Montana. He uses *FORPLAN* to simulate fuel buildup and resulting fire behavior over a 150-year period.

In the next example, manager Melanie realizes that wildfire will undoubtedly strike her 200-year-old Douglas fir forests within the foreseeable future and uses *FORPLAN* both to estimate fire behavior under extreme weather conditions and to predict plant succession following the fire.

In the last example, manager Frank uses *FORPLAN* to display the birds utilizing 150-year-old lodgepole pine communities in the Blue Mountains and then simulates the immediate and 25-year effect of a fire on their habitat utilization.

WELCOME TO THE FOREST PLANNING LANGUAGE SIMULATOR (FORPLAN) UNDER DEVELOPMENT FOR THE USDA FOREST SERVICE,
R.D. AND A. PROGRAM BY
  GRADIENT MODELING, INC.
  BOX 2666
  MISSOULA, MONTANA 59806
PLEASE TYPE IN YOUR NAME.

DENNIS

NOW TYPE TODAYS DATE.

3/17/78

THANK YOU. SOURCE CODE DEVELOPED BY MEREDITH W. POTTER.
PLEASE REFER ALL INQUIRIES TO THE AUTHOR, AND SEE USERS MANUAL FOR JOB INPUT STREAM.
YOU MAY NOW TYPE IN DESIRED SIMULATION RUN.

  AREA MONTANA
  AGE 50
  VEG SPRUCE
  PRINT FUEL
  DISTURB FIRE
  STATISTICS
  AGE 120
  PRINT FUEL
  DISTURB FIRE
  STATISTICS
  AGE 200
  PRINT FUEL
  DISTURB FIRE
  STATISTICS
  END

AREA IS MONTANA.
VEGETATION COMMUNITY IS SPRUCE.
STAND AGE IS CURRENTLY 50 YEARS.

MEDIAN FUEL LOADINGS, PACKING RATIOS, AND PERCENT COVER BY STRATA.

FUEL TYPE	LOADING LB/SQ FT	(MT/ HA)
LITTER	.0410	2.00
GRASS AND FORBS	.0068	.33
1 HOUR DEAD AND DOWN BRANCHWOOD	.0217	1.06
10 HOUR DEAD AND DOWN BRANCHWOOD	.0332	1.62
100 HOUR DEAD AND DOWN BRANCHWOOD	.0692	3.38
1000 HOUR DEAD AND DOWN BRANCHWOOD	.1829	8.93
SHRUB FOLIAGE	.0082	.40
SHRUB BRANCHWOOD	.0276	1.35

PACKING RATIOS (BETA—DIMENSIONLESS)

LITTER STRATUM	.0301
DEAD AND DOWN STRATUM	.0083
SHRUB STRATUM	.0010

PERCENT COVER BY STRATA

LITTER STRATUM	100.00
DEAD AND DOWN STRATUM	85.00
SHRUB STRATUM	40.00

AREA IS MONTANA.
VEGETATION COMMUNITY IS SPRUCE.
STAND AGE IS CURRENTLY 0 YEARS.
WEATHER IS 90 PERCENTILE FROM MOOSE CREEK AFFIRMS STATION.

FIRE BEHAVIOR STATISTICS

RATE OF SPREAD	7.53 FT/MIN	(	2.29 M/MIN)
FIRELINE INTENSITY(BYRAMS)	141.84 BTU/S/FT	(	117.26 KCAL/S/M)
FLAME LENGTH	4.25 FT	(	1.30 M)
SCORCH HEIGHT	30.26 FT	(	9.22M)

TO PRINT FIRE EFFECTS ON FIRE HAZARD, FUELS, PLANTS, AND/OR ANIMALS,
GIVE THE APPROPRIATE AFFECT COMMAND AND/OR ATTRIBUTE AGE COMMAND.

AREA IS MONTANA.
VEGETATION COMMUNITY IS SPRUCE.
STAND AGE IS CURRENTLY 120 YEARS.

MEDIAN FUEL LOADINGS, PACKING RATIOS, AND PERCENT COVER BY STRATA.

FUEL TYPE	LOADING LB/SQ FT	(MT/HA)
LITTER	.0356	1.74
GRASS AND FORBS	.0049	.24
1 HOUR DEAD AND DOWN BRANCHWOOD	.0197	.96
10 HOUR DEAD AND DOWN BRANCHWOOD	.0275	1.35
100 HOUR DEAD AND DOWN BRANCHWOOD	.0702	3.43
1000 HOUR DEAD AND DOWN BRANCHWOOD	.3072	15.00
SHRUB FOLIAGE	.0090	.44
SHRUB BRANCHWOOD	.0328	1.60

PACKING RATIOS (BETA—DIMENSIONLESS)

LITTER STRATUM	.0317
DEAD AND DOWN STRATUM	.0127
SHRUB STRATUM	.0009

PERCENT COVER BY STRATA

LITTER STRATUM	100.00
DEAD AND DOWN STRATUM	85.00
SHRUB STRATUM	40.00

AREA IS MONTANA.
VEGETATION COMMUNITY IS SPRUCE.
STAND AGE IS CURRENTLY 0 YEARS.
WEATHER IS 90 PERCENTILE FROM MOOSE CREEK
AFFIRMS STATION.

FIRE BEHAVIOR STATISTICS

RATE OF SPREAD	7.56 FT/MIN	(	2.30 M/MIN)
FIRELINE INTENSITY (BYRAMS)	314.36 BTU/S/FT	(	259.89 KCAL/S/M)
FLAME LENGTH	6.24 FT	(	1.90 M)
SCORCH HEIGHT	51.45 FT	(	15.68 M)

TO PRINT FIRE EFFECTS ON FIRE HAZARD, FUELS, PLANTS,
AND/OR ANIMALS, GIVE THE APPROPRIATE AFFECT COM-
MAND AND/OR ATTRIBUTE AGE COMMAND.

AREA IS MONTANA.
VEGETATION COMMUNITY IS SPRUCE.
STAND AGE IS CURRENTLY 200 YEARS.

MEDIAN FUEL LOADINGS, PACKING RATIOS, AND PERCENT
COVER BY STRATA.

	LOADING	
	LB/SQ	(MT/
FUEL TYPE	FT	HA)
LITTER	.0440	2.15
GRASS AND FORBS	.0043	.21
1 HOUR DEAD AND DOWN BRANCHWOOD	.0209	1.02
10 HOUR DEAD AND DOWN BRANCHWOOD	.0348	1.70
100 HOUR DEAD AND DOWN BRANCHWOOD	.0348	1.70
1000 HOUR DEAD AND DOWN BRANCHWOOD	.9564	46.70
SHRUB FOLIAGE	.0041	.20
SHRUB BRANCHWOOD	.0139	.68

PACKING RATIOS (BETA—DIMENSIONLESS)

LITTER STRATUM	.0255
DEAD AND DOWN STRATUM	.0124
SHRUB STRATUM	.0005

PERCENT COVER BY STRATA

LITTER STRATUM	100.00
DEAD AND DOWN STRATUM	85.00
SHRUB STRATUM	20.00

AREA IS MONTANA.
VEGETATION COMMUNITY IS SPRUCE.
STAND AGE IS CURRENTLY 0 YEARS.
WEATHER IS 90 PERCENTILE FROM MOOSE CREEK
AFFIRMS STATION.

FIRE BEHAVIOR STATISTICS

RATE OF SPREAD	19.06 FT/MIN	(	5.81 M/MIN)
FIRELINE INTENSITY(BYRAMS)	3904.27 BTU/S/FT	(	3227.73 KCAL/S/M)
FLAME LENGTH	20.02 FT	(	6.10 M)
SCORCH HEIGHT	275.91 FT	(	84.10 M)

TO PRINT FIRE EFFECTS ON FIRE HAZARD, FUELS, PLANTS,
AND/OR ANIMALS,
GIVE THE APPROPRIATE AFFECT COMMAND AND/OR
ATTRIBUTE AGE COMMAND.

THIS ENDS THE SIMULATION FOR DENNIS
REQUESTED ON 3/17/78
THANK YOU FOR SHOPPING USDA FOREST SERVICE.
MAY THE FOREST BE WITH YOU.

WELCOME TO THE FOREST PLANNING LANGUAGE SIMULA-
TOR (FORPLAN) UNDER DEVELOPMENT FOR THE USDA
FOREST SERVICE.
R. D. AND A. PROGRAM BY
    GRADIENT MODELING, INC.
    BOX 2666
    MISSOULA, MONTANA 59805
PLEASE TYPE IN YOUR NAME.

MELANIE

NOW TYPE TODAYS DATE.

3/17/78

THANK YOU. SOURCE CODE DEVELOPED BY MEREDITH W.
POTTER.

PLEASE REFER ALL INQUIRIES TO THE AUTHOR, AND SEE USERS
MANUAL FOR JOB INPUT STREAM.
YOU MAY NOW TYPE IN DESIRED SIMULATION RUN.

    AREA MONTANA
    AGE 200
    VEG DOUGFIR
    PRINT PLANTS
    DISTURB FIRE
    STATISTICS
    AGE 10
    AFFECT PLANTS
    AGE 60
    AFFECT PLANTS
    END

AREA IS MONTANA.
VEGETATION COMMUNITY IS DOUGFIR.
STAND AGE IS CURRENTLY 200 YEARS.

OVERSTORY SPECIES COMPOSITION

SPECIES	RELATIVE DENSITY (%)
SUBALPINE FIR (ABIES LASIOCARPA)	16
WESTERN LARCH (LARIX OCCIDENTALIS)	1
SPRUCE (PICEA)	25
WESTERN WHITE PINE (PINUS MONTICOLA)	2
DOUGLAS FIR (PSEUDOTSUGA MENZIESII)	56

UNDERSTORY SPECIES COMPOSITION

TOTAL HERB AND SHRUB COVER = 65 %
HERB + SHRUB DIVERSITY (NO. SPECIES IN 2 2 × 2 M PLOTS) =
17 SPECIES

SPECIES	ABSOLUTE COVER (%)
GRASS (GRAMINEAE)	4
BEARGRASS (XEROPHYLLUM TENAX)	6
WILLOW SP. (SALIX SP.)	1
STRAWBERRY SP. (FRAGARIA SP.)	3
FIREWEED (EPILOBIUM ANGUSTIFOLIUM)	2
COW-PARSNIP (HERACLEUM LANATUM)	1
HUCKLEBERRY SP. (VACCINIUM SP.)	7
THIMBLEBERRY (RUBUS PARVIFLORUS)	7
MEADOWRUE (THALICTRUM OCCIDENTALE)	2
ARNICA SP. (ARNICA SP.)	6
LADY FERN (ATHYRIUM FELIX-FEMINA)	5
QUEENCUP (CLINTONIA UNIFLORA)	6
FAIRY-BELL SP. (DISPORUM SP.)	1
LACEFLOWER (TIARELLA TRIFOLIATA)	7
SERVICEBERRY (AMELANCHIER ALNIFOLIA)	10
ROSE SP. (ROSA SP.)	2
SHINY-LEAF SPIRAEA (SPIRAEA BETULIFOLIA)	6
MOUNTAIN LOVER (PACHISTIMA MYRSINITES)	6
MOUNTAIN MAPLE (ACER GLABRUM)	2
VIOLET SP. (VIOLA SP.)	3
SARSAPARILLA (ARALIA NUDICAULIS)	7
DOGWOOD SP. (CORNUS SP.)	6
WESTERN TWINFLOWER (LINNAEA BOREALIS)	7
PATHFINDER (ADENOCAULON BICOLOR)	2

AREA IS MONTANA.
VEGETATION COMMUNITY IS DOUGFIR.
STAND AGE IS CURRENTLY 0 YEARS.
WEATHER IS 90 PERCENTILE FROM MOOSE CREEK
AFFIRMS STATION.

FIRE BEHAVIOR STATISTICS

RATE OF SPREAD	19.06 FT/MIN	(	5.81 M/MIN)
FIRELINE INTENSITY (BYRAMS)	3904.27 BTU/S/FT	(	3227.73 KCAL/S/M)
FLAME LENGTH	20.02 FT	(	6.10 M)
SCORCH HEIGHT	275.91 FT	(	84.10 M)

TO PRINT FIRE EFFECTS ON FIRE HAZARD, FUELS, PLANTS,
AND/OR ANIMALS, GIVE THE APPROPRIATE AFFECT COM-
MAND AND/OR ATTRIBUTE AGE COMMAND.

AREA IS MONTANA.
VEGETATION COMMUNITY IS DOUGFIR.
STAND AGE IS CURRENTLY 10 YEARS.

OVERSTORY SPECIES COMPOSITION

SPECIES	RELATIVE DENSITY (%)
WESTERN LARCH (LARIX OCCIDENTALIS)	1
SPRUCE (PICEA)	50
WESTERN WHITE PINE (PINUS MONTICOLA)	4
LODGEPOLE (PINUS CONTORTA)	15
DOUGLAS FIR (PSEUDOTSUSA MENZIESII)	20
QUAKING-ASPEN (POPULUS TREMULOIDES)	10

UNDERSTORY SPECIES COMPOSITION

TOTAL HERB AND SHRUB COVER = 75 %
HERB + SHRUB DIVERSITY (NO. SPECIES IN 2 2 × 2 M PLOTS) =
21 SPECIES

SPECIES	ABSOLUTE COVER (%)
GRASS (GRAMINEAE)	6
WILLOW SP. (SALIX SP.)	2
STRAWBERRY SP. (FRAGARIA SP.)	3
FIREWEED (EPILOBIUM ANGUSTIFOLIUM)	10
COW-PARSNIP (HERACLEUM LANATUM)	1
HUCKLEBERRY SP. (VACCINIUM SP.)	4
THIMBLEBERRY (RUBUS PARVIFLORUS)	10
MEADOWRUE (THALICTRUM OCCIDENTALE)	1
ROSE SP. (ROSA SP.)	1
SHINY-LEAF SPIRAEA (SPIRAEA BETULIFOLIA)	3
MOUNTAIN LOVER (PACHISTIMA MYRSINITES)	1
VIOLET SP. (VIOLA SP.)	1
YARROW (ACHILLEA MILLEOFLIUM)	2

AREA IS MONTANA.
VEGETATION COMMUNITY IS DOUGFIR.
STAND AGE IS CURRENTLY 60 YEARS.

OVERSTORY SPECIES COMPOSITION

SPECIES	RELATIVE DENSITY (%)
SUBALPINE FIR (ABIES LASIOCARPA)	39
WESTERN LARCH (LARIX OCCIDENTALIS)	2
SPRUCE (PICEA)	3
WESTERN WHITE PINE (PINUS MONTICOLA)	5
LODGEPOLE (PINUS CONTORTA)	30
DOUGLAS FIR (PSEUDOTSUGA MENZIESII)	21

UNDERSTORY SPECIES COMPOSITION

TOTAL HERB AND SHRUB COVER = 70 %
HERB + SHRUB DIVERSITY (NO. SPECIES IN 2 2 × 2 M PLOTS) =
16 SPECIES

SPECIES	ABSOLUTE COVER (%)
GRASS (GRAMINEAE)	3
BEARGRASS (XEROPHYLLUM TENAX)	5
STRAWBERRY SP. (FRAGARIA SP.)	4
FIREWEED (EPILOBIUM ANGUSTIFOLIUM)	1
HUCKLEBERRY SP. (VACCINIUM SP.)	6
THIMBLEBERRY (RUBUS PARVIFLORUS)	10
MEADOWRUE (THALICTRUM OCCIDENTALE)	4
ARNICA SP. (ARNICA SP.)	7
SERVICEBERRY (AMELANCHIER ALNIFOLIA)	5
ROSE SP. (ROSA SP.)	4
SHINY-LEAF SPIRAEA (SPIRAEA BETULIFOLIA)	7
MOUNTAIN LOVER (PACHISTIMA MYRSINITES)	3
VIOLET SP. (VIOLA SP.)	1
SARSAPARILLA (ARALIA NUDICAULIS)	1
DOGWOOD SP. (CORNUS SP.)	3
RUSTY MENZIESIA (MENZIESIA FERRUGINEA)	2
WESTERN TWINFLOWER (LINNAEA BOREALIS)	6

THIS ENDS THE SIMULATION FOR MELANIE
REQUESTED ON 3/17/78
THANK YOU FOR SHOPPING USDA FOREST SERVICE.
MAY THE FOREST BE WITH YOU.

WELCOME TO THE FOREST PLANNING LANGUAGE SIMULA-
TOR (FORPLAN) UNDER DEVELOPMENT FOR THE USDA
FOREST SERVICE.
R. D. AND A. PROGRAM BY
    GRADIENT MODELING, INC.
    BOX 2666
    MISSOULA, MONTANA 59806
PLEASE TYPE IN YOUR NAME.

FRANK

NOW TYPE TODAYS DATE.

3/17/78

THANK YOU. SOURCE CODE DEVELOPED BY MEREDITH W.
POTTER.
PLEASE REFER ALL INQUIRIES TO THE AUTHOR, AND SEE USERS
MANUAL FOR JOB INPUT STREAM.
YOU MAY NOW TYPE IN DESIRED SIMULATION RUN.

    AREA BLUE MTNS
    AGE 150
    VEG LODGEPOLE
    PRINT PLANTS
    PRINT ANIMALS BIRDS

DISTURB FIRE
STATISTICS
AFFECT ANIMALS BIRDS
AGE 25
AFFECT ANIMALS BIRDS
END

AREA IS THE BLUE MOUNTAINS.
VEGETATION COMMUNTY IS LODGEPOLE.
STAND AGE IS CURRENTLY 150 YEARS.

LODGEPOLE PINE CORRESPONDS TO THE THREE HALL
(1973) TYPES OF

    LODGEPOLE PINE-PINEGRASS-GROUSE HUCKLE-
    BERRY
    LODGEPOLE PINE-BIG HUCKLEBERRY
    LODGEPOLE PINE-GROUSE HUCKLEBERRY

THESE THREE TYPES FORM A BROADLY OVERLAP-
PING ELEVATION SEQUENCE
RANGING FROM 4000–7500 FEET. COMMON ON
NORTHERLY ASPECTS, SLOPES OF 2–20 PERCENT,
UNDULATING TO STEEP TOPOGRAPHY.

AS AN EXAMPLE, SPECIES COMPOSITION AND SITE
INDICES FOR THE LODGEPOLE
PINE-BIG HUCKLEBERRY TYPE ARE

DOMINANTS	PERCENT COVER	STATUS
LODGEPOLE PINE	40–60	SUCCESSIONAL TO FIRS
WHITE (GRAND) FIR	0–20	CLIMAX
BIG HUCKLEBERRY	20–60	DENSER IN SOUTH BLUES
GROUSE HUCKLEBERRY	0–20	UPPER ELEV, COLDER SITES
PINEGRASS	0–40	LOWER ELEV, WARMER SITES

SITE SPECIES	AGE	MEAN	RANGE
LODGEPOLE PINE	50	31	25–37
WHITE FIR	50	40	37–43

AREA IS THE BLUE MOUNTAINS.
VEGETATION COMMUNITY IS LODGEPOLE.
STAND AGE IS CURRENTLY 150 YEARS.

STATUS OF ALL BIRDS IN THIS HABITAT

	HABITAT UTILIZATION	
SPECIES	FEEDS	REPRO-DUCES
HARLEQUIN DUCK (HISTRIONICUS HISTRIONICUS)		x
WINTER WREN (TROGLODYTES TROGLODYTES)	x	x
PEREGRINE (FALCO PEREGRINUS)	x	x
BLUE GROUSE (DENDRAGAPUS OBSCURUS)	x	
FRANKLINS GROUSE (CANACHITES CANADENSIS)	x	x
RUFFED GROUSE (BONASA UMBELLUS)	x	
HERMIT THRUSH (CATHARUS GUTTATUS)	x	x
DARK-EYED JUNCO (JUNCO HYEMALIS)	x	x
BLACK-BILLED MAGPIE (PICA PICA)	x	
AMERICAN ROBIN (TURDUS MIGRATORIUS)	x	x
MACGILLIVRAYS WARBLER (OPORORNIS TOLMIEI)	x	x
WESTERN FLYCATCHER (EMPIDONAX DIFFICILIS)	x	x
OLIVE-SIDED FLYCATCHER (NUTTALLORNIS BOREALIS)	x	
GOLDEN-CROWNED KINGLET (REGULUS SATRAPA)	x	x
RUBY-CROWNED KINGLET (REGULUS CALENDULA)	x	x
YELLOW-RUMPED WARBLER (DENDROICA CORONATA)	x	x
TOWNSENDS WARBLER (DENDROICA TOWNSENDI)	x	
WESTERN TANAGER (PIRANGA LUDOVICIANA)	x	x
RED CROSSBILL (LOXIA CUCULLATUS)	x	x
GOSHAWK (ACCIPITER GENITILIS)	x	x
SHARP-SHINNED HAWK (ACCIPITER STRIATUS)	x	x
COOPERS HAWK (ACCIPITER COOPERII)	x	
MERLIN (FALCO COLUMBARIUS)	x	x
LONG-EARED OWL (ASIO OTUS)	x	x
RUFOUS HUMMINGBIRD (SELASPHORUS RUFUS)	x	x
HAMMONDS FLYCATCHER (EMPIDONAX HAMMONDII)	x	x
WESTERN WOOD PEEWEE (CONTOPUS SORDIDULUS)	x	x
GRAY JAY (PERISOREUS CANADENSIS)	x	x
VARIED THRUSH (IXOREUS NAEVIUS)	x	x
SOLITARY VIREO (VIREO SOLITARIUS)	x	x
BLACK-HEADED GROSBEAK (PHEUCTICUS)	x	
EVENING GROSBEAK (HESPERIPHONA VESPERTINA)	x	x
PINE GROSBEAK (PINICOLA ENUCLEATOR)	x	x
PINE SISKIN (SPINUS PINUS)	x	x
RED-TAILED HAWK (BUTEO JAMAICENSIS)	x	
GOLDEN EAGLE (AQUILA CHRYSAETOS)	x	

GREAT HORNED OWL (BUBO VIRGINIANUS)	x	
GREAT GREY OWL (STRIX NEBULOSA)	x	
HAIRY WOODPECKER (DENDROCOPOS VILLOSUS)	x	x
BLACK-BACKED THREE-TOED WOODPECKER (PICOIDES ARTICUS)	x	x
NORTHERN THREE-TOED WOODPECKER (PICOIDES TRIDACTYLUS)	x	x
RED-BREASTED NUTHATCH (SITTA CANADENSIS)	x	x
SAW-WHET OWL (AEGOLTUS ACADICUS)	x	
VAUXS SWIFT (CHEATURA VAUXI)	x	
MOUNTAIN CHICKADEE (PARUS GAMBELI)	x	x
CHESTNUT-BACKED CHICKADEE (PARUS RUFFSCENS)	x	x

AREA IS THE BLUE MOUNTAINS.
VEGETATION COMMUNITY IS LODGEPOLE.
STAND AGE IS CURRENTLY 0 YEARS.
WEATHER IS 90 PERCENTILE FROM MEACHAM AFFIRMS STATION.

FIRE BEHAVIOR STATISTICS

RATE OF SPREAD	1.79 FT/MIN	(	.55 M/MIN)
FIRELINE INTENSITY(BYRAMS)	5.79 BTU/S/FT	(	4.78 KCAL/S/M)
FLAME LENGTH	1.01 FT	(	.31 M)
SCORCH HEIGHT	3.59 FT	(	1.09 M)

TO PRINT FIRE EFFECTS ON FIRE HAZARD, FUELS, PLANTS, AND/OR ANIMALS, GIVE THE APPROPRIATE AFFECT COMMAND AND/OR ATTRIBUTE AGE COMMAND.

AREA IS THE BLUE MOUNTAINS.
VEGETATION COMMUNITY IS LODGEPOLE.
STAND AGE IS CURRENTLY 0 YEARS.

STATUS OF ALL BIRDS IN THIS HABITAT

	HABITAT UTILIZATION	
SPECIES	FEEDS	REPRO-DUCES
TURKEY VULTURE (CATHARTES AURA)	x	
PEREGRINE (FALCO PEREGRINUS)	x	x
COMMON RAVEN (CORVUS CORAX)	x	x
BLUE GROUSE (DENDRAGAPUS OBSCURUS)	x	
DARK-EYED JUNCO (JUNCO HYEMALIS)	x	x
BLACK-BILLED MAGPIE (PICA PICA)	x	
AMERICAN ROBIN (TURDUS MIGRATORIUS)	x	
OLIVE-SIDED FLYCATCHER (NUTTALLORNIS BOREALIS)	x	
COOPERS HAWK (ACCIPITER COOPERII)	x	
MERLIN (FALCO COLUMBARTUS)	x	
LONG-EARED OWL (ASIO OTUS)	x	
RUFOUS HUMMINGBIRD (SELASPHORUS RUFUS)	x	
BLACK-HEADED GROSBEAK (PHEUCTICUS)	x	
EVENING GROSBEAK (HESPERIPHONA VESPERTINA)	x	
PINE SISKIN (SPINUS PINUS)	x	
RED-TAILED HAWK (BUTEO JAMAICENSIS)	x	
GOLDEN EAGLE (AQUILA CHRYSAETOS)	x	

GREAT HORNED OWL (BUBO VIRGINIANUS)	x	
GREAT GREY OWL (STRIX NEBULOSA)	x	

AREA IS THE BLUE MOUNTAINS.
VEGETATION COMMUNITY IS LODGEPOLE.
STAND AGE IS CURRENTLY 25 YEARS.

STATUS OF ALL BIRDS IN THIS HABITAT

	HABITAT UTILIZATION	
SPECIES	FEEDS	REPRO-DUCES
WINTER WREN (TROGLODYTES TROGLODYTES)	x	x
PEREGRINE (FALCO PEREGRINUS)	x	x
BLUE GROUSE (DENDRAGAPUS OBSCURUS)	x	
FRANKLINS GROUSE (CANACHITES CANADENSIS)	x	x
HERMIT THRUSH (CATHARUS GUTTATUS)	x	
DARK-EYED JUNCO (JUNCO HYEMALIS)	x	x
WILLOW FLYCATCHER (EMPIDONAX IRAILLII)	x	x
BLACK-BILLED MAGPIE (PICA PICA)	x	
AMERICAN ROBIN (TURDUS MIGRATORIUS)	x	x
MACGILLIVRAYS WARBLER (OPORORNIS TOLMIEI)	x	x
WESTERN FLYCATCHER (EMPIDONAX DIFFICILIS)	x	
OLIVE-SIDED FLYCATCHER (NUTTALLORNIS BOREALIS)	x	
GOLDEN-CROWNED KINGLET (REGULUS SATRAPA)	x	
RUBY-CROWNED KINGLET (REGULUS CALENDULA)	x	
YELLOW-RUMPED WARBLER (DENDROICA CORONATA)	x	
WESTERN TANAGER (PIRANGA LUDOVICIANA)	x	
SHARP-SHINNED HAWK (ACCIPITER STRIATUS)	x	x
COOPERS HAWK (ACCIPITER COOPERII)	x	
MERLIN (FALCO COLUMBARIUS)	x	
LONG-EARED OWL (ASIO OTUS)	x	
RUFOUS HUMMINGBIRD (SELASPHORUS RUFUS)		x
HAMMONDS FLYCATCHER (EMPIDONAX HAMMONDII)	x	
WESTERN WOOD PEEWEE (CONTOPUS SORDIDULUS)	x	
GRAY JAY (PERISOREUS CANDENSIS)	x	x
VARIED THRUSH (IXOREUS NAEVIUS)	x	x
SOLITARY VIREO (VIREO SOLITARIUS)	x	x
BLACK-HEADED GROSBEAK (PHEUCTICUS)	x	
EVENING GROSBEAK (HESPERIPHONA VESPERTINA)	x	
PINE GROSBEAK (PINICOLA ENUCLEATOR)	x	
PINE SISKIN (SPINUS PINUS)	x	x
RED-TAILED HAWK BUTEO JAMAICENSIS)	x	
GOLDEN EAGLE (AQUILA CHRYSAETOS)	x	
GREAT HORNED OWL (BUBO VIRGINIANUS)	x	
MOUNTAIN CHICKADEE (PARUS GAMBELI)	x	x
CHESTNUT-BACKED CHICKADEE (PARUS RUFFSCENS)	x	x

THIS ENDS THE SIMULATION FOR FRANK
REQUESTED ON 3/17/78
THANK YOU FOR SHOPPING USDA FOREST SERVICE.
MAY THE FOREST BE WITH YOU.

20
Conclusions—Technology Transfer Between Researchers and Managers

Throughout this book, I have tried to convey an understanding and appreciation both of the simulation and information needs of land managers and of some of the ways we have gone about meeting those needs. We have the knowledge, the capability, and the technology to make real strides in upgrading resource management and land management planning. Simulation modeling is by no means the sole solution to our resource management problems, but it is a very important and necessary part of the solution.

We have excellent simulation capabilities today, and they improve monthly. Yet there is still so much that we do not understand and cannot predict; every time we answer a question we seem to raise 10 new ones. For example: We can model weather, fuels, and fire behavior—we know that such and such changes in fuel and weather produce a half-meter change in flame length—we know what resolution level of inventory to use to keep the error of predicting that change down to such and such a percent. However, does this error in predicting a half-meter change in flame length make any difference to an elk, an endangered plant, or the ecosystem's nutrient budget?

A major problem for those attempting to improve resource management through simulation is one of communications and information transfer between researchers and managers. As I have noted before (Kessell 1978):

1) Managers often do not communicate their most important problems to researchers.
2) Researchers often conduct projects that are not related to management problems.
3) Research results are frequently presented in a form not interpretable or usable by managers.
4) Managers frequently do not know how to, or are reluctant to, incorporate new information into active management programs.

These problems plague all phases of natural resource management; they are not limited just to the U.S. Department of the Interior National Park Service or to the United States. The seriousness of this problem cannot be overstated as long as land managers act without benefit of the best available information.

This is not meant to detract from some very real strides in technology transfer that have been made in the past few years. The U.S. Department of Agriculture Forest Service, for example, has greatly improved its fire management programs in several areas, including:

1) Availability of computer terminals in field offices for resource management information retrieval and simulation
2) Expanded national training programs, such as the annual Advanced Fire Management course conducted at the National Interagency Training Center
3) Stricter planning requirements that force managers to use existing information and evaluate the effects of all management alternatives
4) Most importantly, significant new research programs to assist managers in integrating fire considerations into land management planning.

Because of the technology transfer problems noted above, new research programs per se often do not help the land manager. The programs that do seem to help are ones that are heavily applications oriented, in which the resources are devoted to the payoffs of state of the art research. These include the R&D (research and development) and RD&A (research, development, and applications) programs described by Lotan (1979). He characterizes them by "a limited life span, an urgent problem assignment, commitment by both research and management, user involvement and commitment, and extra effort on the part of research in transferring technology to user groups." The most valuable features of those programs are, or at least should be, a strong involvment by the user groups and a significant reliance on "outside" research specialists. Let me elaborate on that last statement, as I believe it often determines the success of applications research programs.

Formal user input and feedback are mandatory at all phases of an applications research program. Lack of such input greatly decreases the probability that the results will be useful to managers. Too often, I have seen the development of expensive programs and simulators that later proved to be of marginal value; a user's suggestion of a minor change made early in the study design could have greatly increased their ultimate utility. Throughout the development of the systems described in this book, the constant interactions with the system users, through workshops, training sessions, and field tests, have been one of our greatest assets.

The second point is the need to rely on outside researchers. As Lotan (1979) notes, applications research programs require talents far beyond those readily available within government research organizations. When the problem complexity is coupled with time constraints and the demands of managers (where crises are an everyday event), extraordinary demands are placed on researchers. Too many in-house scientists "are quite adept at not only interpreting the world's problems in terms of personal interest, they quite effectively build cases for continuing ongoing research. There is a great deal of inertia built into any established research organization" (Lotan 1979). By turning to outside groups (universities and research foundations), research administrators are able to tap the talents and facilities required for the desired time frame without incurring the burdens of personnel reassignment (and future placement) and creating additional facilities. Moreover, despite the lack of long-term job security, many researchers find cooperative projects to be very attractive. They are given the opportunity to tackle an interesting problem, free of the red tape surrounding civil servants, without having to make a long-term commitment to an individual program, location, or agency. All of the modeling work described in this book was conducted in a 3-year period by three professionals (and a handful of technicians) under cooperative agreements with government land management agencies. I think the short-term nature of such projects is healthy for all parties concerned. One does not become cavalier with research goals or budgets when operating on a 12-month contact.

Yet despite these accomplishments, like the ones this book documents, we seem still to be following a meandering and rocky road toward sophisticated resource management systems. Activity and accomplishments in this area are sporadic; we achieve real success in one specialty and location, yet plod along in mediocrity in so many other vitally important areas. I believe the kinds of systems described in this book, and the cooperation between such groups as ours and government organizations, have taken us one step closer to the goal of achieving active, professional resource management. But we are still a long way from institutionalizing creativity and innovativeness.

Literature Cited

Aldrich, D. F., and R. W. Mutch. 1972. Ecological interpretations of the White Cap drainage: A basis for wilderness fire management. U.S. Department of Agriculture Forest Service, Intermountain Forest and Range Experiment Station (review draft), 109 p.

Aleksandrova, V. D. 1973. Russian approaches to classification. *In:* R. H. Whittaker (Ed.), Handbook of Vegetation Science 5: Ordination and classification of communities. Junk, The Hague, Netherlands, pp. 493–527.

Angulis, R. P. 1969. Precipitation probabilities in the western United States associated with winter 500-mb map type. U.S. Environmental Science Service Administration Technical Memo. WBTM WR 45-1.

Angulis, R. P. 1970. Precipitation probabilities in the western United States associated with summer 500-mb map type. U.S. Environmental Science Service Administration Technical Memo. WBTM WR 45-3.

Arno, S. F. 1971. Ecology of alpine larch (*Larix lyallii* Parl.) in the Pacific Northwest. Ph.D Dissertation, University of Montana, Missoula. 264 p.

Ayres, H. B. 1900. The Flathead Forest Reserve. 20th Annual Rept., U.S. Geological Survey, Part V (1898–1899), pp. 245–316.

Bailey, V. 1918. Wild animals of Glacier National Park. The mammals, with notes on physiography and life zones. U.S. Department of the Interior National Park Service.

Barkman, J. J. 1958. Phytosociology and ecology of cryptogamic epiphytes, including a taxonomic survey and description of their vegetation units in Europe. Van Gorcum, Assen, 628 p.

Barkman, J. J. 1973. Synusial approaches to classification. *In:* R. H. Whittaker (Ed.), Handbook of Vegetation Science 5: Ordination and classification of communities. Junk, The Hague, Netherlands, pp. 435–491.

Barney, R. J. 1975. Fire management: A definition. J. Forestry **73(8)**:498–519.

Barrows, J. S. 1977. The challenges of forest fire management—A guest editorial. Western Wildlands **4(1)**:55–57.

Batzli, G. O. 1969. Distribution of biomass in rocky intertidal communities on the Pacific Coast of the United States. J. Animal Ecology **38**:531–546.

Beard, J. S. 1955. The classification of tropical American vegetation- types. Ecology **36**:89–100.

Beard, J. S. 1973. The physiognomic approach. *In:* R. H. Whittaker (Ed.), Handbook of Vegetation Science 5: Ordination and classification of communities. Junk, The Hague, Netherlands, pp. 355–386.

Bevins, C. D. 1976. Fire modeling for natural fuel situations in Glacier National Park. First Conference on Scientific Research in the National Parks, New Orleans, November 1976:1225–1229.

Botkin, D. B., J. F. Janak, and J. R. Wallis. 1972. Some ecological consequences of a computer model of forest growth. J. Ecology **60**:849–872.

Bratton, S. P. 1976. Resource division in an understory herb community: responses to temporal and microtopographic gradients. American Naturalist **110**:679–693.

Braun-Blanquet, J. 1913. Die Vegetationsverhältnisse der Schneestufe in den Rätisch-Lepontischen Alpen: Ein Bild des Pflanzenlebens an seinen äussersten Grenzen. Neue Denkschriften Schweizerischen naturforschenden Geselleschaft **48**:1–347.

Braun-Blanquet, J. 1921. Prinzipien einer Systematik der Pflanzengesellschaften auf floristischer Grundlage. Jahrbuch St Gallischen naturwissenschaftlichen Gesellschaft **57**(2):305–351.

Braun-Blanquet, J. 1932. Plant sociology, the study of plant communities. Translated by G. D. Fuller and H. S. Conrad. McGraw Hill, New York. 439 p.

Braun-Blanquet, J. 1951. Pflanzensoziologie: Grundzüge der Vegetationskunde. Springer, Wien, 2nd ed. 631 p; 3rd ed., 1964, 865 p.

Bray, J. R. 1956. A study of the mutual occurrence of plant species. Ecology **37**:21–28.

Bray, J. R. 1960. The composition of the savanna vegetation of Wisconsin. Ecology **41**:721–732.

Bray, J. R. 1961. A test for estimating the relative informativeness of vegetation gradients. J. Ecology **49**:631–642.

Bray, J. R., and J. T. Curtis. 1957. An ordination of the upland forest communities of southern Wisconsin. Ecological Monogr. **27**:325–349.

Brower, L. P., and others. 1972. "The Flooding River: A Study in Riverine Ecology." A 16-mm, color–sound motion picture. Wiley, New York, 34 min.

Brown, J. K. 1971. A planar intersect method for sampling fuel volume and surface area. Forest Sci. **17**(1):96–102.

Brown, J. K. 1974. Handbook for inventorying downed woody material. U.S. Department of Agriculture Forest Service General Technical Rept. INT-16, 24 p.

Cajander, A. K. 1909. Ueber Waldtypen. Acta Forestalia Fennica **1**:1–175.

Cajander, A. K. 1949. Forest types and their significance. Acta Forestalia Fennica **56**(4):1–71.

Cajander, A. K., and Y. Ilvessalo. 1921. Über Waldtypen II. Acta Forestalia Fennica **20**:1–77.

Cattelino, P. J., I. R. Noble, R. O. Slatyer, and S. R. Kessell. 1979. Predicting the multiple pathways of plant succession. Environmental Management **3**(1):41–50.

Clements, F. E. 1905. Research Methods in Ecology. University Publishing Co., Lincoln, Nebraska, 334 p.

Clements, F. E. 1916. Plant succession: an analysis of the development of vegetation. Publications of the Carnegie Instituion, **242**: Washington, D.C., 1–512.

Clements, F. E. 1928. Plant succession and indicators. Wilson, New York, 453 p.

Clements, F. E. 1936. Nature and structure of the climax. J. Ecology **24**:252–284.

Clements, F. E., and V. E. Shelford. 1930. Bio-ecology. Wiley, New York, 425 p.

Colony, W. M. 1974. Some problems arising in the use of maps to maintain basic resource inventory data. U.S. Department of the Interior National Park Service, West Glacier, Montana, 9 p. (mimeo.)

Connell, J. H., and R. O. Slatyer. 1977. Mechanisms of succession in natural communities and their role in community stability and organization. American Naturalist **111**:1119–1144.

Costin, A. B. 1954. A Study of the Ecosystems of the Monaro Region of New South Wales with special reference to soil erosion. Soil Conservation Service of New South Wales, Sydney (A. H. Pettifer, Government Printer), 860 p.

Cottam, G., F. G. Goff, and R. H. Whittaker. 1973. Wisconsin comparative ordination. *In:* R. H. Whittaker (Ed.), Handbook of Vegetation Science 5: Ordination and classification of communities. Junk, The Hague, Netherlands, pp. 193–221.

Cowles, H. C. 1899. The ecological relations of the vegetation on the sand dunes of Lake Michigan, I. Geographical relations of the dune floras. Botanical Gazzette **27**:95–117, 167–202, 281– 308, 361–391.

Cowles, H. C. 1901. The physiographic ecology of Chicago and vicinity: A study of the origin, development, and classification of plant societies. Botanical Gazzette **31**:73–108, 144–182.

Dansereau, P. 1957. Biogeography: An ecological perspective. Ronald Press, New York. 394 p.

Daubenmire, R. F. 1943. Vegetation zonation in the Rocky Mountains. Botanical Rev. **9**:325–393.

Daubenmire, R. F. 1952. Forest vegetation of northern Idaho and adjacent Washington and its bearing on concepts of vegetation classification. Ecological Monogr. **22**:301–330.

Daubenmire, R. F. 1954. Vegetation classification. Veröffentlichungen geobotanischen Instituts, Eidgenössiche Technische Hochschule, Rübel, Zürich **29**:29–34.

Daubenmire, R. F. 1966. Vegetation: Identification of typal communities. Science **151**:291–298.

Daubenmire, R. F. 1968. Plant Communities: A textbook of plant synecology. Harper and Row, New York, 300 p.

Daubenmire, R. F. 1969. Structure and ecology of coniferous forests of the northern Rocky Mountains. *In:* Coniferous Forests of the Northern Rocky Mountains. University of Montana Foundation, Missoula, pp. 25–41.

Deeming, J. E., J. W. Lancaster, M. A. Fosberg, R. W. Furman, and M. J. Schroeder. 1972. The National Fire Danger Rating System. U.S. Department of Agriculture Forest Service Research Paper RM-84; revised 1974. 165 p.

Deeming, J. E., R. E. Burgan, and J. D. Cohen. 1977. The National Fire Danger Rating System—1978. U.S. Department of Agriculture Forest Service General Technical Rept. INT-39, 63 p.

Deevey, E. S. Jr. 1969. Specific diversity in fossil assemblages. Brookhaven Symp. Biology **22**:224–241.

Du Rietz, G. E. 1921. Zur methodologischen Grundlage der modernen Pflanzensoziologie. Holzhausen, Wien. 267 p.

Du Rietz, G. E. 1932. Vegetationsforschung auf soziationsanalytischer Grundlage. Handbuch biologischen Arbeitsmethoden **11(5)**:293–480.

Du Rietz, G. E. 1936. Classification and nomenclature of vegetation units 1930–1935. Svensk Botanisk Tidskrift **30**:580–589.

Dyson, J. L. 1962. The geologic story of Glacier National Park. Glacier Natural History Assoc. Special Bull. No. 3. 24 p.

Eastman Kodak Co. 1970. Basic Scientific Photography. Eastman Kodak Co., Rochester, New York, 40 p.

Eastman Kodak Co. 1971. Photography from Lightplanes and Helicopters. Eastman Kodak Company, Rochester, New York, 24 p.

Eastman Kodak Co. 1972. Applied Infrared Photography. Eastman Kodak Company, Rochester, New York, 88 p.

Egging, L. T. and R. J. Barney. 1979. Fire Management: A Component of land management planning. Environmental Management 3(1):15–20.

Ellenberg, H. 1967. Ecological and pedological methods of forest site mapping (in German with English summaries). Veröffentlichungen geobotanischen Instituts, Eidgenössiche Technische Hochschule, Stiftg. No. 39. Rübel, Zürich, 298 p.

Ellenberg, H., and F. Klotzli. 1967. Vegetation und Bewirtschaftung des Vogelreservates Neeracher Riet. Bericht geobotanischen Forschungsinstitut, Eidgenössiche Technische Hochschule, Stiftg. Rübel, Zürich 37:88–103.

Elton, C. S., and R. S. Miller. 1954. The ecological survey of animal communities; with a practical system of classifying habitats by structural characteristics. J. Ecology 42:460–496.

Emberger, L. 1936. Remarques critizues sur les étages de végétation dans les montagnes marocaines. Bericht Schweizerischen botanischen Gesellschaft 46:614–631.

Emberger, L. 1942. Un projed d'une classification des climate du point de vue phytogéographique. Bull. Société histoire Naturelle Toulouse 77:97–124.

Fosberg, M. 1975. Heat and water vapor flux in coniferous forest litter and duff: A theoretical model. U.S. Department of Agriculture Forest Service Research Paper RM-152. 23 p.

Fosberg, M., and R. W. Furman. 1973. Fire climates of the southwest. Agricultural Meteorology 12:27–34.

Frandsen, W. H. 1971. Fire spread through porous fuels from the conservation of energy. Combustion and Flame 16:9–16.

Frandsen, W. H. 1974. A fire spread model for spatially nonuniform forest floor fuels arrays. Western States Section, The Combustion Institute, Northridge, California. October 1974, 35 p.

Franklin, J. F. 1968. Cone production by upper-slope conifers. U.S. Department of Agriculture Forest Service Research Paper PNW-60. 21 p.

Franklin, J. F. 1979. Simulation modeling and resource management. Environmental Management 3(1):2–5.

Franklin, J. F., and C. T. Dyrness. 1973. Natural vegetation of Oregon and Washington. U.S. Department of Agriculture Forest Service General Technical Rept. PNW-8, 417 p.

Franklin, J. F., and C. E. Smith. 1974a. Seeding habits of upper-slope tree species. II. Dispersal of a mountain hemlock seedcrop on a clearcut. U.S. Department of Agriculture Forest Service Research Note PNW-214, 9 p.

Franklin, J. F., and C. E. Smith. 1974b. Seeding habits of upper-slope tree species. III. Dispersal of white and Shasta red fir seeds on a clearcut. U.S. Department of Agriculture Forest Service Research Note PNW-215, 9 p.

Franklin, J. F., R. Carkin, and J. Booth. 1974. Seeding habits of upper-slope tree species. I. A 12-year record of cone production. U.S. Department of Agriculture Forest Service Research Note PNW-213, 12 p.

Frey, T. E. 1973. The Finnish school and forest site-types. In: R. H. Whittaker (Ed.), Handbook of Vegetation Science 5: Ordination and classification of communities. Junk, The Hague, Netherlands, pp. 403–433.

Fries, T. C. E. 1913. Botanische Untersuchungen im nördlichsten Schweden: Ein Beitrag zur Kenntnis der alpinen und subalpinen Vegetation in Torne Lappmark. Vetensk. och prakt. unders. i Lappland, anordn af Luossavaara-Kiirunavaara Aktiebolag. Flora och Fauna 2:1–361.

Furman, W. R., and G. E. Brink. 1975. The National Fire Weather Data Library: What it is and how to use it. U.S. Department of Agriculture Forest Service General Technical Report RM-19, 8 p.

Gams, H. 1918. Prinzipienfragen der Vegetationsforschung: Ein Beitrag zur Begriffsklärung und Methodik der Biocoenologie. Vierteljahschrift Naturforschenden Gesellschaft, Zürich 63:293–493.

Gams, H. 1927. Von den Follateres zue Dent de Morcles: Vegetationsmonographie aus dem Wallis. Beiträge geobotanischen Landesaufnahme Schweiz 15:1–760.

Gauch, H. G. Jr. 1973a. A quantitative evaluation of the Bray–Curtis ordination. Ecology 54:829–836.

Gauch, H. G. Jr. 1973b. The relationship between sample similarity and ecological distance. Ecology 54:618–622.

Gauch, H. G. Jr. 1973c. The Cornell Ecology Programs Series. Bull. Ecological Soc. America 54(3):10–11.

Gauch, H. G. Jr., and R. H. Whittaker. 1972a. Comparison of ordination techniques. Ecology 53:868–875.

Gauch, H. G. Jr., and R. H. Whittaker. 1972b. Coenocline simulation. Ecology 53:446–451.

Gauch, H. G. Jr., and R. H. Whittaker. 1976. Simulation of community patterns. Vegetatio 33:13–16.

Gauch, H. G. Jr., R. H. Whittaker, and T. R. Wentworth. 1977. A comparative study of reciprocal averaging and other ordination techniques. J. Ecology 65:157–174.

Gauch, H. G. Jr., G. B. Chase, and R. H. Whittaker. 1974. Ordination of vegetation samples by Gaussian species distributions. Ecology 55:1382–1390.

Geiger, R. 1965. The Climate Near the Ground. Harvard University Press, Cambridge, 611 p.

Gill, A. M. 1977. Plant traits adaptive to fires in Meditteranean land ecosystems. In: Proceedings of the Symposium on the Environmental Consequences of Fire and Fuel Management in Mediterranean Climate Ecosystems (Palo Alto, California). U.S. Department of Agriculture Forest Service General Technical Rept. WO-3, pp. 17–26.

Gleason, H. A. 1926. The individualistic concept of the plant association. Bull. Torrey Botanical Club 53:7–26.

Gleason, H. A. 1939. The individualistic concept of the plant association. American Midland Naturalist 21:92–110.

Goodall, D. W. 1954. Vegetation classification and vegetation continua. Angewandte Pflanzensoziologie Wien, Festschr. Aichinger 1:168–182.

Goodall, D. W. 1973. Numerical classification. In: R. H. Whittaker (Ed.), Handbook of Vegetation Science 5: Ordination and classification of communities. Junk, The Hague, Netherlands, pp. 575–615.

Goodall, D. W. 1977. Dynamic changes in ecosystems and their study: the roles of induction and deduction. Journal of Environmental Management 5:309–317.

Goulden, C. E. 1969. Temporal changes in diversity. Brookhaven Symp. Biology 22:96–102.

Habeck, J. R. 1968. Forest succession in the Glacier Park cedar–hemlock forests. Ecology 49:872–880.

Habeck, J. R. 1969. A gradient analysis of a timberline zone at Logan Pass, Glacier National Park, Montana. Northwest Sci. 43:65–73.

Habeck. J. R. 1970a. The vegetation of Glacier National Park, Montana. U.S.

Department of the Interior National Park Service, West Glacier, Montana, 132 p. (mimeo.)

Habeck, J. R. 1970b. Fire ecology investigations in Glacier National Park. University of Montana, Missoula, 80 p. (mimeo.)

Habeck, J. R., and S. F. Arno. 1972. Ecology of alpine larch (*Larix lyallii* Parl.) in the Pacific Northwest. Ecological Monogr. **42**:417–450.

Habeck, J. R., and C. M. Choate. 1963. An analysis of krummholz communities at Logan Pass, Glacier National Park. Northwest Sci. **37**:165–166.

Habeck, J. R., and C. M. Choate. 1967. Alpine plant communities at Logan Pass, Glacier National Park. Proceedings of the Montana Academy of Science **27**:36–54.

Habeck, J. R., and T. Weaver. 1969. A chemsystematic analysis of some hybrid spruce *(Picea)* populations in Montana. Canadian J. Botany **47**:1565–1570.

Hairston, N. G. 1964. Studies on the organization of animal communities. J. Ecology **52**:227–239.

Hall, C. A. S., and J. R. Day, Jr. 1977. Systems and models: terms and basic principles. *In:* C. A. S. Hall and J. W. Day, Jr. (Eds.), Ecosystem Modeling in Theory and Practice: An Introduction with Case Histories. Wiley, New York, pp. 5–36.

Hall, F. C. 1973. Plant communities of the Blue Mountains in eastern Oregon and southeastern Washington. U.S. Department of Agriculture Forest Service Region 6 Area Guide 3-1, 62 p.

Hansen, H. P. 1946. Post glacial forests of the Glacier National Park region. Ecology **29**:146–152.

Hartl, Ph. 1976. Digital picture processing. *In:* E. Schanda (Ed.), Remote Sensing for Environmental Sciences (Ecological Studies 18). Springer-Verlag, New York, pp. 304–350.

Harvey, L. H. 1954a. Checklist of the flora of Glacier National Park. U.S. Department of the Interior National Park Service, West Glacier, Montana. (mimeo.)

Harvey, L. H. 1954b. Additions to the flora of Glacier National Park, Montana. Proceedings of the Montana Academy of Science **14**:23–25.

Hayes, G. L. 1941. Influences of altitude and aspect on daily variations in factors of forest fire danger. U.S. Department of Agriculture Forest Service Circular 591, 39 p.

Heaslip, G. B. 1976. Satellites viewing our world: The NASA LANDSAT and the NOAA SMS/GOES. Environmental Management **1**(1):15–29.

Hill, M. O. 1973a. Reciprocal averaging: An eigenvector method of ordination. J. Ecology **61**:237–249.

Hill, M. O. 1973b. Diversity and evenness: A unifying notation and its consequences. Ecology **54**:427–432.

Hitchcock, C. L., and A. Cronquist. 1973. Flora of the Pacific Northwest. University of Washington Press, Seattle. 730 p.

Holdridge, L. R. 1947. Determination of world plant formations from simple climatic data. Science **105**:367–368.

Holdridge, L. R. 1967. Life Zone Ecology. (Rev. ed.) Tropical Science Center, San Jose, California, 206 p.

Horn, H. S. 1974. Succession. *In:* R. M. May (ed.), Theoretical Ecology: Principles and Applications. Blackwell Scientific Publications, Oxford, pp. 187–204.

James, F. C. 1971. Ordinations of habitat relationships among breeding birds. Wilson Bull. **83**:215–236.

Jeske, B. W., and C. D. Bevins. 1976. Spatial and temporal distribution of natural fuels in Glacier Park. First Conference on Scientific Research in the National Parks, New Orleans, November 1976:1219–1224.

Kalensky, Z., and L. R. Scherk. 1975. Accuracy of forest mapping from LANDSAT computer compatible tapes. Proceedings of the International Symposium on Remote Sensing of the Environment 10:1159–1167.

Kan, E. P., and R. D. Dillman. 1975. Timber type separability in southeastern United States on LANDSAT-1 MSS data. Proceedings of the NASA Earth Resource Survey Symposium 135–157.

Kendeigh, S. C. 1954. History and evaluation of various concepts of plant and animal communities in North America. Ecology 35:152–171.

Kessell, S. R. 1972. A model of eastern and Carolina hemlock (*Tsuga canadensis* and *T. caroliniana*) distribution, productivity, and competition in the Linville Gorge, North Carolina. Honors Thesis, Amherst College, 199 p.

Kessell, S. R. 1973. A model for wilderness fire management. Bull. Ecological Soc. America 54(1):17.

Kessell, S. R. 1974. Checklist of Vascular Plants of Glacier National Park, Montana. Glacier Natural History Assoc., West Glacier, Montana, 79 p.

Kessell, S. R. 1975a. Glacier National Park Basic Resource and Fire Ecology Systems Model. Bull. Ecological Soc. America 56(2):49.

Kessell, S. R. 1975b. Glacier National Park Basic Resource and Fire Ecology Systems Model: User's Manual, (2nd ed.) Gradient Modeling, Inc., West Glacier, Montana, 44 p.

Kessell, S. R. 1976a. Wildland inventories and fire modeling by gradient analysis in Glacier National Park. Proceedings of the 1974 Joint Tall Timbers Fire Ecology Conference and Intermountain Fire Research Council Fire and Land Management Symposium 14:115–162.

Kessell, S. R. 1976b. Gradient Modeling: A new approach to fire modeling and wilderness resource management. Environmental Management 1(1):39–48.

Kessell, S. R. 1976c. Responsible management of biological systems? Environmental Management 1(2):99–100.

Kessell, S. R. 1976d. The Glacier National Park Basic Resource and Fire Ecology Systems Model: Systems Manual. Gradient Modeling, Inc., West Glacier, Montana, 39 p.

Kessell, S. R. 1976e. The Glacier National Park Basic Resource and Fire Ecology Systems Model: Handbook for Coding the Remote Site Inventory. Gradient Modeling, Inc., West Glacier, Montana, 39 p.

Kessell, S. R. 1977. Gradient Modeling: A new approach to fire modeling and resource management. In: C. A. S. Hall and J. W. Day, Jr. (Eds.), Ecosystem Modeling in Theory and Practice: An Introduction with Case Histories. Wiley, New York, pp. 575–605.

Kessell S. R. 1978. Perspectives in fire research. Environmental Management 2(4):291–294.

Kessell, S. R. 1979a. Adaptation and dimorphism in eastern hemlock, *Tsuga canadensis* (L.) Carr. American Naturalist 113:333–350.

Kessell, S. R. 1979b. Phytosociological inference and resource management. Environmental Management 3(1):29–40.

Kessell, S. R. 1979c. Fire modeling, fire management, and land management planning. Environmental Management 3(1):1–2.

Kessell, S. R., and P. J. Cattelino. 1978. Evaluation of a Fire Behavior Information

Integration System for southern California chaparral wildlands. Environmental Management **2(2)**:135–157.

Kessell, S. R., and R. H. Whittaker. 1976. Comparisons of three ordination techniques. Vegetatio **32**:21–29.

Kessell, S. R., D. B. Dwyer, and W. M. Colony. 1975. Gradient analysis and resource management. Bull. Ecological Soc. America **56(2)**:50.

Kessell, S. R., P. J. Cattelino, and M. W. Potter. 1977a. A fire behavior information integration system for southern California chaparral. *In:* Proceedings of the Symposium on the Environmental Consequences of Fire and Fuel Management in Mediterranean Climate Ecosystems (Palo Alto, California,). U.S. Department of Agriculture Forest Service General Technical Rept. WO-3, pp. 354–360.

Kessell, S. R., C. D. Bevins, L. Bradshaw, B. W. Jeske, M. W. Potter, and D. B. Dwyer. 1977b. Final Report to Supplement 3 between USDA Forest Service Intermountain Forest and Range Experiment Station and Gradient Modeling, Inc.: Fire Management Information Integration System *(GANDALF)*. Gradient Modeling, Inc., Missoula, Montana, 248 p.

Kessell, S. R., M. W. Potter, and L. Bradshaw. 1977c. Final Report to Supplement 2 between USDA Forest Service Intermountain Forest and Range Experiment Station and Gradient Modeling, Inc.: Evaluation of Fire Management Information Integration System in Operational Land Management Situations. Gradient Modeling, Inc., Missoula, Montana, 106 p.

Kessell, S. R., M. W. Potter, C. D. Bevins, L. Bradshaw, and B. W. Jeske. 1978. Analysis and application of forest fuels data. Environmental Management **2(4)**:347–363.

Kessell, S. R., P. J. Cattelino, and M. W. Potter. 1979. Guidelines for implementing *FORPLAN* (The *FORest Planning LANguage* and Simulator) on any selected national forest. Gradient Modeling, Inc., Missoula, Montana, 48 p.

King, C. E. 1964. Relative abundance of species and MacArthur's model. Ecology **45**:716–727.

Koterba, W. D. 1968. An analysis of the North Fork grasslands in Glacier National Park, Montana. Masters Thesis, University of Montana, Missoula, 81 p.

Kourtz, P. H. 1977. An application of LANDSAT digital technology to forest fire fuel type mapping. Proceedings of the International Symposium on Remote Sensing of the Environment **11**:1111–1115.

Küchler, A. W. 1964. Potential natural vegetation of the conterminous United States. American Geographical Soc., Special Publication No. 36, 116 p. (Rev. ed. of map, 1965.)

Küchler, A. W. 1966. International bibliography of vegetation maps. Vol. 2, Europe. University of Kansas Library Series, 584 p.

Küchler, A. W. 1967. Vegetation Mapping. Ronald Press, New York, 472 p.

Küchler, A. W. 1968. International bibliography of vegetation maps. Vol. 3, USSR, Asia and Australia. University of Kansas Library Series, 389 p.

Küchler, A. W. 1970. International bibliography of vegetation maps. Vol. 4, Africa, South America and the World (General). University of Kansas Library Series, 561 p.

Küchler, A. W., and J. McCormick. 1965. International bibliography of vegetation maps. Vol. 1, North America. University of Kansas Library Series, 453 p.

Küchler, A. W. and J. O. Sawyer, Jr. 1967. A study of the vegetation near Chiengmai, Thailand. Kansas Academy of Science **70**:281–348.

Lippmaa, T. 1933. Aperçu général sur la végétation autochtone du Lautaret

(Hautes-Alpes) avec des remarques critiques sur quelques notions phytosociolo-giques. (Estonian summary). Acta Instituti horti botanici Tartuenis 3(3):1–108.

Lippmaa, T. 1935. Une analyse des forêts de l'île estonienne d'Abruka (Abro) sur la base des associations unistrates. Acta Instituti horti Botanici Tartuenis 4(no. 1/2, art. 5):1–97.

Lippmaa, T. 1939. The unistratal concept of plant communities (the unions). American Midland Naturalist 21:111–145.

Lloyd, M., and R. J. Ghelardi. 1964. A table for calculating the "equitability" component of species diversity. J. Animal Ecology 33:217–225.

Lotan, J. E. 1979. Integrating fire management into land management planning: A multiple-use management research, development, and applications program. Environmental Management 3(1):7–14.

Lund, I. P. 1963. Map-pattern classification by statistical methods. J. Applied Meteorology 2:56–65.

MacArthur, R. H. 1957. On the relative abundance of bird species. Proceedings of the National Academy of Sciences (U.S.A.) 43:293–295.

MacArthur, R. H. 1960. On the relative abundance of species. American Naturalist 94:25–36.

Maki, D. P., and M. Thompson. 1973. Mathematical Models and Applications. Prentice-Hall, Englewood Cliffs, N.J., 492 p.

Mason, D. L., and S. R. Kessell. 1975. A fuel moisture and microclimate model for Glacier National Park. Bull. Ecological Soc. America 56(2):50.

McArthur, A. G. 1977. Bush fires in Australia. Australian Government Publishing Service, Canberra, 359 p.

Merriam, C. H. 1890. Results of a biological study of the San Francisco Mountain region and desert of the Little Colorado, Arizona. North American Fauna 3:1–136.

Merrian, C. H. 1894. Laws of temperature control of the geographic distribution of terrestrial animals and plants. National Geographic Magazine 6:229–238.

Merrian, C. H. 1898. Life zones and crop zones of the United States. Bull. Bureau Biological Surveys (USDA) 10:1–79.

Merrian, C. H. 1899. Results of a biological survey of Mount Shasta, California. North American Fauna 16:1–179.

Motomura, I. 1932. A statistical treatment of associations. Japanese J. Zoology 44:379–383.

Mueller-Dombois, D. 1972. Crown distortion and elephant distribution in the woody vegetations of Ruhuna National Park, Ceylon. Ecology 53:208–226.

Mueller-Dombois, D., and H. Ellenberg. 1974. Aims and Methods of Vegetation Ecology. Wiley, New York, 547 p.

Nichols, J. D. 1975. Mapping of the wildland fuel characteristics of the Santa Monica Mountains of southern California. Proceedings of the NASA Earth Research Survey Symposium 159–166.

Niering, W. A., and R. H. Goodwin. 1973. Creation of relatively stable shrublands with herbicides: arresting succession on rights of way and pastureland. Bull. Ecological Soc. America 54(1):17.

Noble, I. R., and R. O. Slatyer. 1977. Post-fire succession of plants in Mediterra-nean ecosystem. In: Proceedings of the Symposium on the Environmental Consequences of Fire and Fuel Management in Mediterranean Climate Ecosys-tems (Palo Alto, California). U.S. Department of Agriculture Forest Service General Technical Rept. WO-3, pp. 27–36.

Noble, I. R., and R. O. Slatyer, 1978. Recent developments in ecological succession theory. *In:* A. O. Nicholls (Ed.), Report on a Workshop on Succession and Disturbance in Australia held at Shortwater Bay, Queensland, 26 November to 2 December 1977. Commonwealth Scientific and Industrial Research Organization Division of Land Use Research (Canberra) Technical Memorandum 78/26. pp. 14–24.

Noble, I. R., and R. O. Slatyer. 1979. The use of vital attributes to predict successional changes in plant communities subject to recurring disturbance. Vegetatio: in press.

Noy-Mier, I., and R. H. Whittaker. 1977. Continuous multivariate methods in community analysis: some problems and developments. Vegetatio 33:79–98.

O'Brien, D. M. 1969. A report on occurrence of lightning caused fires in Glacier National Park (1910–1968). U.S. Department of the Interior National Park Service, West Glacier, Montana, 28 p. (mimeo.)

Odum, E. P. 1969. The strategy of ecosystem development. Science 164:262–270.

Orloci, L. 1966. Geometric models in ecology. I. The theory and application of some ordination methods. J. Ecology 54:193–215.

Orloci, L. 1973. Ordination by resemblance matrices. *In:* R. H. Whittaker (Ed.), Handbook of Vegetation Science 5: Ordination and classification of communities. Junk, The Hague, Netherlands, pp. 249–286.

Orloci, L. 1975. Multivariate Analysis in Vegetation Research. Junk, The Hague, Netherlands, 276 p.

Patrick, R., M. H. Hohn, and J. H. Wallace. 1954. A new method for determining the pattern of the diatom flora. Notulae Naturae (Academy Natural Sciences of Philadelphia) 259:1–12.

Paegle, J. N. 1974. Prediction of precipitation probability based on 500-mb flow types. J. Applied Meteorology 13:213–220.

Paegle, J. N., and R. P. Wright. 1975. Forecast of precipitation probability based on a pattern recognition algorithm. J. Applied Meteorology 14:180–188.

Peet, R. K. 1974. The measurement of species diversity. Annual Rev. Ecology and Systematics 5:285–307.

Pfister, R. D., B. L. Kovalchik, S. F. Arno, and R. C. Presby. 1977. Forest habitat types of Montana. U.S. Department of Agriculture Forest Service General Technical Rept. INT-34, 174 p.

Phillips, C. B. 1977. Fire protection and fuel management on privately-owned wildlands in California. *In:* Proceedings of the Symposium on the Environmental Consequences of Fire and Fuel Management in Mediterranean Climate Ecosystems (Palo Alto, California). U.S. Department of Agriculture Forest Service General Technical Rept. WO-3, pp. 348–353.

Pielou, E. C. 1966. The measurement of diversity in different types of biological collections. J. Theoretical Biology 13:131–144.

Potter, M. W., S. R. Kessell, and P. J. Cattelino. 1978. *FORPLAN (FORest Planning LANguage and Simulator)* User's Guide. (2nd Ed.) Gradient Modeling, Inc., Missoula, Montana, 75 p.

Potter, M. W., S. R. Kessell, and P. J. Cattelino. 1979. *FORPLAN—A FORest Planning LANguage and Simulator.* Environmental Management 3(1):59–72.

Preston, F. W. 1948. The commonness and rarity of species. Ecology 29:254–283.

Preston, F. W. 1962. The canonical distribution of commonness and rarity. Ecology 43:185–215, 410–432.

Ramensky, L. G. 1926. Die Grundgesetzmassigkeiten im Aufbau der Vegetationsdecke. (Abstract from Vêstnik opỹtnago dela, Voronezh 1924; 37–73). Bot. Zbl. N. F. **7**:453–455.

Ramensky, L. G. 1930. Zur Methodik der vergleischended Bearbeitung und Ordnung von Pflanzenlisten und anderen Objekten, die durch mehrere, verschiedenartig wirkende Faktoren bestimmt werden. Beiträge Biologie Pflanzen **18**:269–304.

Rasch, G. E., and A. E. MacDonald. 1975. Map type precipitation probabilities for the Western Region. NOAA Technical Memo. NWS WR-96.

Reifsnyder, W. E. 1976. Development of a fire danger rating system for universal application. Proceedings of the Fourth Conference on Fire and Forest Meteorology (St. Louis, Missouri), November 1976. U.S. Department of Agriculture Forest Service General Technical Rept. RM-32.

Reiners, W. A., I. A. Worley, and D. B. Lawrence. 1970. Plant diversity in a chronosequence at Glacier Bay, Alaska. Ecology **52**:55–69.

Rothermel, R. C. 1972. A mathematical model for predicting fire spread rate and intensity in wildland fuels. U.S. Department of Agriculture Forest Service Research Paper INT-115, 40 p.

Rowe, J. S. 1959. (3rd ed., 1972.) Forest regions of Canada. Canadian Department of Northern Affairs and Natural Resources, Forestry Branch Bull. **123**:1–71.

Salerno, A. E. 1976. Aerospace photography. *In:* E. Schanda (Ed.), Remote Sensing for Environmental Sciences (Ecological Studies 18). Springer-Verlag, New York, pp. 11–83.

Schanda, E. 1976. Introductory remarks on remote sensing. *In:* E. Schanda (Ed.), Remote Sensing for Environmental Sciences (Ecological Studies 18). Springer-Verlag, New York, pp. 1–10.

Schmithusen, J. 1968. Vegetation maps at 1:25 million of Europe, North Asia, South Asia, SW Asia, Australia, N Africa, S Africa, N America, Central America, South America (north part), South America (south part). **XVIII**:321–346. *In:* Grosses Duden-Lexikon (Bibliographisches Institut A. G., Mannheim).

Schroeder, M. J., and C. C. Buck. 1970. Fire Weather. U.S. Department of Agriculture Forest Service Agricultural Handbook No. 360, 229 p.

Singer, F. J. 1975a. Wildfire and ungulates in the Glacier National Park area, northwestern Montana. Masters Thesis, University of Idaho, Moscow, 53 p.

Singer, F. J. 1975b. Behavior of mountain goats, elk, and other wildlife in relation to U.S. Highway 2, Glacier National Park. U.S. Department of the Interior National Park Service, West Glacier, Montana, 96 p.

Stage, A. R. 1973. Prognosis model for stand development. U.S. Department of Agriculture Forest Service Research Paper INT-137, 32 p.

Standley, P. C. 1921. Flora of Glacier National Park. U.S. National Herbarium **22**:235–438.

Standley, P. C. 1926. Plants of Glacier National Park. U.S. Department of the Interior National Park Service.

Storey, T. G. 1972. FOCUS: A computer simulation model for fire control planning. Fire Technology **8(2)**:91–103.

Tansley, A. G. (Ed.) 1911. Types of British Vegetation. Cambridge Univ. Press, Cambridge, 416 p.

Tansley, A. G. 1920. The classification of vegetation and the concept of development. J. Ecology **8**:118–149.

Tansley, A. G. 1939. The British Islands and their Vegetation. Cambridge University Press, Cambridge, 930 p.

Thomas, J. W., R. Miller, C. Maser, R. Anderson, and B. Carter. 1977. The relationship of terrestrial vertebrates to plant communities and their successional stages. Proposed final draft, for Forest–Wildlife Relationships in the Blue Mountains of Washington and Oregon, U.S. Department of Agriculture Forest Service.

Trass, H., and N. Malmer. 1973. North European approach to classification. *In:* R. H. Whittaker (Ed.), Handbook of Vegetation Science **5**: Ordination and classification of communities. Junk, The Hague, Netherlands, pp. 529–574.

Vandermeer, J. H., and R. H. MacArthur. 1966. A reformulation of alternative (b) of the broken stick model of species abundance. Ecology **47**:139–140.

Vogl, R. J. 1977. Fire frequency and site degradation. *In:* Proceedings of the Symposium on the Environmental Consequences of Fire and Fuel Management in Mediterranean Climate Ecosystems (Palo Alto, California). U.S. Department of Agriculture Forest Service General Technical Rept. WO-3, pp. 193–201.

Walter, H. 1977. Effects of fire on wildlife communities. *In:* Proceedings of the Symposium on the Environmental Consequences of Fire and Fuel Management in Mediterranean Climate Ecosystems (Palo Alto, California). U.S. Department of Agriculture Forest Service General Technical Rept. WO-3, pp. 183–192.

Weaver, J. E., and F. E. Clements. 1929. Plant Ecology. McGraw Hill, New York, 520 p. 3rd ed., 1938, 601 p.

Westhoff, V., and E. van der Maarel. 1973. The Braun-Blanquet approach. *In:* R. H. Whittaker (Ed.), Handbook of Vegetation Science **5**: Ordination and Classification of Communities. Junk, The Hague, Netherlands, pp. 617–726.

Whittaker, R. H. 1953. A consideration of climax theory: the climax as a population and pattern. Ecological Monogr. **23**:41–78.

Whittaker, R. H. 1956. Vegetation of the Great Smoky Mountains. Ecological Monogr. **26**:1–80.

Whittaker, R. H. 1960. Vegetation of the Siskiyou Mountains, Oregon and California. Ecological Monogr. **30**:279–338.

Whittaker, R. H. 1962. Classification of natural communities. Botanical Rev. **28**:1–239.

Whittaker, R. H. 1965. Dominance and diversity in land plant communities. Science **147**:250–260.

Whittaker, R. H. 1967. Gradient analysis of vegetation. Biological Rev. **42**:207–264.

Whittaker, R. H. 1969. Evolution of diversity in plant communities. Brookhaven Symp. Biology **22**:173–196.

Whittaker, R. H. 1970. The population structure of vegetation. *In:* R. Tuxen, (Ed.), Gesellschaftsmorphologie. Bericht Internationalen Symposium Rinteln, 1966. Junk, The Hague, Netherlands, pp. 39–62.

Whittaker, R. H. 1972. Evolution and measurement of species diversity. Taxon **21**:213–251.

Whittaker, R. H. 1973a. Introduction. *In:* R. H. Whittaker (Ed.), Handbook of Vegetation Science **5**: Ordination and classification of communities. Junk, The Hague, Netherlands, pp. 3–6.

Whittaker, R. H. 1973b. Approaches to classifying vegetation. *In:* R. H. Whittaker (Ed.), Handbook of Vegetation Science **5**: Ordination and classification of communities. Junk, The Hague, Netherlands, pp. 323–354.

Whittaker, R. H. 1973c. Dominance-types. *In:* R. H. Whittaker (Ed.), Handbook of Vegetation Science **5**: Ordination and classification of communities. Junk, The Hague, Netherlands, pp. 387–402.

Whittaker, R. H. 1973d. Direct gradient analysis: techniques. *In:* R. H. Whittaker (Ed.), Handbook of Vegetation Science 5: Ordination and classification of communities. Junk, The Hague, Netherlands, pp. 7–31.

Whittaker, R. H. 1973e. Direct gradient analysis: results. *In:* R. H. Whittaker (Ed.), Handbook of Vegetation Science 5: Ordination and classification of communities. Junk, The Hague, Netherlands, pp. 33–51.

Whittaker, R. H. 1974. Climax concepts and recognition. *In:* R. Knapp (Ed.), Handbook of Vegetation Science 8: Vegetation dynamics. Junk, The Hague, Netherlands, pp. 137–154.

Whittaker, R. H. 1975a. Functional aspects of succession in deciduous forests. *In:* R. Tuxen (Ed.), Sukzessionsforschung. Bericht Internationalen Symposium Rinteln, 1973. Junk, The Hague, Netherlands, pp. 377–405.

Whittaker, R. H. 1975b. Communities and Ecosystems. (2nd ed.) MacMillan, New York. 385 p.

Whittaker, R. H., and H. G. Gauch, Jr. 1973. Evaluation of ordination techniques. *In:* R. H. Whittaker (ed.), Handbook of Vegetation Science 5: Ordination and classification of communities. Junk, The Hague, Netherlands, pp. 287–321.

Whittaker, R. H., and W. A. Niering. 1964. Vegetation of the Santa Catalina Mountains, Arizona. I. Ecological classification and distribution of species. J. Arizona Academy of Science 3:9–34.

Whittaker, R. H., and W. A. Niering. 1965. Vegetation of the Santa Catalina Mountains. Arizona: a gradient analysis of the south slope. Ecology 46:429–452.

Whittaker, R. H., and G. M. Woodwell. 1969. Structure, production, and diversity of the oak–pine forest at Brookhaven, New York. J. Ecology 57:155–174.

Williams, C. B. 1953. The relative abundance of different species in a wild animal population. J. Animal Ecology 22:14–31.

Williams, C. B. 1964. Patterns in the Balance of Nature, and Related Problems of Quantitative Ecology. Academic Press, New York, 324 p.

Williams, D. L. 1976. A canopy-related stratification of a southern pine forest using LANDSAT digital data. *In:* Proceedings of the American Society of Foresters Fall Convention (Seattle, Washington). pp. 231–239.

APPENDIX 1

SCIENTIFIC AND COMMON NAMES FOR SPECIES MENTIONED IN THE TEXT*

Abies Mill. (Fir) (Pinaceae)
   A. grandis (Dougl.) Forbes
   (Grand fir)
   A. lasiocarpa (Hook.) Nutt.
   (Subalpine fir)
Acer L. (Maple) (Aceraceae)
   A. glabrum Torr. (Mountain
   maple)
Achillea L. (Yarrow)
   (Compositae)
   A. millefolium L. (Yarrow)
Adenocaulon Hook. (Pathfinder)
   (Compositae)
   A. biocolor Hook.
   (Pathfinder)
Agropyron Gaertn. (Wheatgrass)
   (Gramineae)
   A. canium (L.) Beauv.
   (Awned wheatgrass)

A. spicatum (Pursh) Scrign. &
   Smith (Bluebunch
   wheatgrass)
Alnus Hill (Alder) (Betulaceae)
   A. incana (L.) Moench
   (Mountain alder)
   A. sinuata (Regel) Rybd.
   (Green alder)
Amelanchier Medic.
   (Serviceberry) (Rosaceae)
   A. alnifolia Nutt. (Western
   serviceberry)
Aralia L. (Sarsaparilla)
   (Araliaceae)
   A. nudicaulis L. (Wild
   sarsaparilla)
Arnica L. (Arnica) (Compositae)
   A. alpina (L.) Olin (Alpine
   arnica)

*Based on Hitchcock and Cronquist (1973) and Kessell (1974).

A. diversifolia Greene (Sticky
arnica)
(probably = *A. mollis* Hook.
or *A. amplexicaulis* Nutt. X
A. cordifolia Hook. or *A.
latifolia* Bong.)
A. latifolia Bong. (Mountain
arnica)
A. longifolia D. C. Eat. (Seep-
spring arnica)
A. mollis Hook. (Hairy
arnica)
A. parryi Gray (Nodding
arnica)
Artemesia L. (Sagebrush)
(Compositae)
A. michauxiana Bess.
(Michaux mugwort)
A. tridentata Nutt. (Big
sagebrush)
Athyrium Roth (Lady fern)
(Polypodiaceae)
A. filix-femina (L.) Roth
(Lady fern)

Betula L. (Birch) (Betulaceae)
B. occidentalis Hook. (Water
birch)
B. papyrifera Marsh.
(Western paper birch)

Carex. L. (Sedge) (Cyperaceae)
C. geyeri Booth (Geyer's
sedge)
C. nigricans Retz. (Black
alpine sedge)
C. spectabilis Dewey (Showy
sedge)
Castilleja Motis ex L.f. (Indian-
paintbrush) (Scrophulariaceae)
C. rhexifolia Rybd. (Red
indian-paintbrush)
C. sulphurea Rybd. (Sulfur
indian-paintbrush)

Ceanothus L. (Ceanothus)
(Rhamnaceae)
C. sanguineus Pursh
(Redstem ceanothus)
C. velutinus Dougl.
(Deerbrush)
Chimaphila Pursh (Pipsissewa)
(Ericaceae)
C. umbellata (L.) Bart.
(Pipsissewa)
Clintonia Raf. (Queencup)
(Liliaceae)
C. uniflora (Schult.) Kunth.
(Queencup)
Cornus L. (Dogwood)
(Cornaceae)
C. canadensis L.
(Bunchberry)
C. stolonifera Michx. (Red-
osier dogwood)
Cryptogramma R. Br. (Rock-
brake) (Polypodiaceae)
C. crispa (L.) R. Br. (Parsley-
fern)

Disporum Salisb. (Fairy-bell)
(Liliaceae)
D. hookeri (Torr.) Nicholson
(Hooker fairy-bell)
D. trachycarpum (Wats.)
Benth. & Hook. (Sierra
fairy-bell)

Eleocharis R. Br. (Spike-rush)
(Cyperaceae)
E. palustris (L.) R.&S.
(Common spike-rush)
Epilobium L. (Fireweed)
(Onagraceae)
E. alpinum L. (Talus
fireweed)
E. angustifolium L. (Common
fireweed)
E. latifolium L. (Alpine
fireweed)

Erigeron L. (Fleabane, daisy)
(Compositae)
    E. acris L. (Bitter fleabane)
    E. leiomerus Gray (Smooth
    daisy)
    E. peregrinus (Pursh) Greene
    (Showy fleabane)
    E. simplex L. (Alpine daisy)
    E. speciosus (Lindl.) DC.
    (Showy fleabane)
Erigonum Michx. (Buckwheat)
(Polygonaceae)
    E. flavum Nutt. (Sulphur-
    plant)

Festuca L. (Fescue) (Gramineae)
    F. idahoensis Elmer (Idaho
    fescue)
    F. scabrella Torr. (Rough
    fescue)
Fragaria L. (Strawberry)
(Rosaceae)
    F. vesca L. (Woods
    strawberry)
    F. virginiana Duchesne
    (Blueleaf strawberry)

Galium L. (Bedstraw)
(Rubiaceae)
    G. boreale L. (Baby's-breath)
    G. triflorum Michx.
    (Sweetscented bedstraw)
Goodyera R. Br. (Rattlesnake-
plantain) (Orchidaceae)
    G. oblongifolia Raf. (Western
    rattlesnake-plantain)

Heracleum L. (Cow-parsnip)
(Umbelliferae)
    H. lanatum Michx. (Cow-
    parsnip)
Hypericum L. (St. John's-wort)
(Hypericaceae)
    H. formosum H.B.K.
    (Western St. John's-wort)

Juncus L. (Rush) (Juncaceae)
    J. drummondii E. Meyer
    (Drummond's rush)
Juniperis L. (Juniper)
(Cupressaceae)
    J. communis L. (Common
    juniper)
    J. horizontalis Moench.
    (Creeping juniper)
    J. scopulorum Sarg. (Rocky
    Mountain juniper)

Larix Adans. (Larch) (Pinaceae)
    L. lyallii Parl. (Alpine larch)
    L. occidentalis Nutt.
    (Western larch)
Lemna L. (Duckweed)
(Lemnaceae)
    L. minor L. (Water lentil)
    L. trisulca L. (Star duckweed)
Linnaea L. (Twinflower)
(Caprifoliaceae)
    L. borealis L. (Western
    twinflower)
Luzula DC. (Woodrush)
(Juncaceae)
    L. hitchcockii Hamet-Ahti
    (Smooth woodrush)
    L. piperi (Cov.) Jones (Piper's
    woodrush)
    L. spicata (L.) DC. (Spiked
    woodrush)
Lycopodium L. (Clubmoss)
(Lycopodiaceae)
    L. alpinum L. (Alpine
    clubmoss)
    L. selago L. (Fir clubmoss)

Menziesia Smith (Menziesia)
(Ericaceae)
    M. ferruginea Smith (Busty
    menziesia)
Mimulus L. (Monkey-flower)
(Scrophulariaceae)
    M. lewisii Pursh (Purple
    monkey-flower)

M. tilingii Regel (Yellow
monkey-flower)

Osmorhiza Raf. (Sweet-cicely)
(Umbelliferae)
  O. chilensis H.&A.
  (Mountain sweet-cicely)
  O. occidentalis (Nutt.) Torr.
  (Western sweet-cicely)
Oxyria Hill (Mountain sorrel)
(Polygonaceae)
  O. digyna (L.) Hill (Mountain
  sorrel)

Pachistima Raf. (Mountain lover)
(Celastraceae)
  P. myrsinites (Pursh) Raf.
  (Mountain lover)
Penstemon Mitch. (Beardtongue)
(Scrophulariaceae)
  P. lyallii Gray (Lyall's
  beardtongue)
  P. nitidus Dougl. (Shining
  penstemon)
  P. wilcoxii Rybd. (Wilcox's
  penstemon)
Picea A. Dietr. (Spruce)
(Pinaceae)
  P. engelmannii Parry
  (Engelmann spruce)
  P. glauca (moench) Voss
  (White spruce)
  (also widespread *P.
  engelmannii* X *P. glauca*)
Pinus L. (Pine) (Pinaceae)
  P. albicaulis Engelm.
  (Whitebark pine)
  P. contorta Dougl.
  (Lodgepole pine)
  P. flexilis James (Limber pine)
  P. monticola Dougl. (Western
  white pine)
  P. ponderosa Dougl.
  (Ponderosa pine)

(also *P. albicaulis* X *P.
monticola*)
Poa L. (Bluegrass) (Gramineae)
  P. alpina L. (Alpine
  Bluegrass)
  P. gracillima Vasey (Pacific
  bluegrass)
  P. leptocoma Trin. (Bog
  bluegrass)
Polystichum Roth (Holly fern)
(Polypodiaceae)
  P. lonchitis (L.) Roth (Holly
  fern)
Populus L. (Aspen) (Salicaceae)
  P. tremuloides Michx.
  (Quaking-aspen)
  P. trichocarpa T.&G.
  (Cottonwood)
Potamogeton L. (Pondweed)
(Potamogetonaceae)
  P. alpinus Balbis (Northern
  pondweed)
  P. natans L. (Broad-leaved
  pondweed)
Pseudotsuga Carr. (Douglas-fir)
(Pinaceae)
  P. menziesii (Mirbel) Franco.
  (Douglas-fir)
Pyrola L. (Pyrola) (Ericaceae)
  P. assarifolia Michx. (Pink
  pyrola)
  P. secunda L. (Sidebells
  pyrola)

Ribes L. (Currant)
(Grossulariaceae)
  R. cereum Dougl. (Squaw
  currant)
  R. inerme Rybd.
  (Gooseberry)
  R. lacustre (Pers.) Poir.
  (Spiny currant)
  R. viscosissimum Pursh
  (Sticky currant)

Rosa L. (Rose) (Rosaceae)
  R. acicularis Lindl. (Prickly
    rose)
  R. gymnocarpum Nutt.
    (Little wild rose)
  R. woodsii Lindl. (Pearhip
    rose)
Rubus L. (Blackberry, raspberry)
  (Rosaceae)
  R. idaeus L. (Red raspberry)
  R. parviflorus Nutt.
    (Thimbleberry)

Salix L. (Willow) (Salicaceae)
  S. arctica Pall. (Arctic
    willow)
  S. candida Fluegge (Hoary
    willow)
  S. drummondiana Barratt
    (Drummond willow)
  S. glauca L. (Glaucous
    willow)
  S. nivalis Hook. (Snow
    willow)
  S. scouleriana Barratt (Rock
    willow)
Sedum L. (Stonecrop)
  (Crassulaceae)
  S. lanceolatum Torr. (Yellow
    stonecrop)
  S. rosea (L.) Scop. (Red
    orpine)
  S. stenopetalum Pursh
    (Wormleaf sedum)
Senecio L. (Ragwort,
  butterweed) (Compositae)
  S. foetidus Howell (Sweet-
    marsh butterweed)
  S. fremontii T.&G. (Dwaft
    mountain butterweed)
  S. megacephalus Nutt.
    (Large-headed butterweed)
  S. resedifolius Less. (Dwaft
    arctic butterweed)
  S. triangularis Hook. (Tall
    ragwort)

Sorbus L. (Mountain-ash)
  (Rosaceae)
  S. scopulina Greene (Cascade
    mountain-ash)
Sparganium L. (Bur-reed)
  (Sparganiaceae)
  S. angustifolium Michx.
    (Narrowleaf bur-reed)
Spiraea L. (Spiraea) (Rosaceae)
  S. betulifolia Pall. (Shiny-leaf
    spiraea)
  S. densiflora Nutt. (Subalpine
    spiraea)
Stipa L. (Needlegrass)
  (Gramineae)
  S. richardsonii Link
    (Richardson's needlegrass)

Taxus L. (Yew) (Taxaceae)
  T. brevifolia Nutt. (Western
    yew)
Thalictrum L. (Meadowrue)
  (Ranunculaceae)
  T. occidentale Gray (Western
    meadowrue)
Thuja L. (Arbovitae, redcedar)
  (Cupressaceae)
  T. plicata Donn (Western
    redcedar)
Tiarella L. (Laceflower)
  (Saxifragaceae)
  T. trifoliata L. (Laceflower,
    foamflower)
Tsuga L. (Hemlock) (Pinaceae)
  T. heterophylla (Raf.) Sarg.
    (Western hemlock)
Typha L. (Cat-tail) (Typhaceae)
  T. latifolia L. (Common cat-
    tail)

Urtica L. (Nettle) (Urticaceae)
  U. dioica L. (Stinging nettle)

Vaccinium L. (Blueberry,
  huckleberry) (Ericaceae)
  V. caespitosum Michx.
    (Dwaft huckleberry)

V. membranaceum Dougl.
(Tall huckleberry)
V. scoparium Leiberg (Red
huckleberry)
Veratrum L. (False helleborne)
(Liliaceae)
   V. viride Ait. (False
   helleborne)
Viola L. (Violet) (Violaceae)
   V. adunca Sm. (Early blue
   violet)
   V. canadensis L. (Canadian
   violet)
   V. glabella Nutt. (Yellow
   violet)
   V. nephrophylla Greene
   (Purple violet)

V. nuttallii Pursh (Nuttall
violet)
V. orbiculata Geyer
(Evergreen violet)
V. palustris L. (March violet)

Woodsia R. Br. (Woodsia)
(Polypodiaceae)
   W. scopulina D. C. Eat.
   (Rocky Mountain woodsia)

Xerophyllum Michx. (Beargrass)
(Liliaceae)
   X. tenax (Pursh) Nutt.
   (Beargrass)

APPENDIX 2

GRADIENT POPULATION
NOMOGRAMS FOR
GLACIER NATIONAL PARK

(112 Figures)

The non-linear responses of species populations to the elevation and time since burn gradients and the overlap among topographic-moisture categories are important to the interpretation of these nomograms. These considerations are discussed in detail in Chapter 8, while the populations and communities that the nomograms represent are described in Chapters 9, 10, and 11.

Figures 1–28. Gradient population nomograms for overstory species in the McDonald drainage of Glacier National Park plot elevation vs. topographic-moisture for four stand age classes, and elevation vs. stand age (time since the last canopy fire) for five elevation classes. Contour lines (isodens) connect areas of equal species relative density; all importance values are relative density (%). Species include:

Figures   1 and   2   *Abies grandis*
Figures   3 and   4   *Abies lasiocarpa*
Figure   5   *Larix lyallii*
Figures   6 and   7   *Larix occidentalis*
Figures   8 and   9   *Picea glauca* × *engelmannii* complex
Figures 10 and 11   *Pinus monticola* × *albicaulis* complex
Figures 12 and 13   *Pinus contorta*
Figures 14 and 15   *Pinus ponderosa*
Figures 16 and 17   *Pseudotsuga menziesii*
Figures 18 and 19   *Tsuga heterophylla*
Figures 20 and 21   *Thuja plicata*
Figures 22 and 23   *Populus tremuloides*
Figures 24 and 25   *Populus trichocarpa*
Figures 26 and 27   *Betula papyrifera*
Figure   28   *Betula occidentalis*

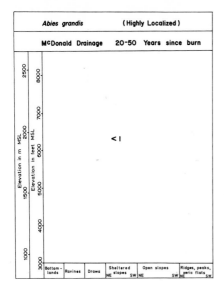

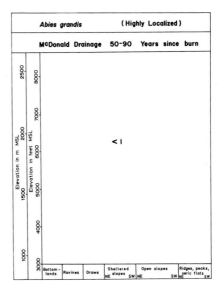

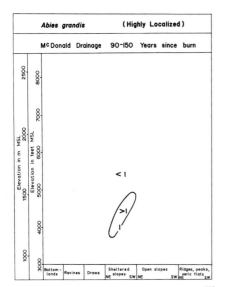

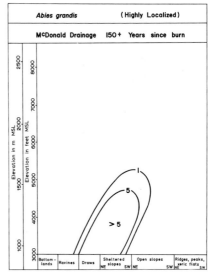

Figure 1

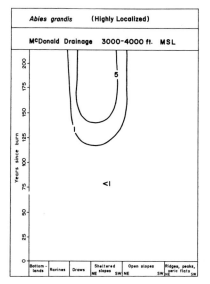
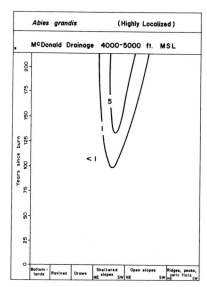

Figure 2

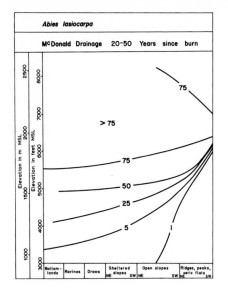

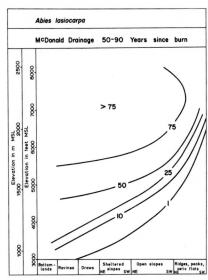

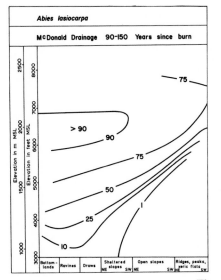

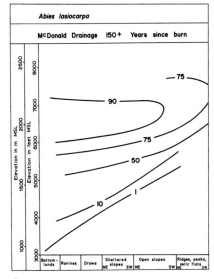

Figure 3

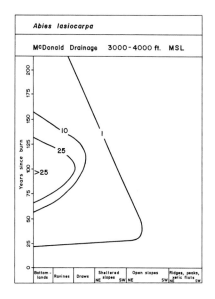

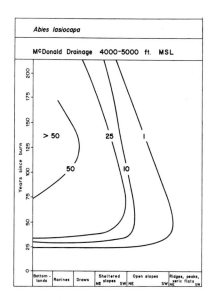

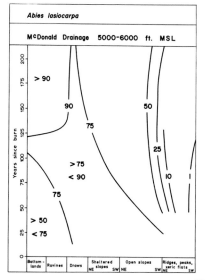

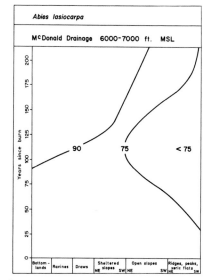

Figure 4

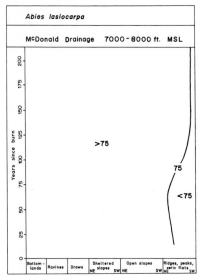

Figure 4 (*Continued.*)

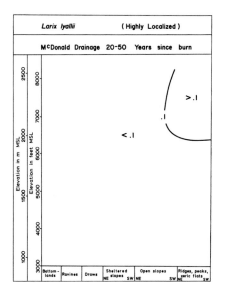

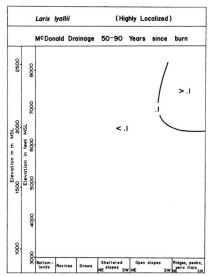

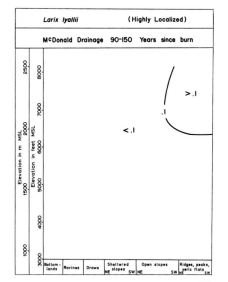

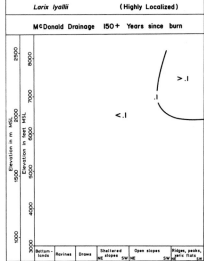

Figure 5

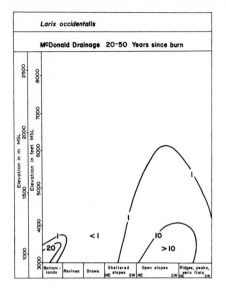

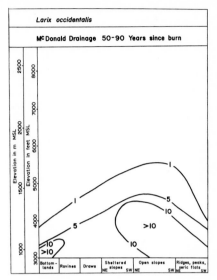

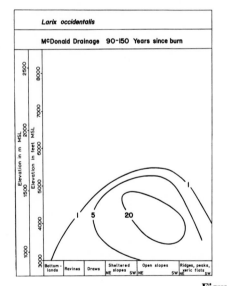

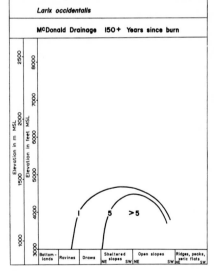

Figure 6

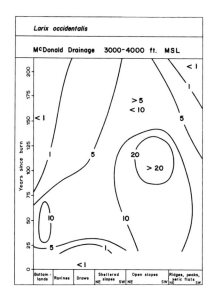

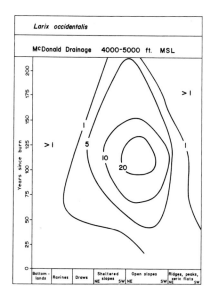

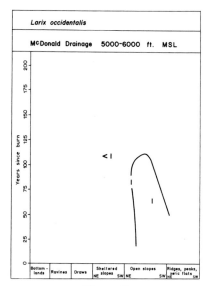

Figure 7

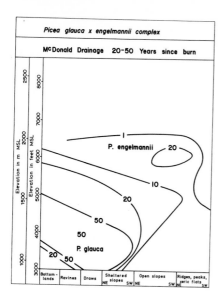

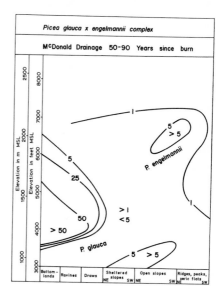

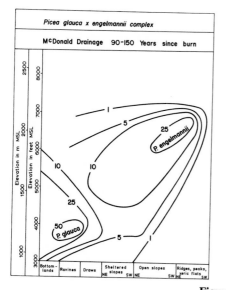

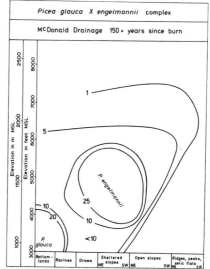

Figure 8

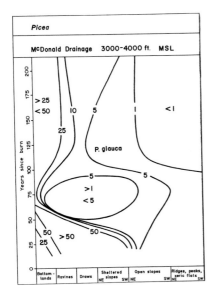

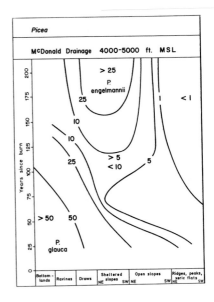

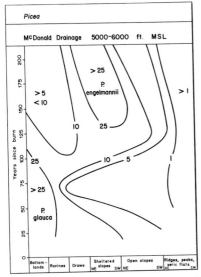

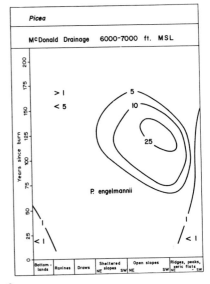

Figure 9

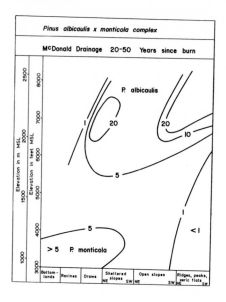

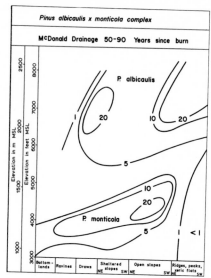

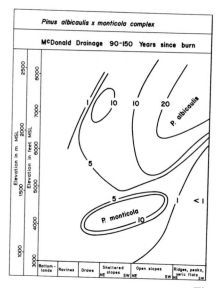

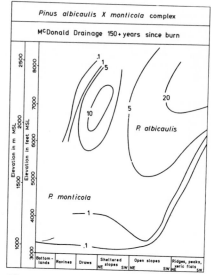

Figure 10

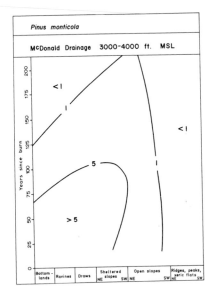

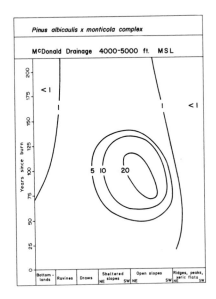

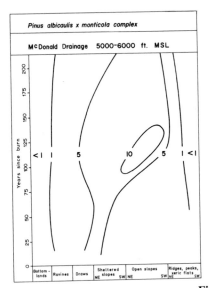

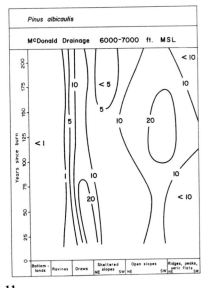

Figure 11

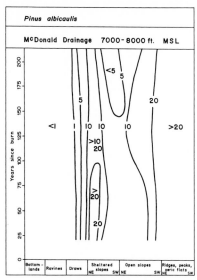

Figure 11 (*Continued.*)

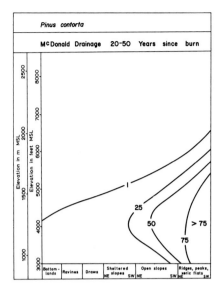

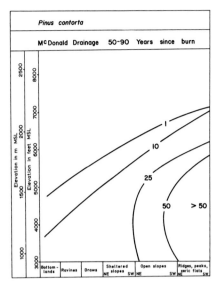

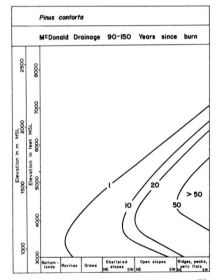

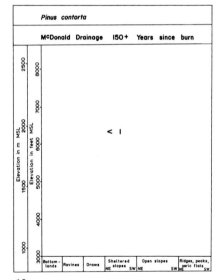

Figure 12

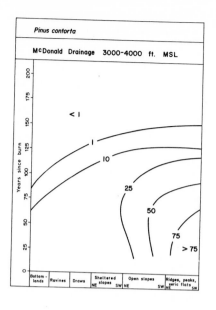

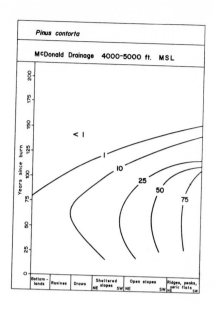

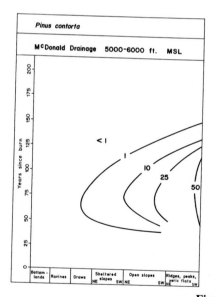

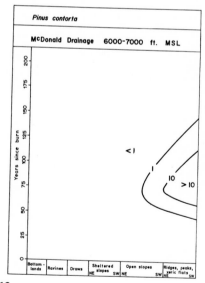

Figure 13

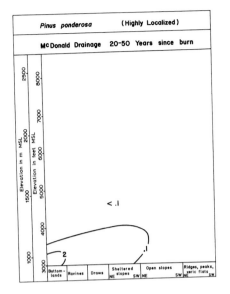

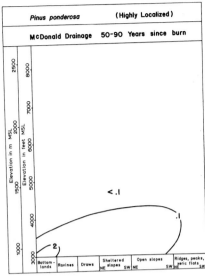

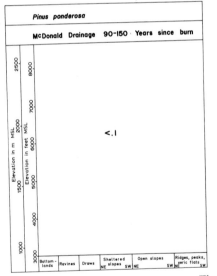

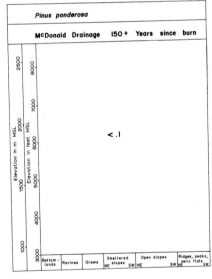

Figure 14

APPENDIX 2

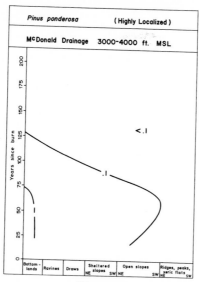

Figure 15

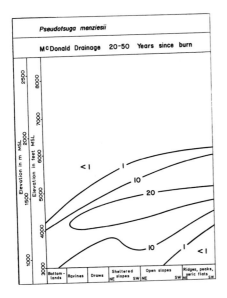

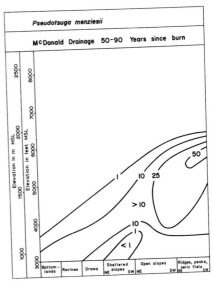

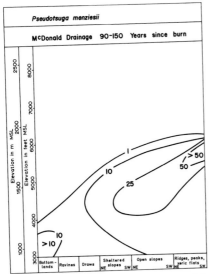

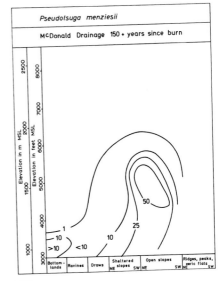

Figure 16

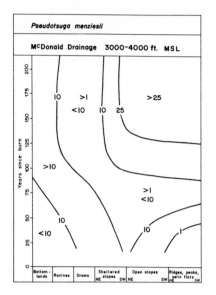

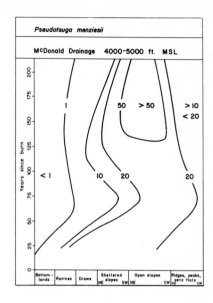

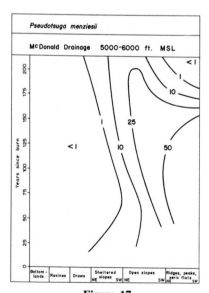

Figure 17

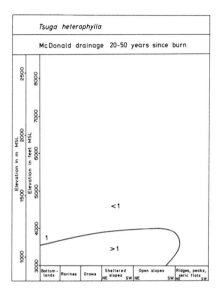

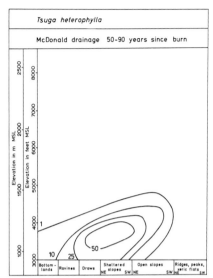

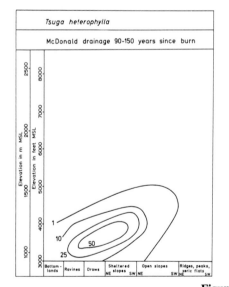

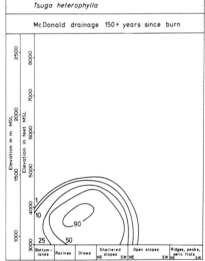

Figure 18

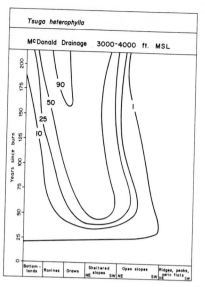

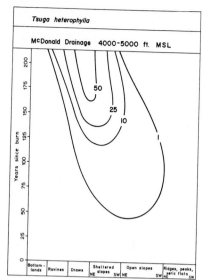

Figure 19

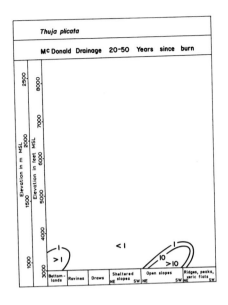

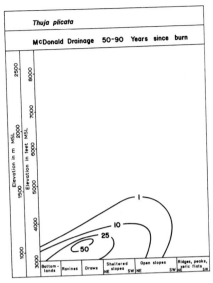

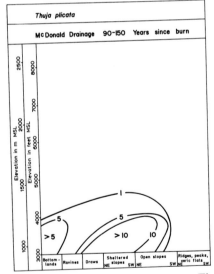

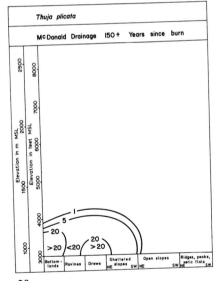

Figure 20

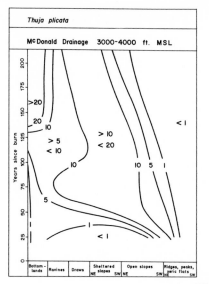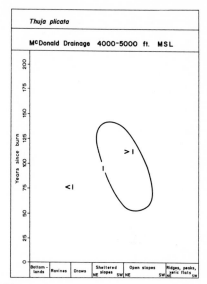

Figure 21

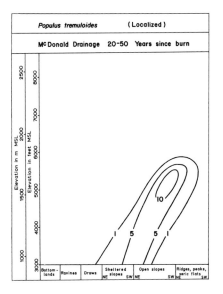

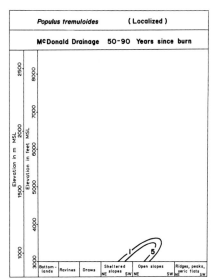

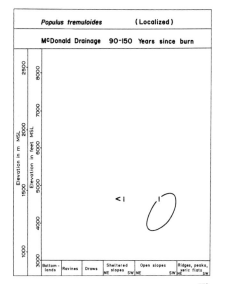

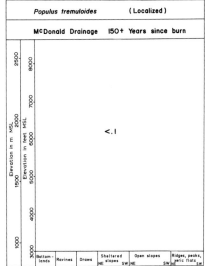

Figure 22

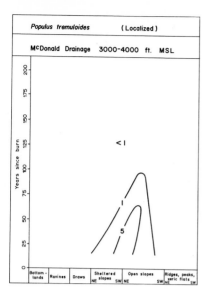

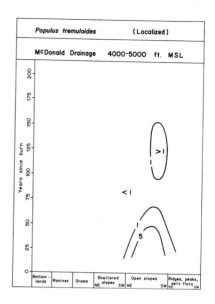

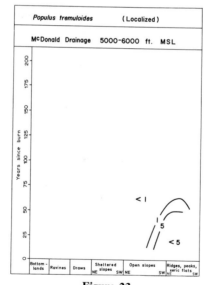

Figure 23

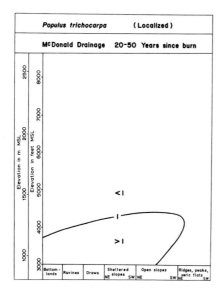

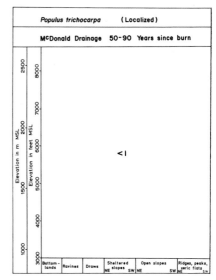

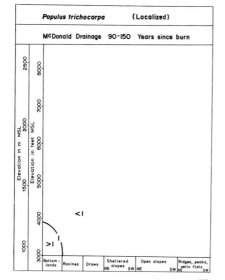

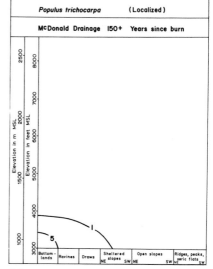

Figure 24

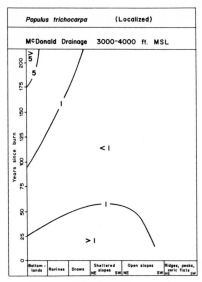

Figure 25

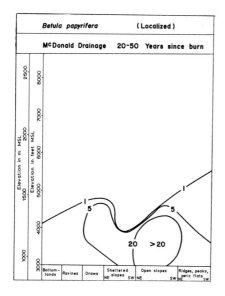

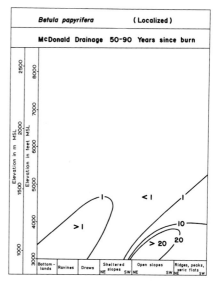

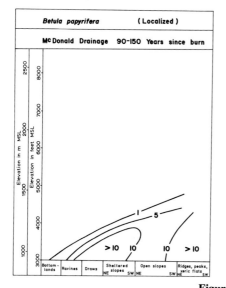

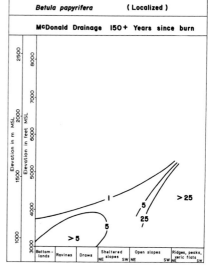

Figure 26

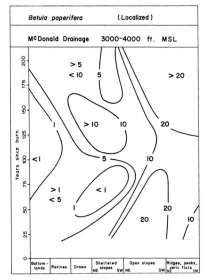
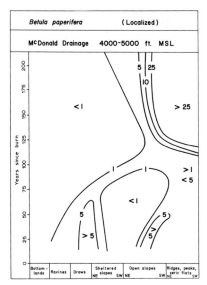

Figure 27

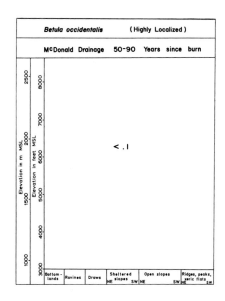

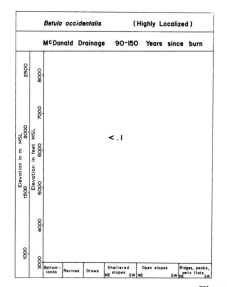

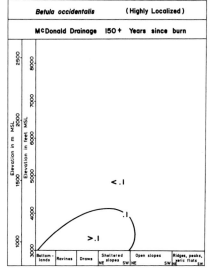

Figure 28

Figures 29–43. Gradient population nomograms for overstory species in the Camas drainage of Glacier National Park plot elevation vs. topographic-moisture for four stand age classes. Contour lines (isodens) connect areas of equal species relative density; all importance values are relative density (%). Species include:

Figure 29   *Abies grandis*
Figure 30   *Abies lasiocarpa*
Figure 31   *Larix lyallii*
Figure 32   *Larix occidentalis*
Figure 33   *Picea glauca* × *engelmannii* complex
Figure 34   *Pinus monticola* × *albicaulis* complex
Figure 35   *Pinus contorta*
Figure 36   *Pinus ponderosa*
Figure 37   *Pseudotsuga menziesii*
Figure 38   *Tsuga heterophylla*
Figure 39   *Thuja plicata*
Figure 40   *Populus tremuloides*
Figure 41   *Populus trichocarpa*
Figure 42   *Betula papyrifera*
Figure 43   *Betula occidentalis*

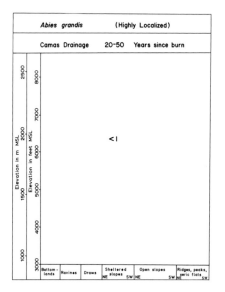

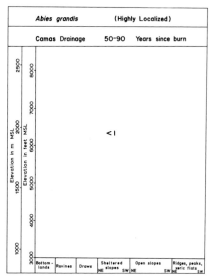

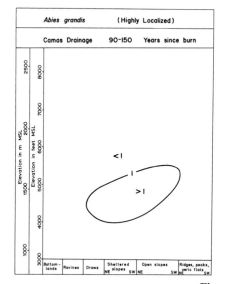

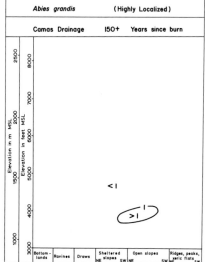

Figure 29

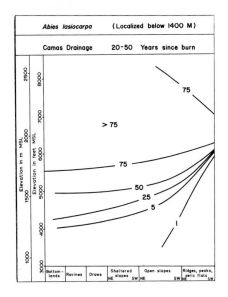

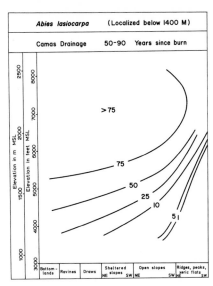

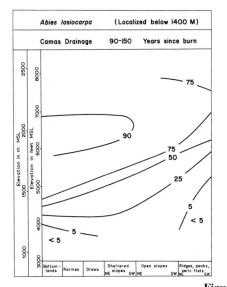

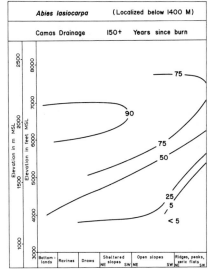

Figure 30

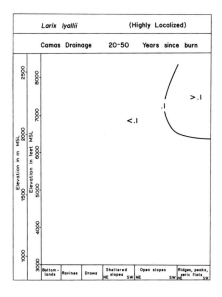

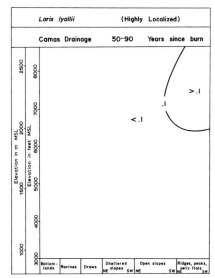

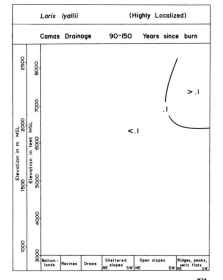

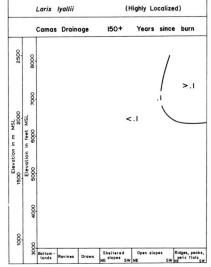

Figure 31

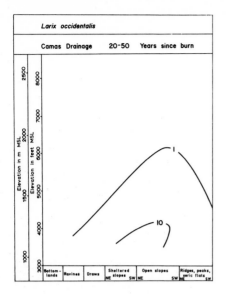

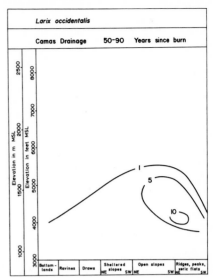

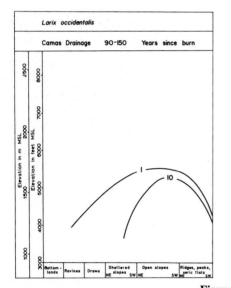

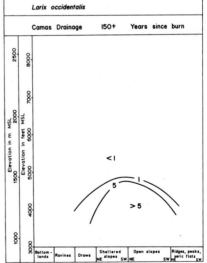

Figure 32

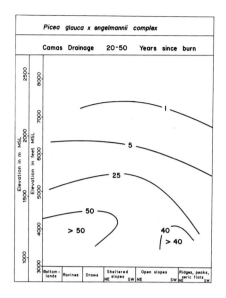

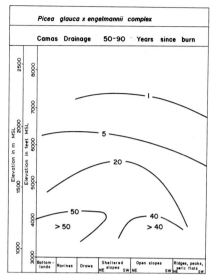

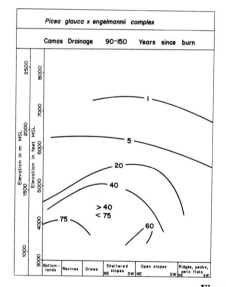

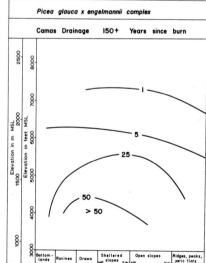

Figure 33

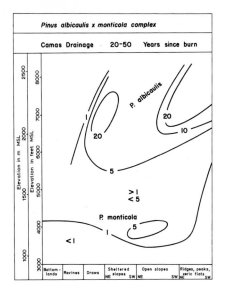

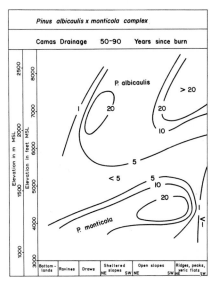

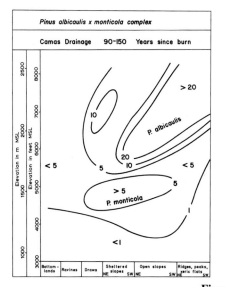

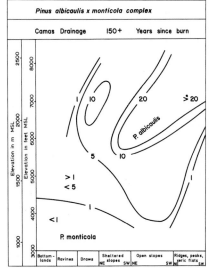

Figure 34

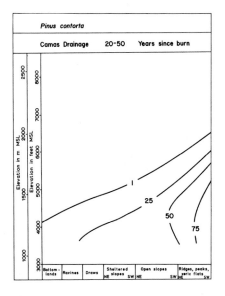

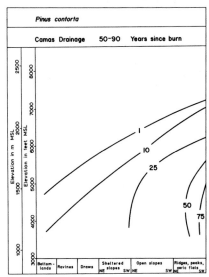

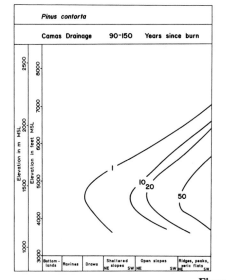

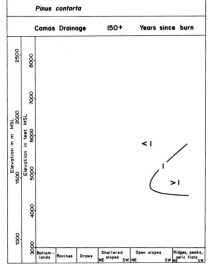

Figure 35

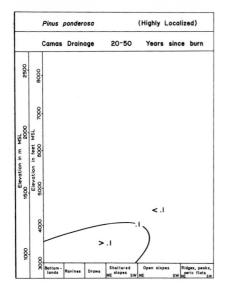

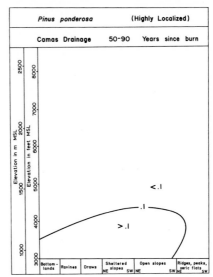

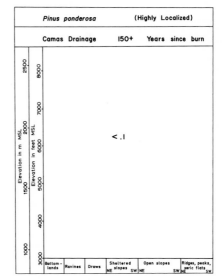

Figure 36

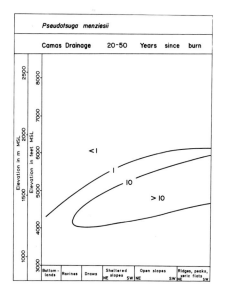

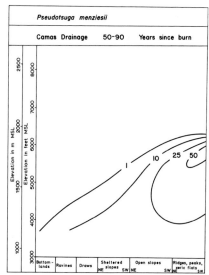

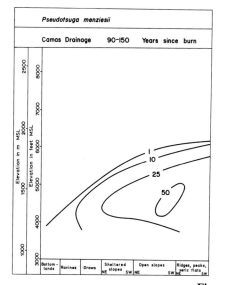

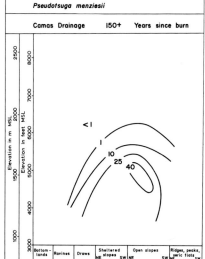

Figure 37

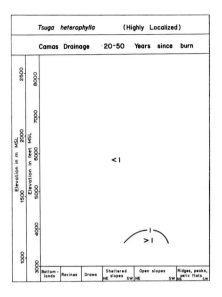

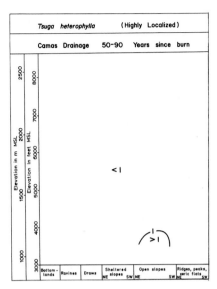

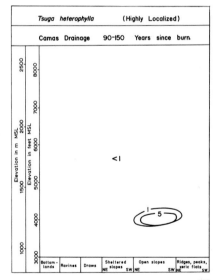

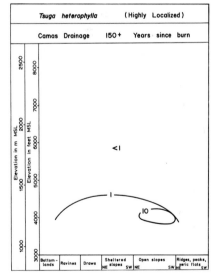

Figure 38

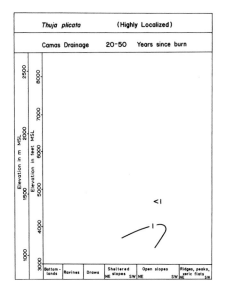

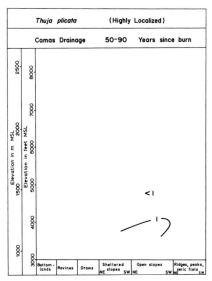

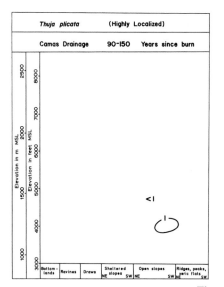

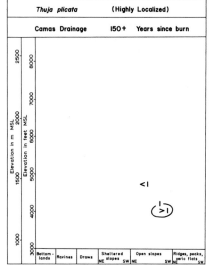

Figure 39

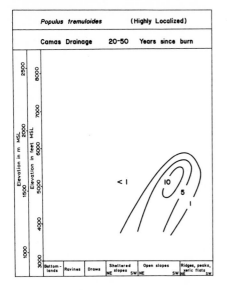

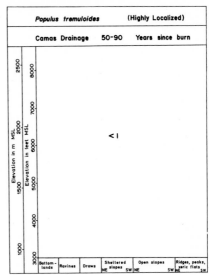

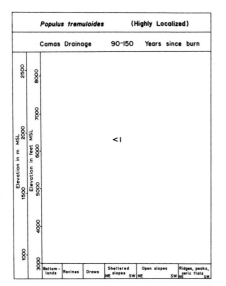

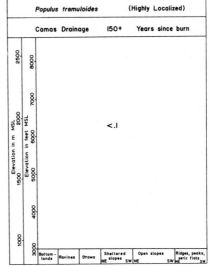

Figure 40

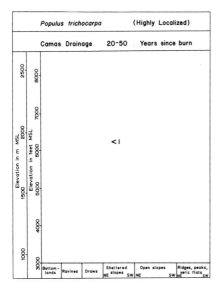

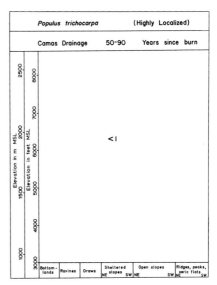

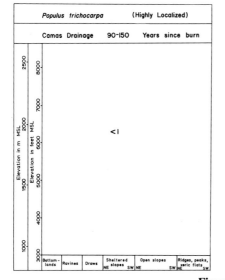

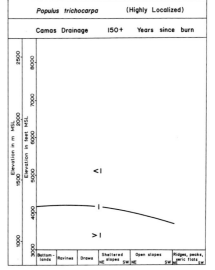

Figure 41

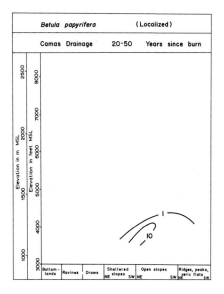

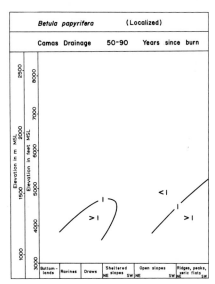

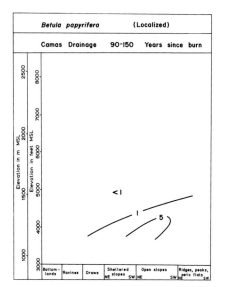

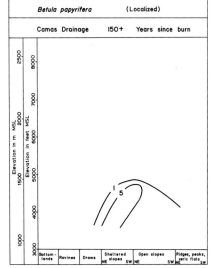

Figure 42

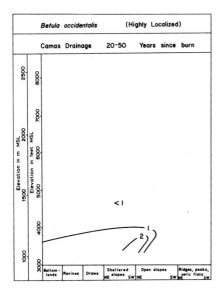

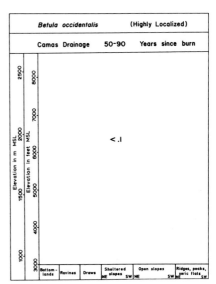

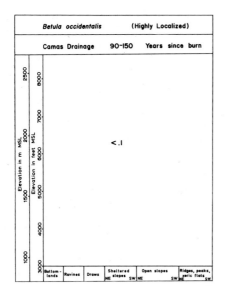

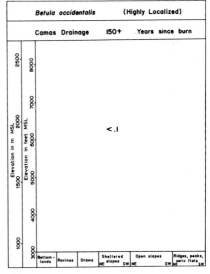

Figure 43

Figures 44–54. Gradient population nomograms for overstory species in the North Fork of the Flathead River drainage in Glacier National Park plot elevation vs. topographic-moisture for four stand age classes. Contour lines (isodens) connect areas of equal species relative density; all importance values are relative density (%). Species include:

Figure 44 *Abies lasiocarpa*
Figure 45 *Larix lyallii*
Figure 46 *Larix occidentalis*
Figure 47 *Picea glauca* × *engelmannii* complex
Figure 48 *Pinus monticola* × *albicaulis* complex
Figure 49 *Pinus contorta*
Figure 50 *Pinus ponderosa*
Figure 51 *Pseudotsuga menziesii*
Figure 52 *Populus tremuloides*
Figure 53 *Populus trichocarpa*
Figure 54 *Betula papyrifera*

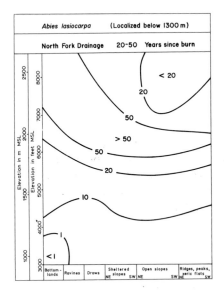

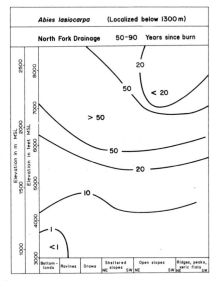

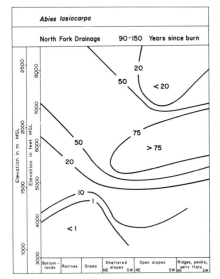

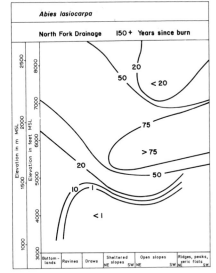

Figure 44

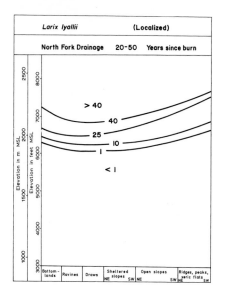

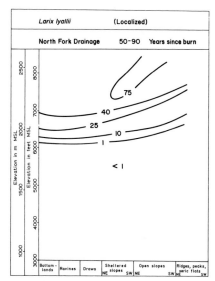

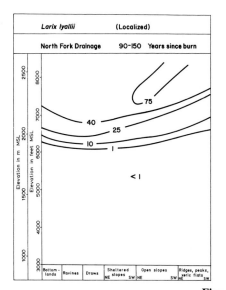

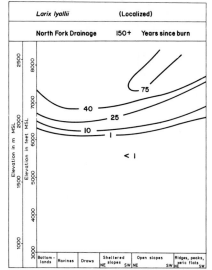

Figure 45

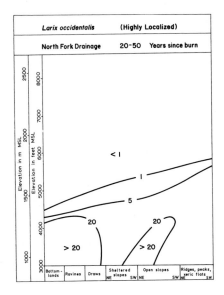

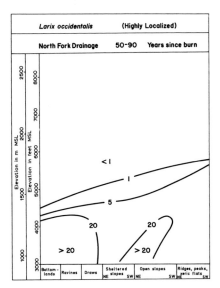

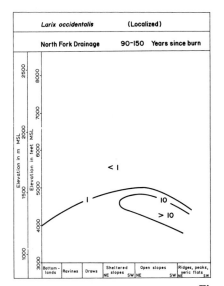

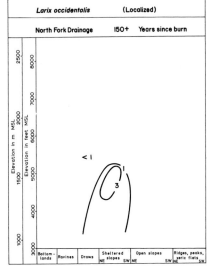

Figure 46

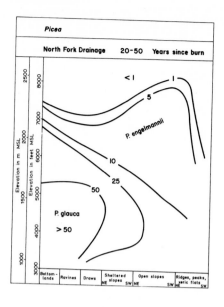

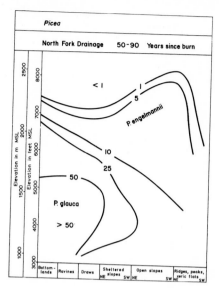

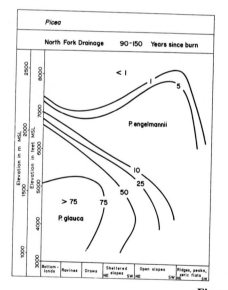

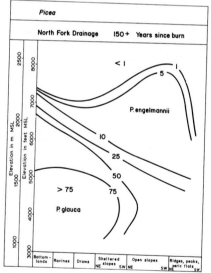

Figure 47

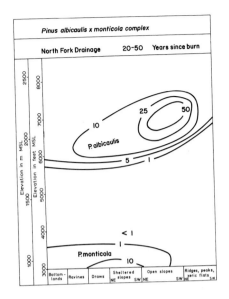

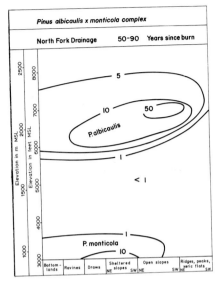

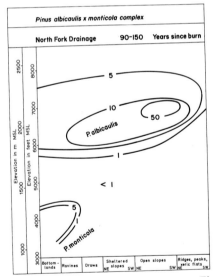

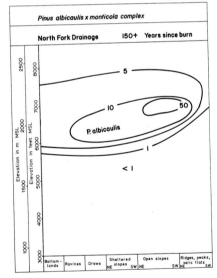

Figure 48

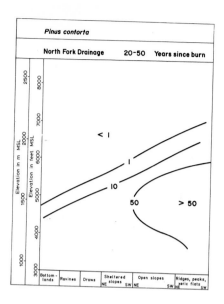

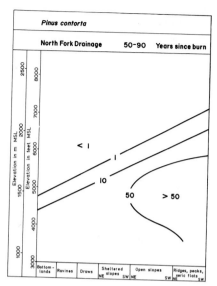

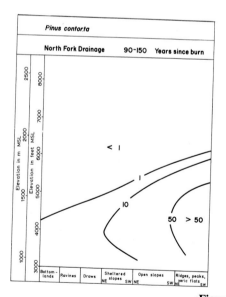

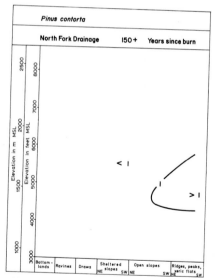

Figure 49

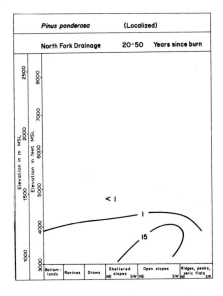

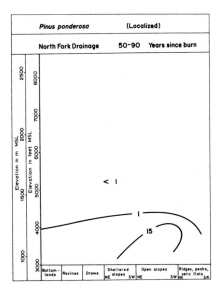

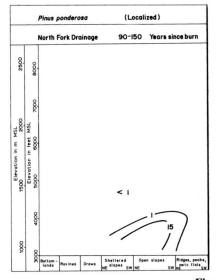

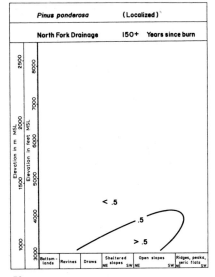

Figure 50

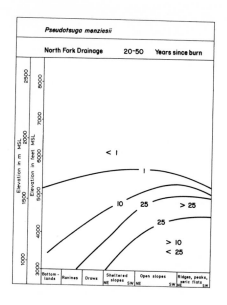

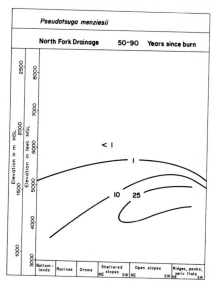

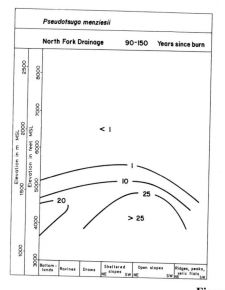

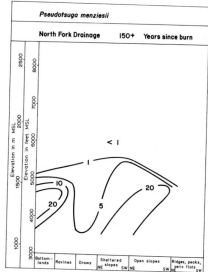

Figure 51

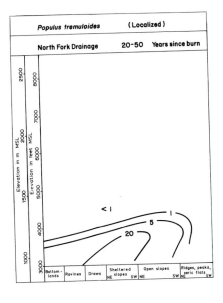

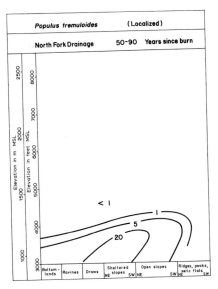

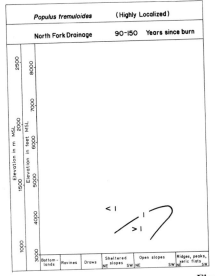

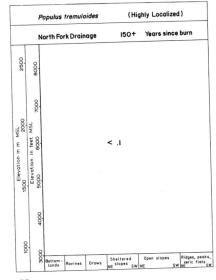

Figure 52

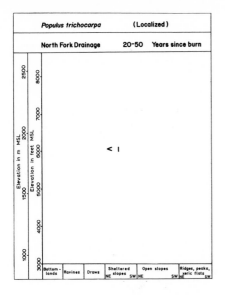

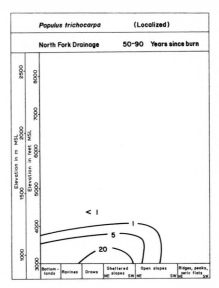

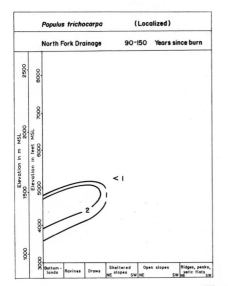

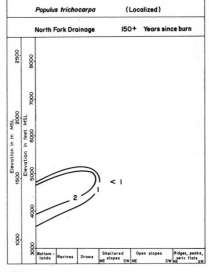

Figure 53

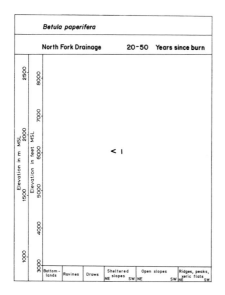

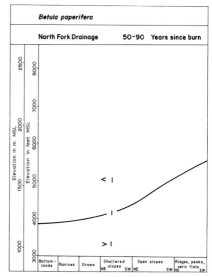

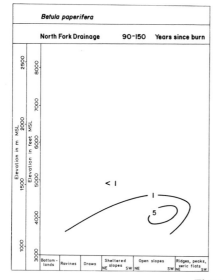

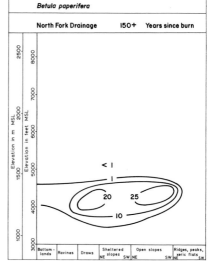

Figure 54

Figures 55–87. Gradient population nomograms for common understory herb and shrub species in the combined West Lakes drainages of
Glacier National Park plot elevation vs. topographic-moisture for five stand
age classes. **Figure 55** plots total herb and shrub cover in percent. **Figure 56**
plots herb and shrub species diversity defined as the number of species in a
pair of random 2 × 2 m plots. **Figures 57–87** plot absolute cover (%) for
individual species using the following cover code: 0, (no cover); T, (trace—
less than 1%); 1, 1%–5%; 2, 6%–25%; 3, 26%–50%; 4, 51%–75%; and 5,
76%–95%; contour lines connect areas of equal absolute cover. All of these
stands support a forest overstory by an age of 30 years after disturbance.
Species include:

Figure 57   *Athyrium filix-femina*
Figure 58   Gramineae family
Figure 59   *Xerophyllum tenax*
Figure 60   *Clintonia uniflora*
Figure 61   *Disporum* spp.
Figure 62   *Goodyera oblongifolia*
Figure 63   *Salix* spp.
Figure 64   *Thalictrum occidentale*
Figure 65   *Tiarella trifoliata*
Figure 66   *Ribes* spp.
Figure 67   *Fragaria* spp.
Figure 68   *Amelanchier alnifolia*
Figure 69   *Rosa* spp.
Figure 70   *Rubus parviflorus*
Figure 71   *Sorbus scopulina*
Figure 72   *Spiraea betulifolia*
Figure 73   *Pachistima myrsinites*
Figure 74   *Viola* spp.
Figure 75   *Epilobium angustifolium*
Figure 76   *Aralia nudicaulis*
Figure 77   *Heracleum lanatum*
Figure 78   *Osmorhiza* spp.
Figure 79   *Cornus* spp.
Figure 80   *Menziesia ferruginea*

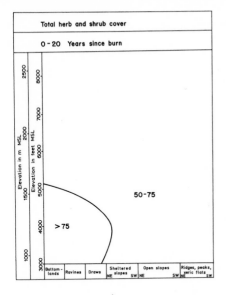

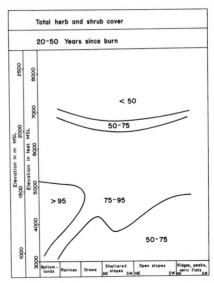

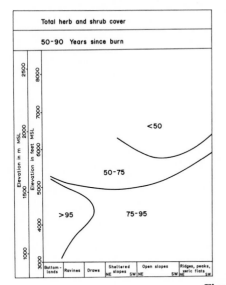

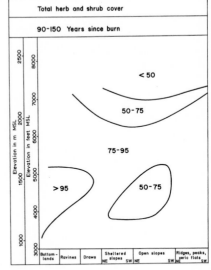

Figure 55

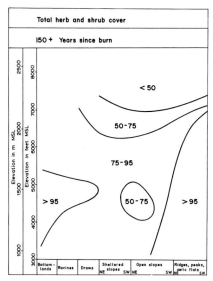

Figure 55 (*Continued.*)

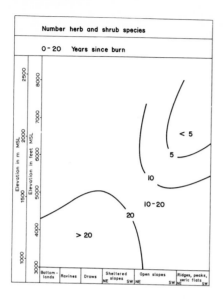

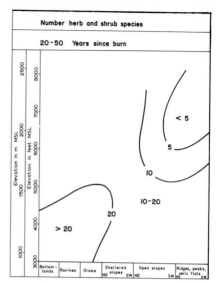

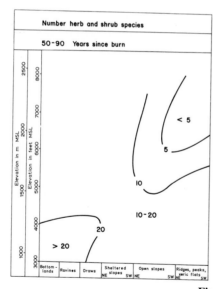

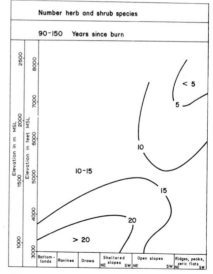

Figure 56

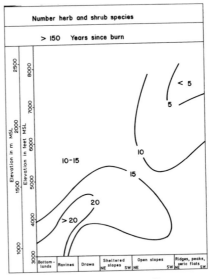

Figure 56 (*Continued.*)

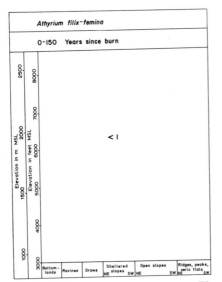

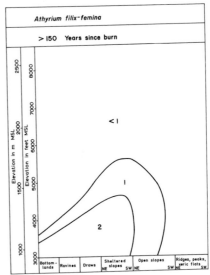

Figure 57

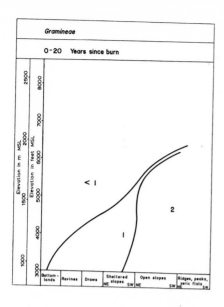

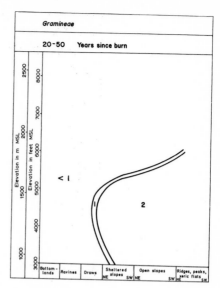

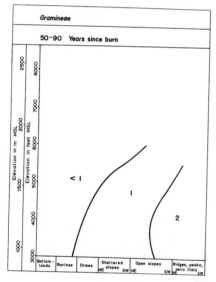

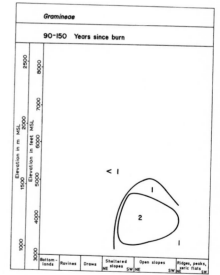

Figure 58

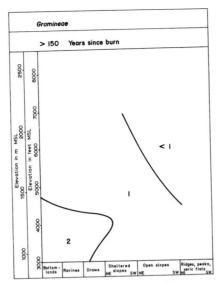

Figure 58 (*Continued.*)

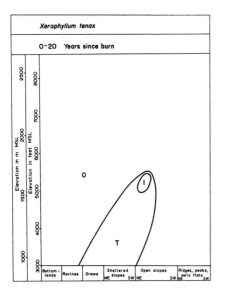

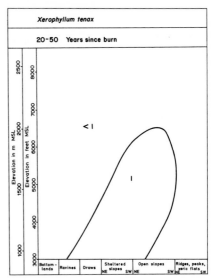

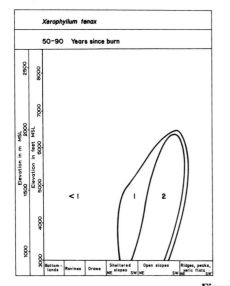

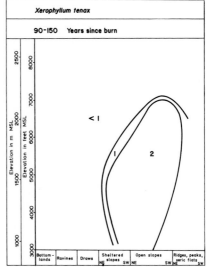

Figure 59

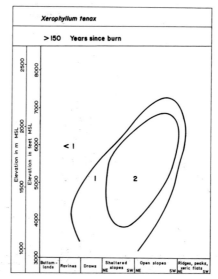

Figure 59 (*Continued*.)

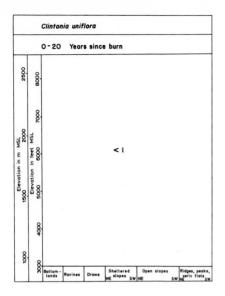

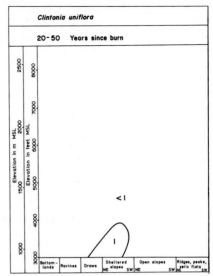

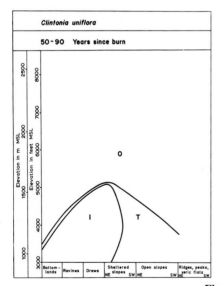

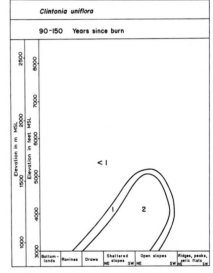

Figure 60

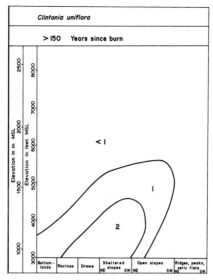

Figure 60 (*Continued.*)

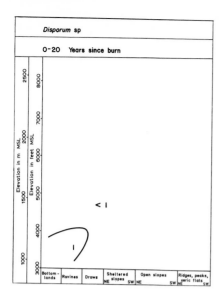

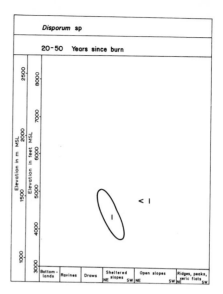

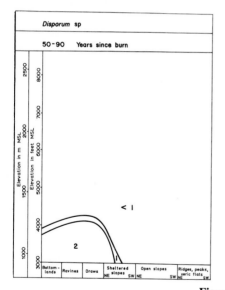

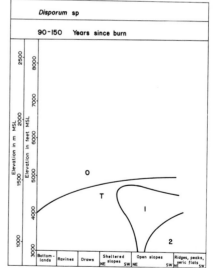

Figure 61

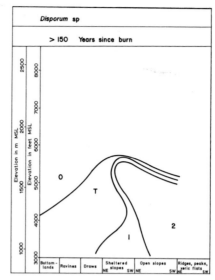

Figure 61 (*Continued.*)

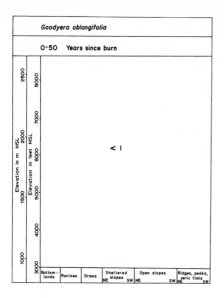

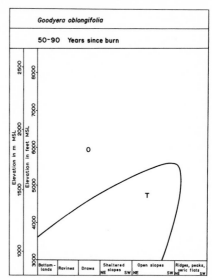

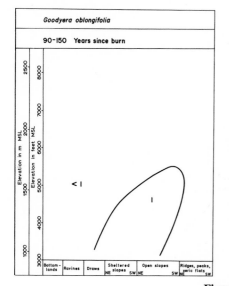

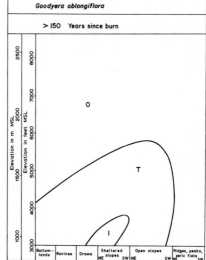

Figure 62

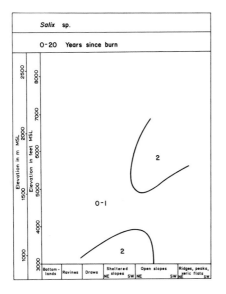

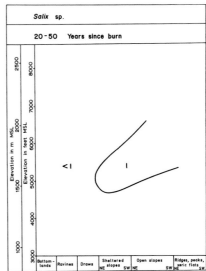

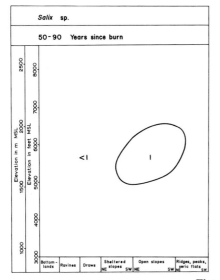

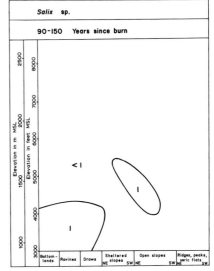

Figure 63

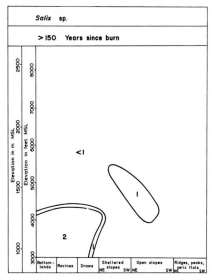

Figure 63 (*Continued.*)

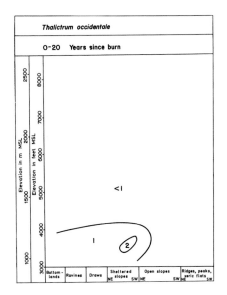

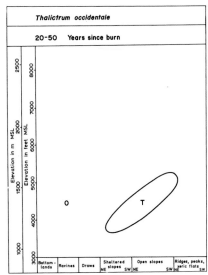

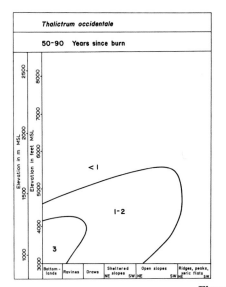

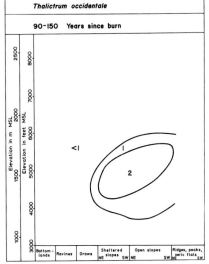

Figure 64

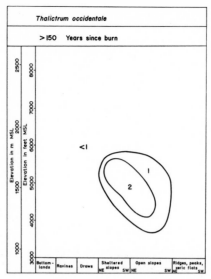

Figure 64 (*Continued.*)

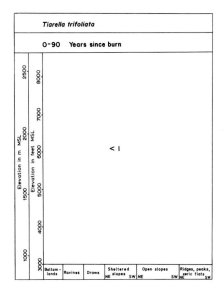

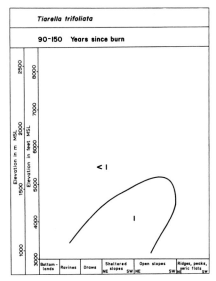

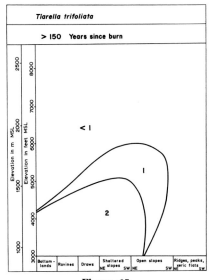

Figure 65

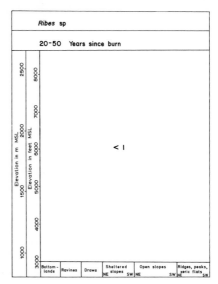

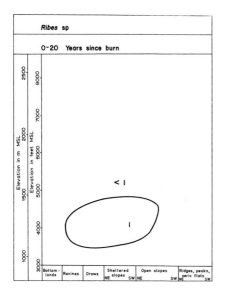

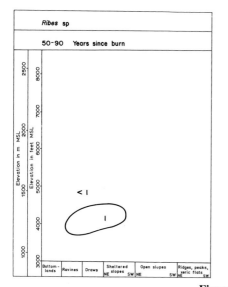

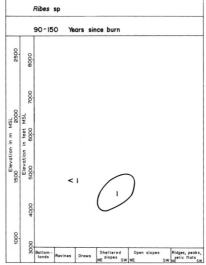

Figure 66

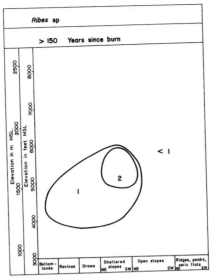

Figure 66 (*Continued.*)

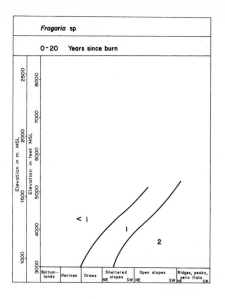

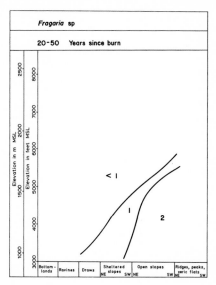

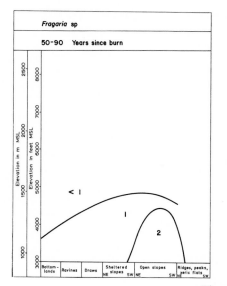

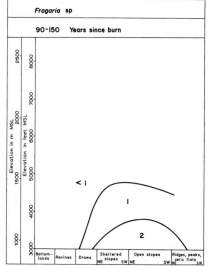

Figure 67

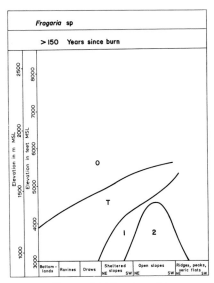

Figure 67 (*Continued*.)

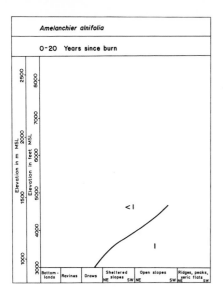

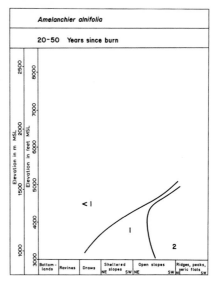

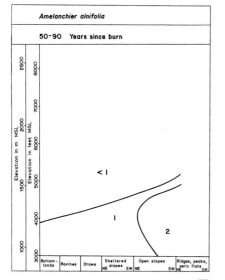

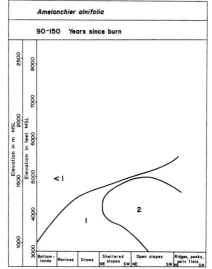

Figure 68

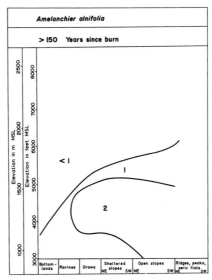

Figure 68 (*Continued.*)

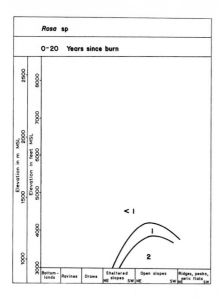

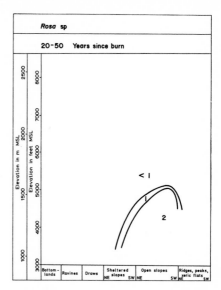

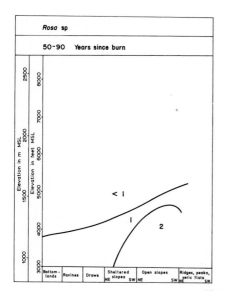

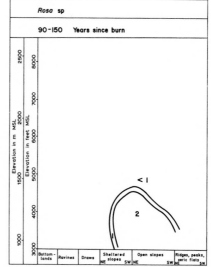

Figure 69

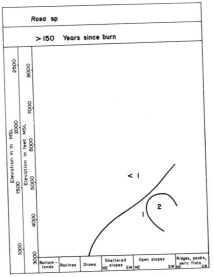

Figure 69 (*Continued.*)

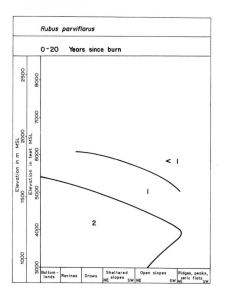

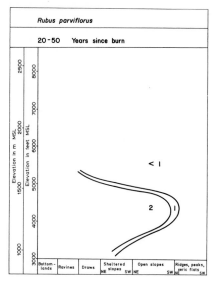

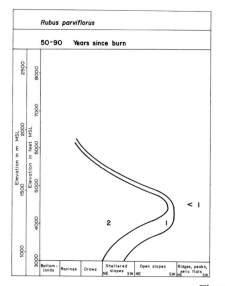

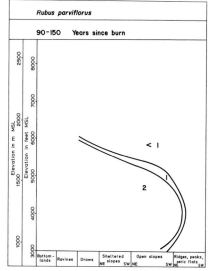

Figure 70

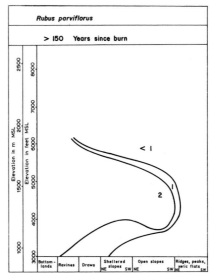

Figure 70 (*Continued.*)

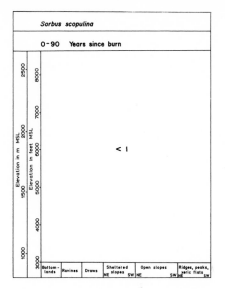

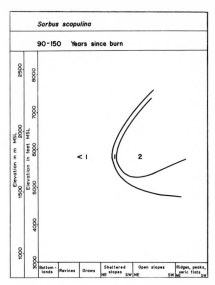

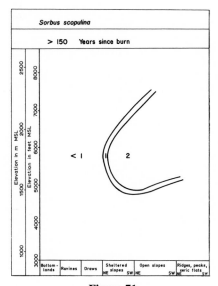

Figure 71

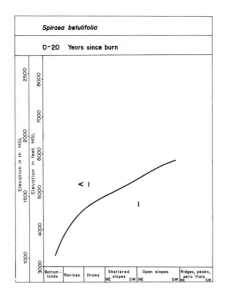

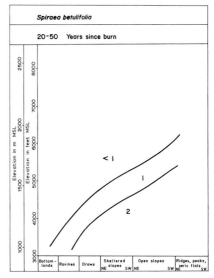

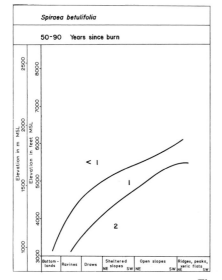

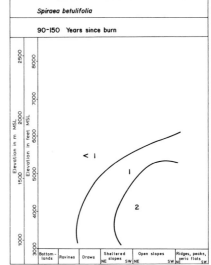

Figure 72

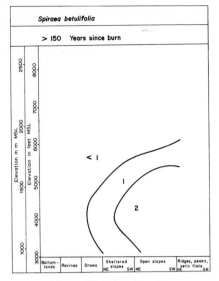

Figure 72 (*Continued*.)

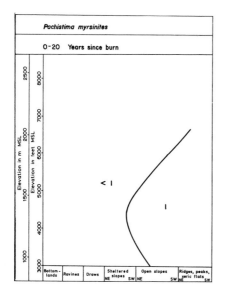
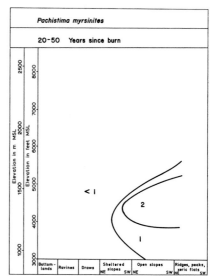
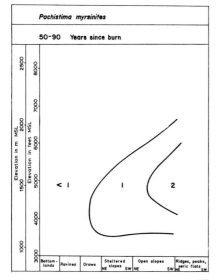
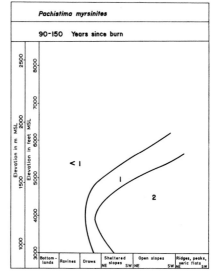

Figure 73

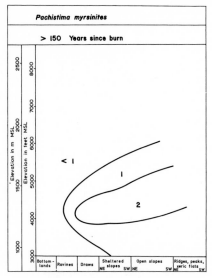

Figure 73 (*Continued.*)

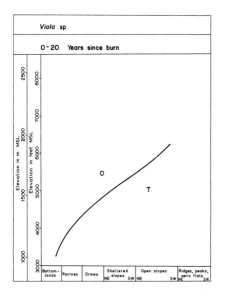

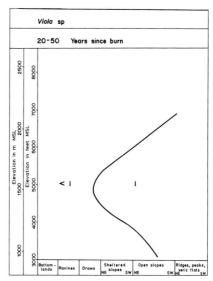

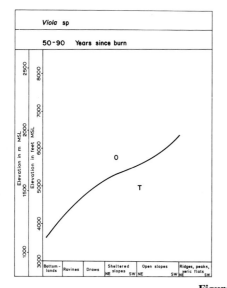

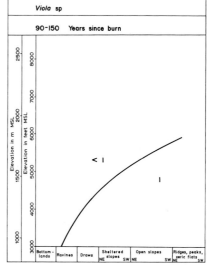

Figure 74

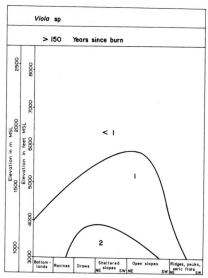

Figure 74 (*Continued.*)

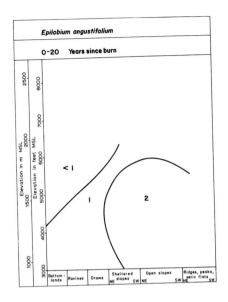

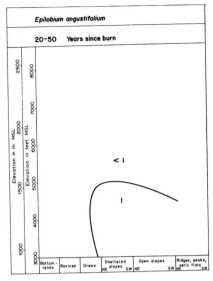

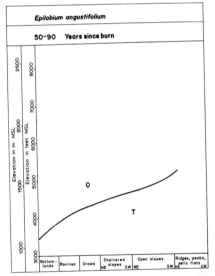

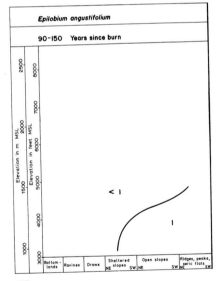

Figure 75

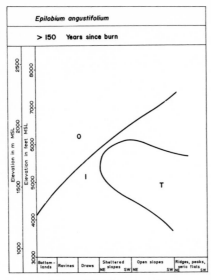

Figure 75 (*Continued.*)

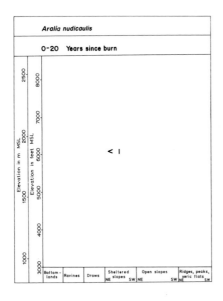

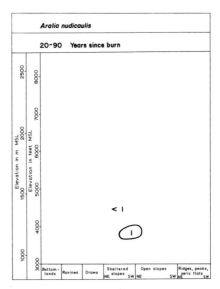

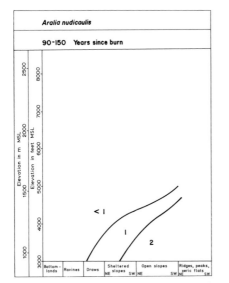

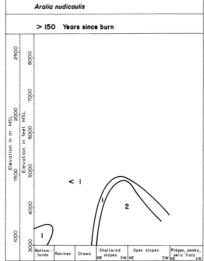

Figure 76

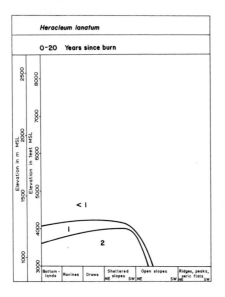

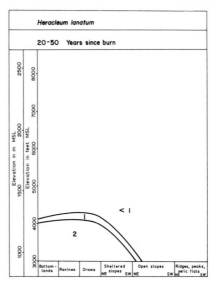

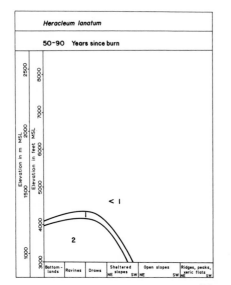

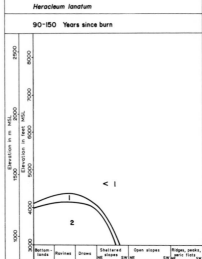

Figure 77

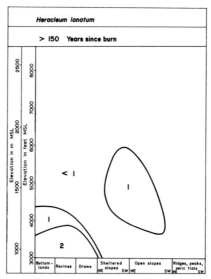

Figure 77 (*Continued.*)

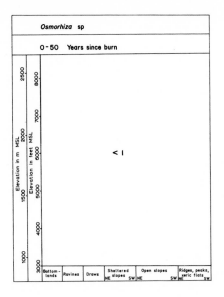

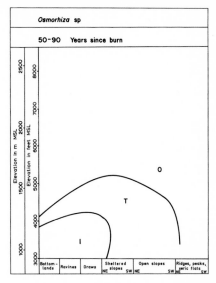

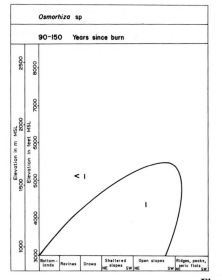

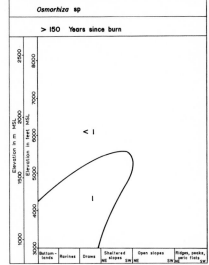

Figure 78

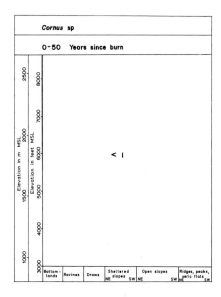

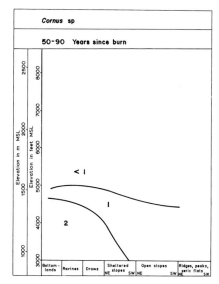

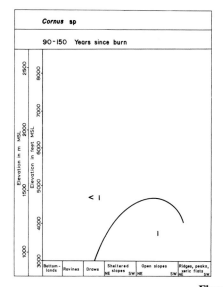

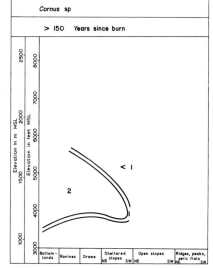

Figure 79

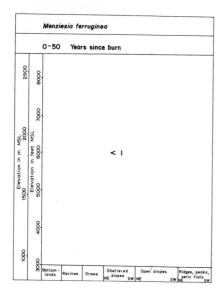

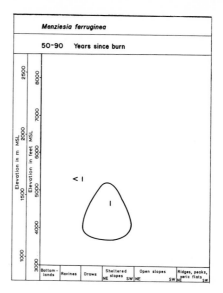

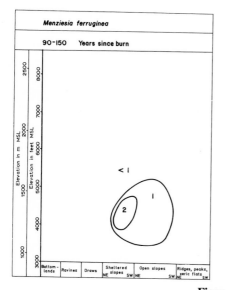

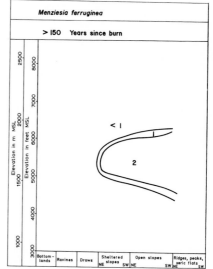

Figure 80

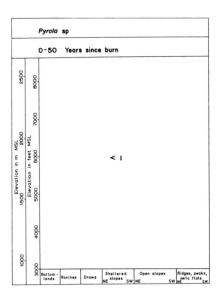

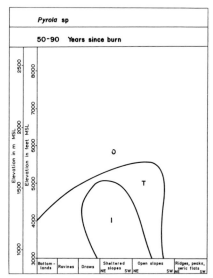

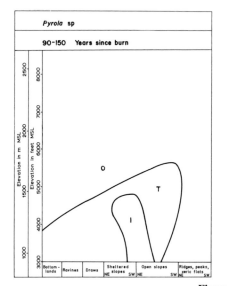

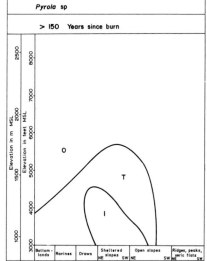

Figure 81

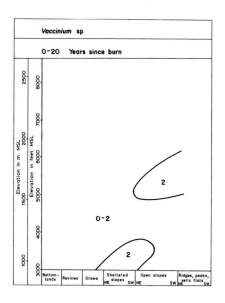

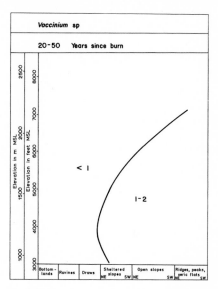

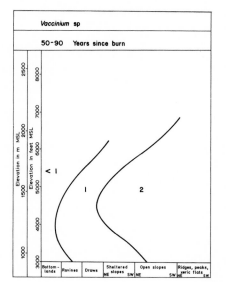

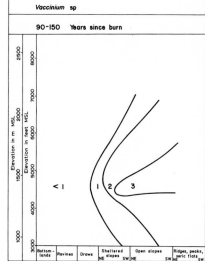

Figure 82

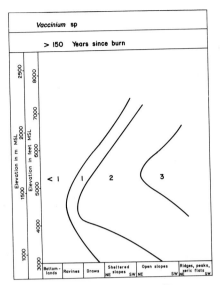

Figure 82 (*Continued.*)

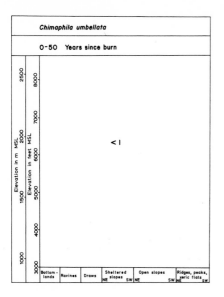

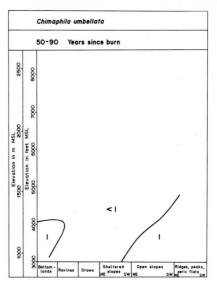

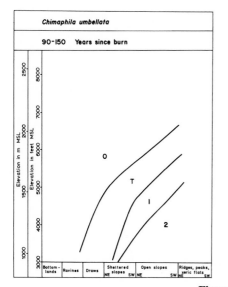

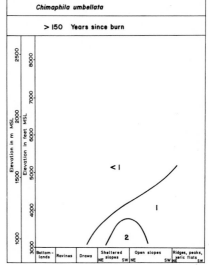

Figure 83

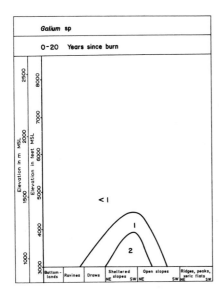

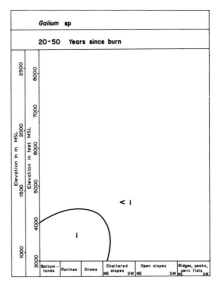

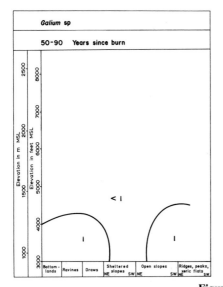

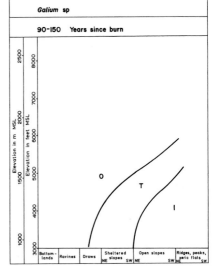

Figure 84

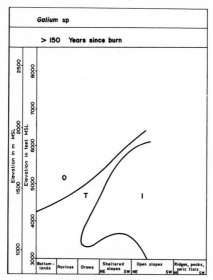

Figure 84 (*Continued.*)

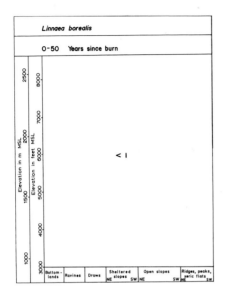

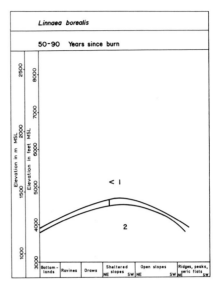

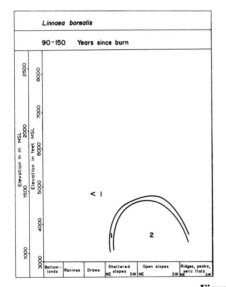

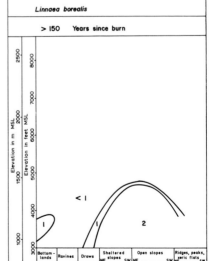

Figure 85

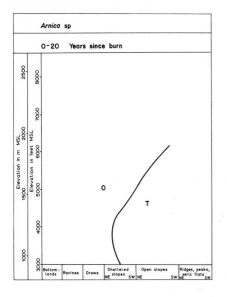

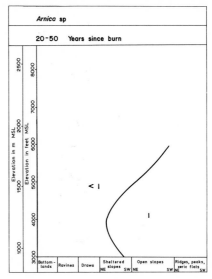

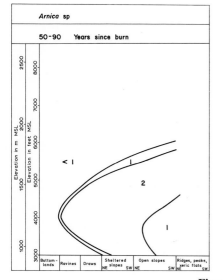

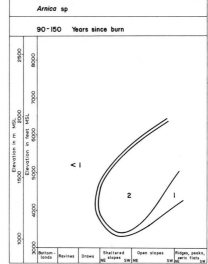

Figure 86

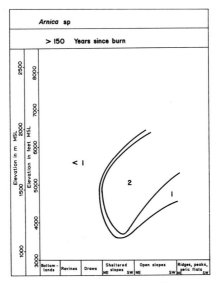

Figure 86 (*Continued.*)

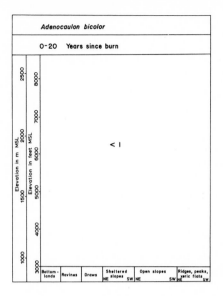

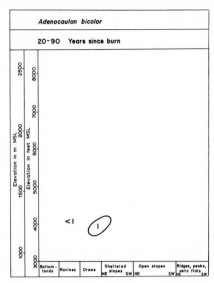

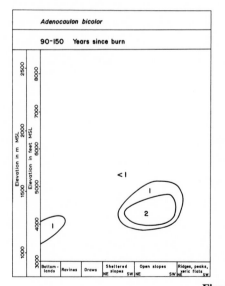

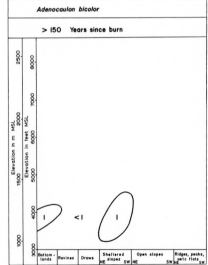

Figure 87

Figures 88–112. Gradient population nomograms for common herb and shrub species in the combined West Lakes drainages of Glacier National Park plot elevation vs. the primary succession gradient for up to five elevation classes of 305 m (1000 ft). **Figure 88** plots total herb and shrub cover in percent. **Figure 89** plots herb and shrub species diversity defined as the number of species in a pair of random 2 × 2 m plots. **Figures 90–112** plot absolute cover (%) for individual species using the following cover code: 0, (no cover); T, (trace—less than 1%); 1, 1%–5%; 2, 6%–25%; 3, 26%–50%; 4, 51%–75%; and 5, 76%–95%; contour lines connect areas of equal absolute cover. Species include:

Figure	90	*Carex* spp.
Figure	91	Gramineae family
Figure	92	*Poa* spp.
Figure	93	*Agropyron* spp.
Figure	94	*Xerophyllum tenax*
Figure	95	*Salix* spp.
Figure	96	*Alnus sinuata*
Figure	97	*Veratrum viride*
Figure	98	*Ribes* spp.
Figure	99	*Fragaria* spp.
Figure	100	*Epilobium angustifolium*
Figure	101	*Heracleum lanatum*
Figure	102	*Vaccinium* spp.
Figure	103	*Sorbus* spp.
Figure	104	*Rubus parviflorus*
Figure	105	*Penstemon* spp.
Figure	106	*Achillea millefolium*
Figure	107	*Thalictrum occidentale*
Figure	108	*Arnica* spp.
Figure	109	*Erigeron* spp.
Figure	110	*Urtica dioica*
Figure	111	*Senecio* spp.
Figure	112	*Spiraea* spp.

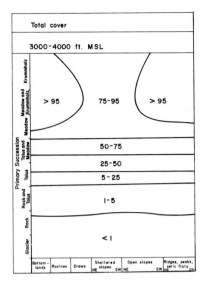

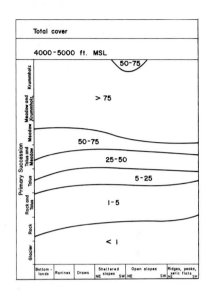

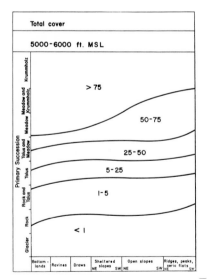

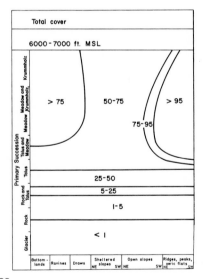

Figure 88

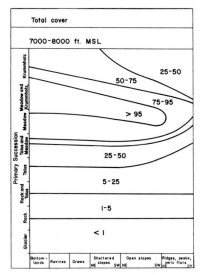

Figure 88 (*Continued.*)

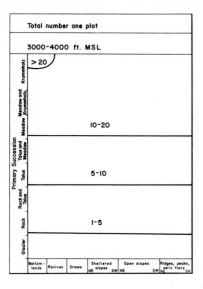

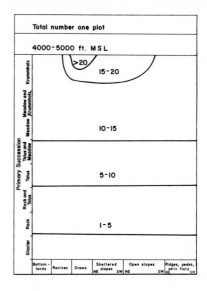

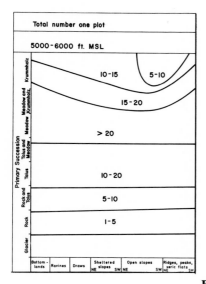

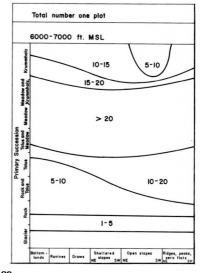

Figure 89

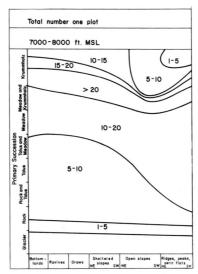

Figure 89 (*Continued.*)

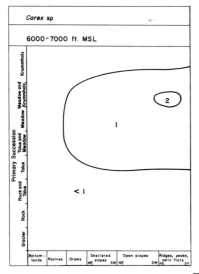

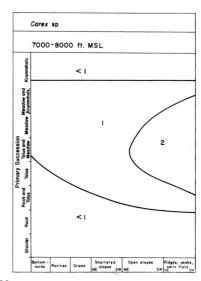

Figure 90

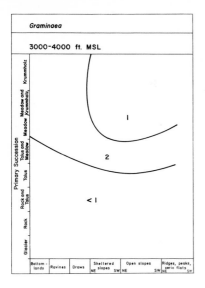

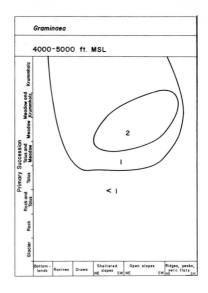

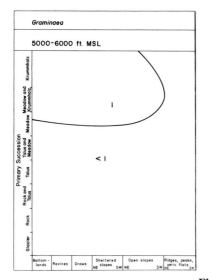

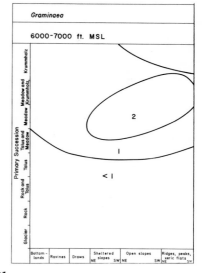

Figure 91

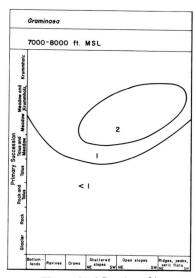

Figure 91 (*Continued.*)

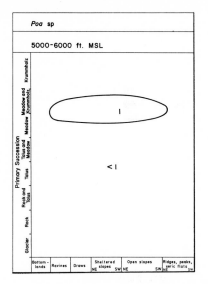

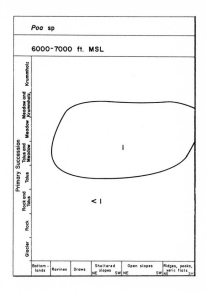

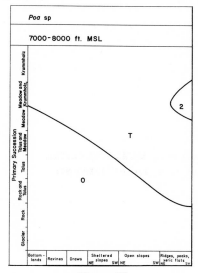

Figure 92

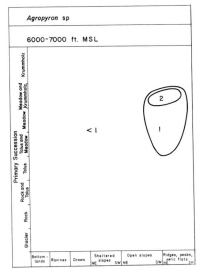

Figure 93

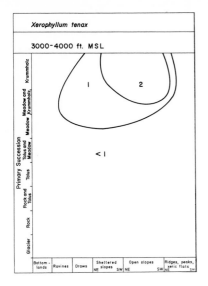

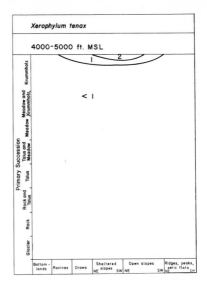

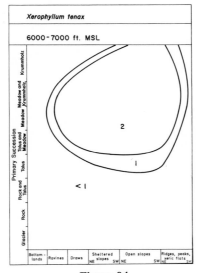

Figure 94

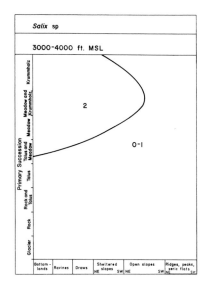

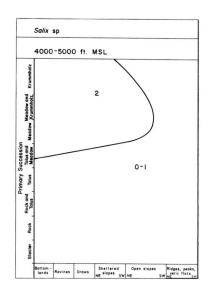

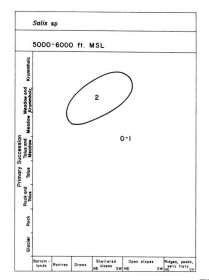

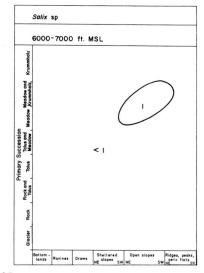

Figure 95

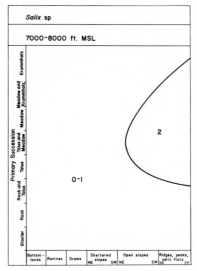

Figure 95 (*Continued.*)

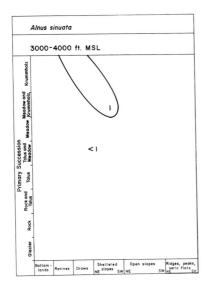

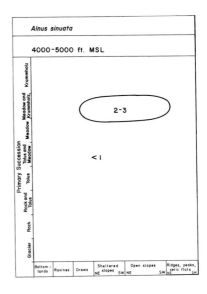

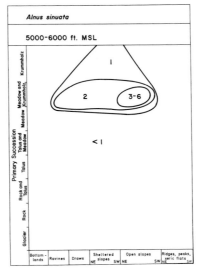

Figure 96

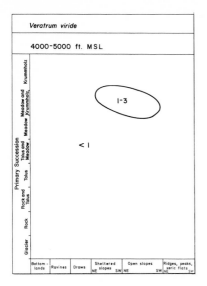

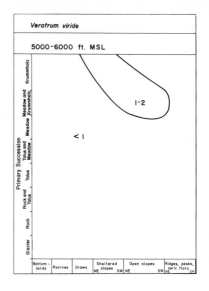

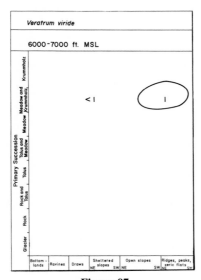

Figure 97

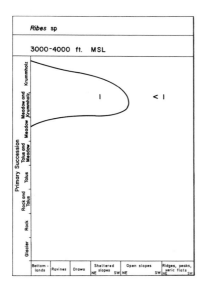

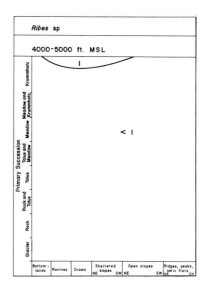

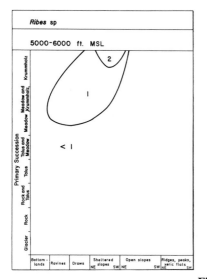

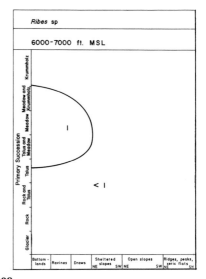

Figure 98

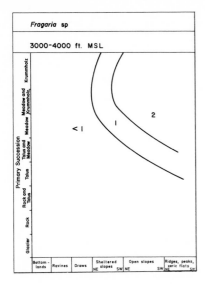

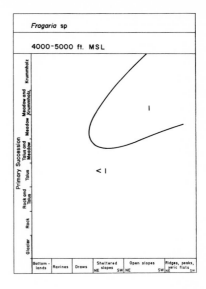

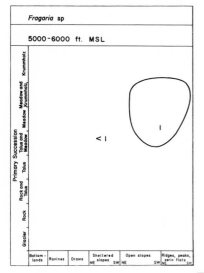

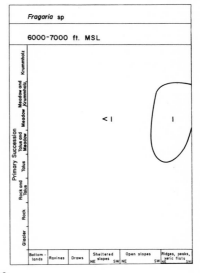

Figure 99

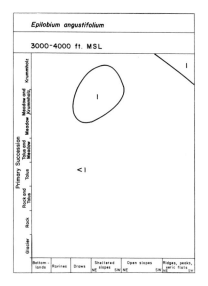

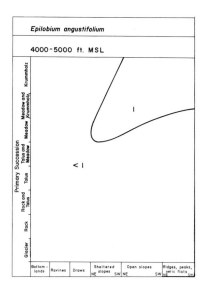

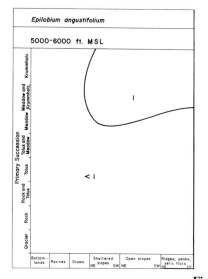

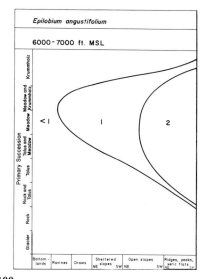

Figure 100

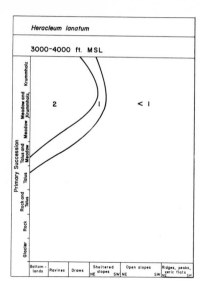

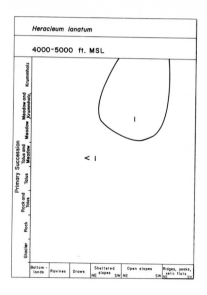

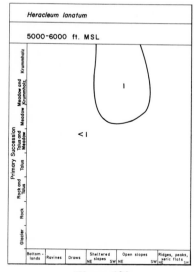

Figure 101

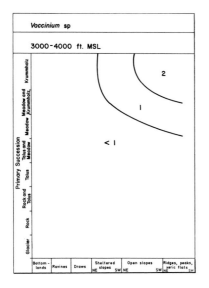

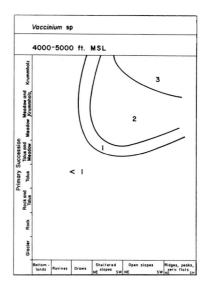

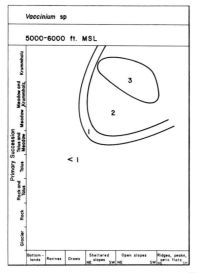

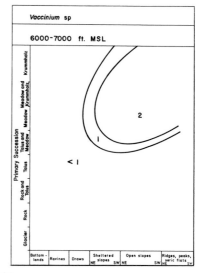

Figure 102

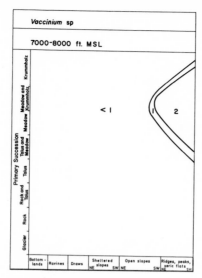

Figure 102 (*Continued.*)

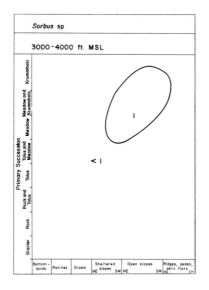

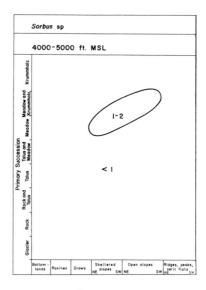

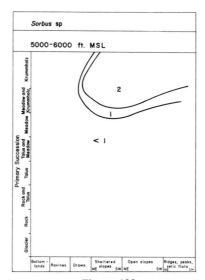

Figure 103

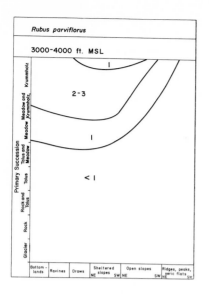

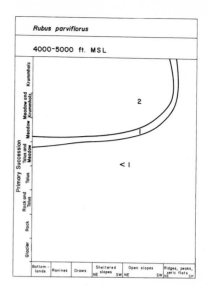

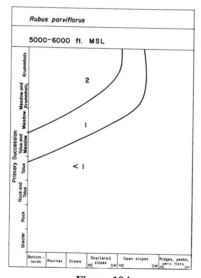

Figure 104

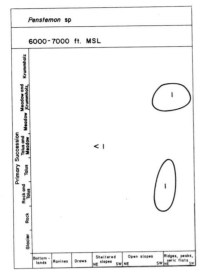
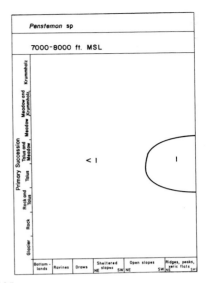

Figure 105

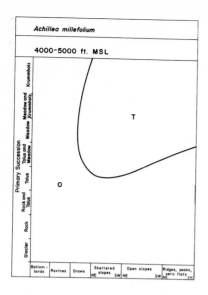

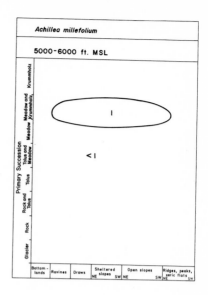

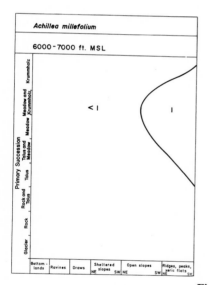

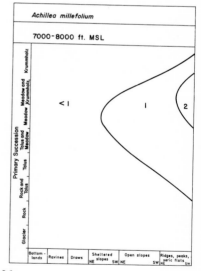

Figure 106

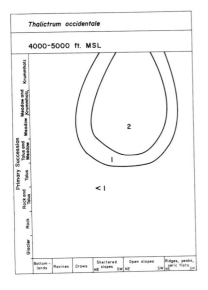

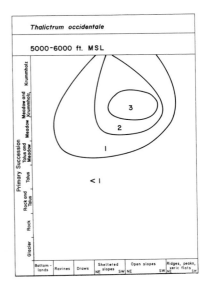

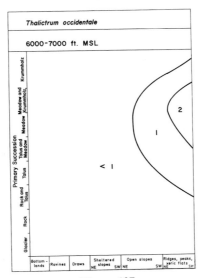

Figure 107

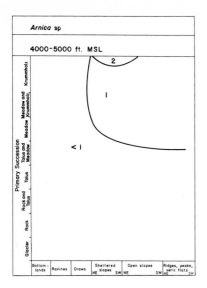

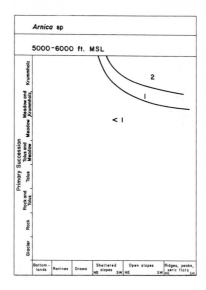

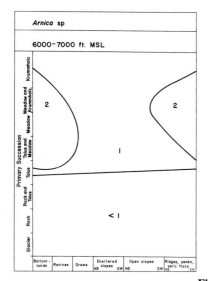

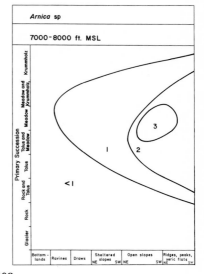

Figure 108

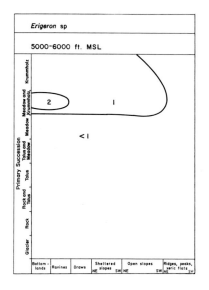

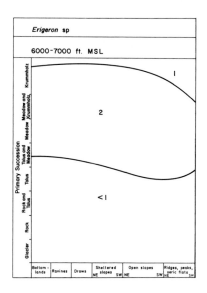

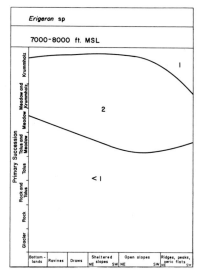

Figure 109

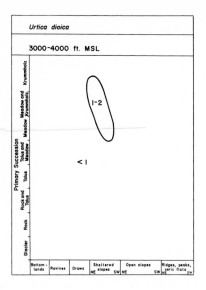

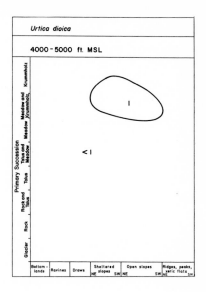

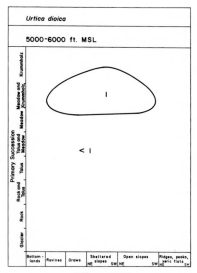

Figure 110

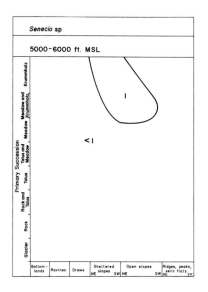

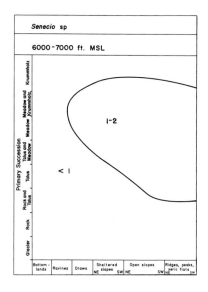

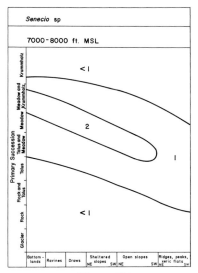

Figure 111

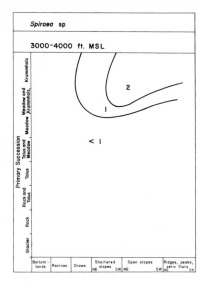

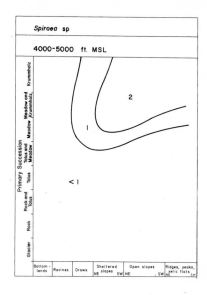

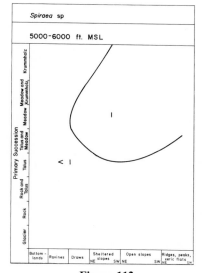

Figure 112

Index